Risks and Decisions for Conservation and Environmental Management

This book outlines how to conduct a complete environmental risk assessment. The first part documents the psychology and philosophy of risk perception and assessment, introducing a taxonomy of uncertainty and the importance of context; it provides a critical examination of the use and abuse of expert judgement and goes on to outline approaches to hazard identification and subjective ranking that account for uncertainty and context. The second part of the book describes technical tools that can help risk assessments to be transparent and internally consistent; these include interval arithmetic, ecotoxicological methods, logic trees and Monte Carlo simulation. These methods have an established place in risk assessments in many disciplines and their strengths and weaknesses are explored. The last part of the book outlines some new approaches, including p-bounds and information-gap theory, and describes how quantitative and subjective assessments can be used to make transparent decisions.

MARK BURGMAN is Professor of Environmental Science in the School of Botany at the University of Melbourne, Australia. He teaches environmental risk assessment, conservation biology and ecology. His research covers habitat models, population ecology and viability analysis, decision making and community ecology.

To Terry

Mark

ECOLOGY, BIODIVERSITY AND CONSERVATION

The world's biological diversity faces unprecedented threats. The urgent challenge facing the concerned biologist is to understand ecological processes well enough to maintain their functioning in the face of the pressures resulting from human population growth. Those concerned with the conservation of biodiversity and with restoration also need to be acquainted with the political, social, historical, economic and legal frameworks within which ecological and conservation practice must be developed. This series will present balanced, comprehensive, up-to-date and critical reviews of selected topics within the sciences of ecology and conservation biology, both botanical and zoological, and both 'pure' and 'applied'. It is aimed at advanced final-year undergraduates, graduate students, researchers and university teachers, as well as ecologists and conservationists in industry, government and the voluntary sectors. The series encompasses a wide range of approaches and scales (spatial, temporal and taxonomic), including quantitative, theoretical, population, community, ecosystem, landscape, historical, experimental, behavioural and evolutionary studies. The emphasis is on science related to the real world of plants and animals, rather than on purely theoretical abstractions and mathematical models. Books in this series will, wherever possible, consider issues from a broad perspective. Some books will challenge existing paradigms and present new ecological concepts, empirical or theoretical models, and testable hypotheses. Other books will explore new approaches and present syntheses on topics of ecological importance.

Risks and Decisions for Conservation and Environmental Management

MARK BURGMAN

School of Botany
University of Melbourne

CAMBRIDGE UNIVERSITY PRESS
Cambridge, New York, Melbourne, Madrid, Cape Town, Singapore, São Paulo

Cambridge University Press
The Edinburgh Building, Cambridge CB2 2RU, UK

www.cambridge.org
Information on this title: www.cambridge.org/9780521835348

First published 2005
Reprinted 2006

Printed in the United Kingdom at the University Press, Cambridge

A catalogue record for this book is available from the British Library

Library of Congress Cataloguing in Publication data

Burgman, Mark A.
Risks and decisions for conservation and environmental management / Mark Burgman.
    p.   cm. – (Ecology, biodiversity, and conservation)
Includes bibliographical references and index.
ISBN 0 521 83534 8 (hardback : alk. paper) – ISBN 0 521 54301 0 (pbk. : alk. paper)
1. Environmental risk assessment.   2. Environmental management.   I. Title.   II. Series.
GE145.B78   2005
333.71′4–dc22   2004054564

ISBN-13   978-0-521-83534-8 hardback
ISBN-10   0-521-83534-8 hardback

ISBN-13   978-0-521-54301-0 paperback
ISBN-10   0-521-54301-0 paperback

Contents

Preface

This book intends to create a professional standard for 'honest and complete' environmental risk assessments.

Complete risk assessments are defined as those that undertake all stages of the risk management cycle guided by stakeholders, a marriage of risk analysis methods, adaptive management, decision tools, monitoring and validation.

Honest risk assessments are defined as those that are faithful to assumptions about the kinds of uncertainties embedded in an assessment, that carry these uncertainties through chains of calculations and judgements, and that represent and communicate them reliably and transparently.

The philosophy of this book is that it is incumbent on risk analysts to make all relevant uncertainties, and the sensitivity of decisions to these uncertainties, as plain and as accessible as possible. This book treats both the qualitative and quantitative aspects of risk assessment. It takes the position that in most circumstances, the best use of models is to interrogate options, resulting in choices that are robust to a range of assumptions and uncertainties.

Reconciling the dispassionate and personal elements is the essence of creating an honest and complete environmental risk assessment.

The book explores a variety of approaches to risk assessment relevant to the management and conservation of the environment without providing full coverage. Thus, it does not explore the full details of ecotoxicology, but focuses instead on the kinds of models used by ecotoxicologists to solve environmental problems, and on the conventions used to represent uncertainty. It does not provide a complete introduction to intervals, Monte Carlo or logic trees for environmental risk assessment. Other books provide extensive details on these topics.

Here, there is enough background so that readers will be able to come to terms with the things that methods such as Monte Carlo and intervals do, how they can be critically interpreted and used to assist people to explore options. The book outlines their assumptions and weaknesses, and the kinds of uncertainties the methods can and cannot accommodate.

The US EPA (1992) used a technical perspective when it defined ecological risk assessment as a process that evaluates the likelihood that adverse ecological effects may occur or are occurring as a result of exposure to one or more stressors. Uncertainty derived from different perceptions and values will always be present in risk assessments (Jasanoff 1993). The results of this dichotomy are approaches to risk assessment that range from those that emphasize the sociological context (Adams 1995, O'Brien 2000) to those that focus primarily on the technical and probabilistic nature of risks (Vose 1996, Stewart and Melchers 1997), with almost no overlap between them.

The book assumes the reader has completed a first year university course in statistics, or its equivalent. That is, it assumes the reader is familiar with concepts such as probability distributions (density functions), probability theory, confidence intervals, the normal distribution, the binomial distribution, data transformations, linear regression, null hypothesis tests and analysis of variance. It assumes high school calculus and linear algebra, but no more than that.

An opportunity exists to create a role and a framework for environmental risk assessments so that they represent the ideas and priorities of stakeholders and are internally consistent, transparent and relatively free from the conceptual and linguistic ambiguities that plague less formal attempts to evaluate and manage human impacts on the environment. The opportunity exists because the theoretical and philosophical foundations have matured to a point where generalizations are possible, though not to the point at which conventions have become immutably entrenched.

The flaws in this book are entirely my responsibility. I will appreciate any corrections or suggestions.

Acknowledgements

I am indebted to Helen Regan and Mark Colyvan for allowing me to use our paper (Regan et al. 2002a) as the substantial basis for Chapter 2. Parts of the section on monitoring are drawn from Burgman *et al.* (1997) and Burgman and Lindenmayer (1998). I thank Keith Hayes for explaining hazard assessments to me. I thank Nick Linacre for introducing me to the way actuaries work. I am grateful to Brendan Wintle, Louisa Flander, Bernd Rohrmann, Rosemary Robins, Nick Linacre, Helen Regan, Tracey Regan and Jan Carey for sharing with me their notes and the ideas they use in teaching.

Michelle Zaunbrecher was generous in explaining how to operationalize ecological risk assessments in a corporate setting. David Fox, Peter Walley, Deborah Mayo, Michael McCarthy, Yakov Ben-Haim and Scott Ferson were patient in explaining, and explaining how to explain, statistical process control, imprecise probabilities, severity, maximum likelihood, infomation-gap theory and p-boxes, respectively. Barry Hart and Mike Grace tested ideas and forged a workable model for risk assessments for freshwater systems. I thank Nick Caputi for giving me permission to cite the IRC report on the Western Rock Lobster fishery. I thank Will Wilson, Helen Regan, Sandy Andelman, Per Lundberg, Bill Langford, Neil Thomason, Tracey Regan, Natalie Baran, Fiona Fidler, Rob Buttrose and Geoff Cumming for allowing me to use elements and ideas from our collaborative works-in-progress.

Much of this work was conducted while I was a Sabbatical Fellow at the US National Center for Ecological Analysis and Synthesis, a Center funded by NSF (Grant no. DEB-0072909), the University of California, and the Santa Barbara campus.

I am grateful to Claire Drill who created the glossary and most of the figures. Claire Layman, Wendy Layman, Barry Hart, Per Lundberg, Nick Caputi, Warren Powell, Nick Linacre, Michael McCarthy, Terry Fernandez, Keith Hayes, Kevin Korb, Jan Carey, Genevieve Hamilton, Bill Dixon, Resit Akçakaya, Fiona Fidler, Deborah Mayo, Aris Spanos,

Will Wilson, Brendan Wintle, Jane Elith, Bill Langford, Sarah Bekessy, Helen Regan, Elisabeth Bui, David Fox, Mark Colyvan, Jason Grossman, Anne Findlay, Christine Croydon, Mark Borsuk, Rick van Dam, Peter Bayliss and Robert Maillardet read drafts of the book and made numerous helpful suggestions that greatly improved it.

I thank Michael Usher for taking this on and for his careful advice. I am indebted to the editorial team at Cambridge for their efficient and professional work.

I thank the students in the University of Melbourne Environmental Risk Assessment classes and the people in government agencies, universities and businesses in Australia, the United States, Pakistan, Papua New Guinea, Thailand and Canada on whom I experimented before this made sense.

1 · *Values, history and perception*

Risk is the chance, within a time frame, of an adverse event with specific consequences. The person abseiling down the front of the city building next to the giant, hostile ear of corn is protesting about the risks of eating genetically modified food (Figure 1.1). The same person is willing to accept the risks associated with descending from a building suspended by a rope. This book explores the risk assessments that we perform every day and introduces some tools to make environmental risk assessments reliable, transparent and consistent.

Risk assessments help us to make decisions when we are uncertain about future events. Environmental risk assessments evaluate risks to species (including people), natural communities and ecosystem processes. Whatever the focus, the risk analyst's job is to evaluate and communicate the nature and extent of uncertainty. To discharge this duty diligently, we need professional standards against which to assess our performance.

Epidemiologists, toxicologists, engineers, ecologists, geologists, chemists, sociologists, economists, foresters and others conduct environmental risk assessments routinely. Yet analysts often select methods for their convenience or familiarity. Choices should be determined by data, questions and analytical needs rather than by professional convention.

Different philosophies of risk influence how risk assessments are conducted and communicated. Societies and science have evolved conventions for communicating uncertainty that do not acknowledge the full extent of uncertainty. This book explores different paradigms for risk assessment. It evaluates how well different tools for risk assessment serve different needs. It recommends a broad framework that encourages honest and complete environmental risk assessments.

1.1 Uncertainty and denial

Scientific training leaves us with an unreasonable preoccupation with best estimates of variables. We focus on means, medians and central tendencies

Figure 1.1. Risks of different kinds: abseiling and eating genetically modified corn. Photo by Shawn Best. © Reuters 2000.

of other kinds. When we report an estimate, we are not obliged to report its reliability.

We rarely, if ever, think about the tails, the extremes, of a distribution. Usually, we tacitly deny the tails exist. Risk assessment differs from mainstream science because it obliges us to think about the tails of distributions, to account for the full extent of possibilities.

Many risk assessments communicate ideas about risk with words. For example, ADD (1995) argued in an environmental effects assessment for a new port facility that oil spills from ships were a 'low risk'. To justify this assessment, they said 'Between 1986 and 1994 there has been, on average, one ship grounding per annum. Despite the restricted waterways, ... the probability of a ship grounding is low. This is due to the high standard of training given to ... officers, risk control measures ...'.

The words 'low' and 'risk' were not formally defined, but were used in context with other words such as 'moderate risk' and 'high risk'. The adjectives have a natural order but do not communicate the magnitude of the risk or the extent of consequences. To resolve such difficulties, we need a description of the kinds of uncertainty and some indication of how they should be expressed, combined and managed.

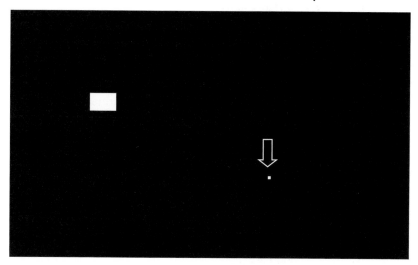

Figure 1.2. The US NRC (1983) reported five million known chemicals (represented by the black area). Of these, 7000 were tested for carcinogenicity (the larger white box). A total of 30 (the small white box, indicated by the arrow) were definitely linked to cancer in humans (from Adams 1995).

All risk assessments involve a mixture of kinds of uncertainties, only some of which may be quantified. Even quantified uncertainty comes in different forms. For example, we may measure directly the strength of a wall, the flow rate of a chemical through soil, or the birth rate of a threatened species. We may estimate them from scientific theory. If theory or direct measurements are unavailable, we may extrapolate from data from similar systems or circumstances and assume our system behaves the same way. Extrapolations are dangerous but may be better than nothing. We may turn to what is often the weakest kind of evidence, expert opinion.

Too often, different kinds of information are treated as though they were equivalent. For instance, expert opinion is usually available, but reliable expert opinion is deceptively difficult to obtain. Chapter 4 is devoted to exploring its weaknesses and usefulness.

Direct measurements are almost always missing but the lack of information is rarely communicated. Take the case of human health risks posed by chemicals. Adams (1995) gave a visual interpretation of uncertainty by shading a square black in proportion to the number of chemicals in use that had not been tested for their effects (Figure 1.2). But even this underestimates the extent of uncertainty. If the toxicities of all the chemicals were thoroughly tested in laboratories, there would still be uncertainty about interactions between chemicals, exposure and susceptibility among individuals, the effects of the chemicals in field conditions and so on (see Suter 1993, Silbergeld 1994).

In an example from an entirely different field, Briggs and Leigh (1996) created a list of 'threatened' Australian plants. They estimated there were 4955 species of vascular plants considered to be 'at risk (i.e., endangered, vulnerable, rare or poorly known and thought to be threatened)'. Are such lists trustworthy? Should they be used to design conservation reserves, set priorities for recovery, or constrain trade or development (Possingham et al. 2002a)?

Briggs and Leigh's (1996) list was modified as new information came to hand, as people and agencies responsible for the list changed, and as taxonomists revised species descriptions. Five years later, only 65% of the species in the earlier list remained on the official government list (Burgman 2002). Most species were removed because the taxonomy was revised, or because additional populations were discovered, reducing the threat classification for the species. In the meantime, the presence of about 2000 species that were not, in fact, threatened affected planning decisions and diverted resources for environmental protection. The costs may be worth the benefits of protecting the species on the list that turn out to need protection, but the considerable uncertainty in the status of threatened species lists is not communicated.

Science creates for itself a mantle of objective certainty. This impression often is unjustified and misleading. For example, Table 1.1 gives the numbers of species thought to be at risk in several countries. Such lists are published and used routinely by governments throughout the world.

The mantle of certainty is reflected in the assumptions that untrained people make when they see the lists. For example, I have shown lists like Table 1.1 to several thousand high school and first year undergraduate students over the last 10 years. I asked them 'why are there no officially listed endangered fungi, and so few listed endangered invertebrates in Australia, China and the United States?' They have given some wonderfully inventive and plausible explanations: fungi have a resistant life stage that makes them immune to environmental stress; insects produce large numbers of eggs and can breed their way out of trouble; insects and fungi can disperse long distances, thereby avoiding trouble. Almost none considered the fact that scientists haven't looked. The vast majority of nonvascular plants, insects and fungi have never been assessed.

Table 1.1. *The number of species listed as endangered in Australia, USA and China (each of these countries has relatively large numbers of endemic species) compared to the number of species listed as endangered in the UK and the number thought to exist globally (data from Groombridge 1994, Anonymous 1995, Burgman 2002)*

Taxon	Estimated total number of species in the world[a]	Number listed as 'Threatened'			
		Australia	USA	China	UK[b]
Fungi	500 000	0	0	0	46
Nonvascular plants	100 000	0	0	0	70
Vascular plants	250 000	1597	1845	343	61
Invertebrates	3 000 000	372	860	13	171
Fish	40 000	54	174	16	7
Amphibians	4 000	20	16	0	3
Reptiles	6 000	42	23	8	6
Birds	9 500	51	46	86	25
Mammals	4 500	43	22	42	18

[a] Estimates of the global numbers of species in each taxon were estimated crudely from numbers of currently described species and expert judgements of the proportion remaining to be described (see May *et al.* 1995, Burgman 2002).

[b] Numbers from the UK represent all those species in the UK's Biodiversity Action Plan for which conservation action plans were written (Anonymous 1995).

The list from the UK has relatively large numbers of fungi, nonvascular plants and invertebrates. The total number of plants and animals in the UK is modest. Survey and taxonomic effort per species has been high.

Most lists from most countries are uncertain and biased. May *et al.* (1995) called them popularity polls. Scientists are guided by funding opportunities and personal interests. To most, mammals and birds are more interesting than insects and fungi. The distribution of research and taxonomic effort among taxa reflects the preferences of scientists as well as the interests of broader society. Science is motivated by interest so the bias is acceptable. But the failure to communicate uncertainty and bias is professionally negligent, despite its conventional acceptability.

This chapter lays a foundation for improving communication about uncertainty by exploring concepts of uncertainty, the history of thinking about risk and the cognitive foundations of risk perception. Later chapters offer strategies for dealing with the context of a problem and the values and preferences of individuals involved in risk assessments.

1.2 Chance and belief

Concepts of probability influence risk measurement, interpretation and communication. Mostly, a probability is assumed to be a relative frequency. In practice, other definitions are often applied. For instance, in a paper on estimating parameters in engineering systems, Stephens *et al.* (1993, p. 272) suggested probability be defined as, 'the degree of belief, held by the person providing the information, that a quantitative system characteristic will have a value in a particular interval under specified conditions of measurement'. It is remarkable that, in an early application of probabilistic risk assessment, Kaplan and Garrick (1981, p. 17) defined probability as '. . . a numerical measure of a state of knowledge, a degree of belief, a state of confidence'. This section explores the foundations of a curious ambiguity about the word 'probability'.

1.2.1 Two kinds of probability

Probability has two dimensions. It can be viewed as the statistical frequency (or relative frequency) with which an event is expected to occur, and it can be viewed as the degree of belief warranted by evidence. The former definition is in most standard statistical textbooks (e.g. Casella and Berger 1990, p. 5). The latter perspective is associated with 'Bayesian' statistics.

The dual nature of probability was recognized in texts dating from the 1660s including the Port Royal Logic (1662, the first influential text on probability; see Hacking 1975) but it is rarely distinguished in modern applications. Hacking (1975, p. 143) called it the 'squirming duality'.

Good (1959) defined 'classical' probability as 'the proportion of equally probable cases', a concept he attributes to Candan, a sixteenth century gambler. For example, of the 36 possible results of throwing two dice, 3 give a total of 11 or more points, so the probability is defined as 1/12. The purpose of the definition was to explain the observed, long-run frequencies of particular events.

When the concept of probability relates to an event that, when repeated, occurs with a certain frequency, it is a statistical concept that describes a chance process. It exists, independent of our knowledge of it. For example, the probability of failure of a dam wall may be known (within some confidence limits) because there have been many such walls built and because physical characteristics and failure rates are known from theory and measurement. The concept is equivalent to the chance of rolling a given number when playing a dice game.

The other side of the probabilistic coin is concerned with reasonable degrees of belief. It applies when a probability is unknown or unknowable. Despite their subjective origin, in most applications they are expected to be rational in the sense that they follow the rules of probability (Cooke and Kraaikamp 2000). For example, if I believe that the chance that an oil spill will eliminate a bird rookery is 70%, then the chance that it will not eliminate the rookery should be 30%.

We will return to the dual meaning of probability later in this chapter and in subsequent chapters.

1.2.2   Two kinds of subjective probability

The term 'subjective probability' is in widespread use. Like the word 'probability' alone, the phrase also has two meanings. The first meaning is a lack of knowledge about a process or bias. The second meaning is that it indicates purely personal degrees of belief.

To illustrate the term when it refers to a lack of knowledge, Hacking (1975) used an example of a sack full of biased coins. Each coin turns to either heads or tails more frequently than 1/2 of the time. In this sack, on average, the biases cancel. If we select a single coin, the probability of getting heads on the next toss is 1/2. The probability is composed of two components: the probability that this particular coin will result in a head (the objective element); and our ignorance about which coin it is (the subjective element of incomplete knowledge). In this view, the subjective uncertainty may be resolved by repeating the experiment (tossing the coin) many times.

In contrast, when subjective probabilities indicate personal belief, probabilities are unknown only insofar as a person doesn't know their own mind (Hacking 1975). Usually, we expect personal beliefs to be rational in the sense that subjective probabilities should coincide with relative frequencies, when frequency data are available. While they don't need to be exactly the same, they need to be *answerable* to frequencies in the sense that frequency data should influence judgements when they are known (Colyvan 2004).

1.2.3 Disentangling meanings

Understanding statements about probability is complicated. A proposition may be stated in probabilistic terms, but there may be no underlying fact. Language may allow borderline cases or ambiguities so that it is hard to know what the statement means. People use gaming analogies to represent nonprobabilistic uncertainties on a numerical scale (even the Port Royal Logic used this analogy in 1662; see Hacking 1975, Walley and DeCooman 2001). We will return to this topic in Chapter 2.

In yet another twist, statements may have frequency interpretations, but the assignment of a probability may be subjective. Understanding what is meant may not depend on repeated trials. For example, the following circumstances do not require repetition to be understood:

> The Tasmanian Tiger is probably extinct.
> It will probably rain tomorrow.
> Living in this city, you will probably learn to like football.
> My brother is probably sleeping at this moment.

For instance, I can say that my brother is probably sleeping just because I know he is lazy. Equally reasonably, the statements may have a frequency interpretation and be amenable to measurement. For instance, in support of the assertion that my brother is sleeping at this time, I could sample days and times randomly, record my brother's behaviour and, eventually, build up a reliable picture of his sleeping habits. I could then, after making some assumptions, report a probability that he is, in fact, sleeping.

'Bayesian' statisticians use subjective probabilities (degrees of belief, such as, 'I believe that my brother is asleep') whereas 'frequentist' statisticians see probabilities as relative frequencies (such as, 'at this time of day, he is asleep 9 times out of 10'). Subjective probabilities can be updated (via Bayes' theorem) when new data come to hand. In most cases, repeated application of Bayes' theorem results in subjective probabilities that converge on objective chance. And there are objective ways to arrive at subjective probabilities, including the use of betting behaviour (see Regan et al. 2002a for more details, and Anderson 1998b, Carpenter et al. 1999 and Wade 2000 for application of Bayesian methods to environmental problems).

1.2.4 Probability words

Many words cluster around the concept of probability. They are used to capture a component of the broader concept. But they are used carelessly and somewhat interchangeably. The following list describes perhaps their most common current uses in risk assessments (see also Good 1959, Hacking 1975):

- *Chance*: the frequency of a given outcome, among all possible outcomes of a random process, or within a given time frame.
- *Belief*: the degree to which a proposition is judged to be true, often reported on an interval (0,1) or per cent scale creating an analogy with 'chance'.
- *Tendency*: the physical properties or traits of a system or a test that result in stable long-run frequencies with repeated trials, or that yield one of several possible outcomes.
- *Credibility*: the believability of detail in a narrative or model (acceptance of ideas based on the skill of the communicator, the trust placed in a proponent).
- *Possibility*: the set of things (events, outcomes, states, propositions) that could be true, to which some (nonzero) degree of belief or relative frequency might be assigned.
- *Plausibility*: the relative chance or relative degree of belief (the rank) assigned to elements of a set of possible outcomes.
- *Confidence*: the degree to which we are sure that an estimate lies within some distance of the truth.
- *Bounds*: limits within which we are sure (to some extent) the truth lies.
- *Likelihood*: how well a proposition or model fits available data.
- *Risk*: the chance (within a time frame) of an adverse event with specific consequences.

Hacking (1975) pointed out that common usage of the word 'probable' prior to the 1660s was evaluative. If a thing was probable, it was worth doing (my sleeping brother would see this definition clearly). It also meant trustworthy. A probable doctor was a trusted one. Jesuit theologians in the period before 1660 used the term probable to mean 'approved by the wise', propositions supported by testimony or some authorized opinion. These meanings are like the terms belief and credibility above, and allow that a proposition could be both probable (because it is made by a trustworthy person) and false.

Today, the words under the umbrella of probability make up a dynamic linguistic landscape. People interpret probability intuitively, leading to different interpretations of risk (Anderson 1998b). There is no doubt that the meanings of these words will shift in the future as circumstances change.

This makes it important to be clear about the use and interpretation of words.

1.2.5 Probability and inference

Scientists meet the word 'probability' as a 'p-value' most commonly in the context of statistical hypothesis tests. In this context, usually probability is thought to mean chance. More exactly, the p-value of a test is the probability of the observed data assuming the null hypothesis is true. It deals with chance conceptually. If the null hypothesis is true, and if an experiment is repeated many times, the p-value is the proportion of experiments that would give less support to the null than the experiment that was performed.

In so-called null hypothesis significance testing, frozen and rigidly interpreted in introductory textbooks, the scientist sets a threshold such as 0.05, and rejects the null hypothesis whenever the p-value is less than or equal to 0.05. In this way, in the long run if all sources of error are taken into account, the scientist will reject a true null hypothesis 5% of the time. The p-value is a measure of the plausibility of the assertion that the null is true (Salsburg 2001).

Null hypothesis tests are routinely misinterpreted by scientists in many disciplines including ecology (see Anderson *et al.* 2000). The standard model is particularly vulnerable to psychological frailties (see Chapters 4 and 11). Despite these difficulties, food and drug regulatory authorities, environmental protection agencies, law courts and medical trials all accept null hypothesis testing as an appropriate method of proof. Chapter 11 explores some ways of fixing the problem.

1.3 The origin of ideas about risk

The history of ideas about risk gives insight into the dual nature of probability. It also illustrates the long history of people making poor judgements about risks; people are not moved to 'rational' responses by empirical evidence alone.

Bernstein (1996) noted that the lack of analytical tools in the Greek, Hebrew and Roman systems led people to make bets on knucklebones (one of the earliest common randomizers; Hacking 1975) that today would be considered irrational. When thrown, a knucklebone can come to rest in only 1 of 4 ways. The values for narrow faces were 1 and 6 and for wide faces were 3 and 4. Throwing a sequence of 1, 3, 4, 6 earned more than throwing 6, 6, 6, 6 or 1, 1, 1, 1, even though the latter sequences are less likely. But the wisdom of betting on the wide faces is only apparent because we have been trained to think in terms of frequencies, at least to some extent. Things we take as common sense are, in fact, learned.

Numbers provide a symbolism that makes arithmetic easy because each digit denotes a power, usually of 10 (Gigerenzer 2002). The Hindus created a number system that allowed calculations (Hacking 1975, Bernstein 1996). Quantitative probabilistic reasoning can be found in Indian literature that dates to at least the ninth century, 800 years before it emerged in Europe (Hacking 1975). al-Khowârizmî, an Arab mathematician living around 825, established the rules for adding, subtracting, multiplying and dividing (Bernstein 1996). Most cultures did not adopt Hindu-Arabic notation for numbers until well into the second millennium.

Around 1200, Fibonacci, an Italian mathematician, visited Algeria and learned al-Khowârizmî's system. He wrote a book called *Liber Abaci* in which he documented the mechanics of calculations for whole numbers, fractions, proportions and roots. The centrepiece was the invention of zero. The system provided solutions to linear and quadratic equations. It allowed people to calculate profit margins and interest rates, and to convert weights and measures for arbitrarily large numbers. The Arabic ancestry of mathematical ideas lives on in words with Arabic roots such as algorithm, algebra and hazard (from *al zahr*, the Arabic word for dice; Bernstein 1996).

When we flip two coins, there are three partitions (HT, TT, HH) and four permutations (HT, TH, TT, HH). Partitions record only the kinds of outcomes, ignoring the different ways of getting them. Gambling on combinations of three 6-sided dice dates to at least AD 1200. Rolling three dice can generate a sum of 4 with only one partition, 1–1–2. But a 4 can be obtained with three different permutations (1–1–2, 1–2–1, 2–1–1) and is (we know now) three times more likely than rolling a 3, which can be generated in only one way (with one partition), namely, 1–1–1. It was not obvious to the people of AD 1200 if permutations or partitions were equally probable. In 1477, a commentary was written that gave the probabilities of various totals when three dice are thrown (Good 1959). Hacking (1975) suggested that the truth could only have been determined by observation.

Theories of frequency, randomness and probability coalesced in Europe around the 1660s and there was a suite of spectacular coincidences (see Hacking 1975 and Bernstein 1996). The theory of probability usually is attributed to Pascal who, in correspondence with Fermat, solved problems for games of chance in 1654 (Good 1959). Huygens published the first textbook on probability in 1657. At about the same time and seemingly independently, Hobbes wrote that while experience allows no universal conclusions, 'if the signs hit twenty times for one missing, a man may lay a wager of twenty to one of the event; but may not conclude it for a truth' (in Hacking 1975, p. 48).

The first application of probabilistic reasoning to something other than a game of chance appeared in the Port Royal Logic in 1662. In the same year, Gaunt published statistical inferences drawn from mortality records. In 1665, Leibniz applied probabilistic measurements to legal reasoning, using numbers to represent 'degrees of probability', which we would now call degrees of belief. His 'natural jurisprudence' encompassed conditional probabilities and 'mixed' or conflicting evidence. In the late 1660s, Hudde and de Wit developed systems that put annuities on a sound actuarial footing. Many of these people were unaware of the work of the others.

Why did all these ideas arise simultaneously? The explanation is related to the spread of the new number system and to the evolution of ideas about two kinds of probability in scientific reasoning. Hacking (1975) argued that the concept of inductive evidence was an additional, necessary precursor to the development of theories of probability. Prior to the 1600s, evidence included concepts of testimony (supported by witnesses) and authority (supported by religious elders, experts and books). Evidence 'of the senses' comprised information that was gained first-hand, sometimes called the evidence of things, or internal evidence. Lastly, evidence sometimes referred to the demonstration of effects from knowledge of underlying causes.

Inference requires the formulation of explanations from observations. For a long time, there was no accepted basis for inductive reasoning. Hacking (1975) argued that inductive reasoning developed from medical diagnosis and the related 'low' sciences of alchemists, astrologers and miners who relied on empirical evidence to guide them to explanations. Symptoms were used as evidence of the state of the system. It led to the rise of reasoning from observed effects to hypothetical causes. The concept of diagnosis sits more comfortably with belief than it does with chance. One can have partial belief in several different explanations but only one of them will be true.

The merger of probability (as both belief and chance) with the machinery of arithmetic led to the notion of expected utility, one of the foundations of decision theory. Expected utility is the magnitude of an anticipated gain, discounted by the chance that the outcome will be achieved. For instance, the Port Royal Logic (1662) stated that 'Fear of harm ought to be proportional not merely to the gravity of the harm, but also to the probability of the event' (in Bernstein 1996). Bernoulli (1713, in Bernstein 1996) outlined the law of large numbers, utility and diminishing returns explicitly, '[The] utility resulting from any small increase in wealth will be inversely proportionate to the quantity of goods already previously possessed.'

During the period from the late 1700s to the late 1800s, concepts such as conditional probabilities, statistical power, the central limit theorem and the standard deviation were discovered by Bayes, de Moivre, Laplace, Bernoulli and Gauss. Many discoveries were in response to practical issues such as dividing wagers before a game was complete and dividing resources fairly between merchants (Stigler 1986).

The evolution of ideas about risk occurred hand in hand with the development of ideas about probability and decision theory. Pascal discussed the decision about the costs and benefits of believing in God in a decision-theoretic framework (Hacking 1975). This thinking led, eventually, to the formalization of statistical inference and null hypothesis tests by R. A. Fisher and Neyman and Pearson between about 1920 and 1935.

1.4 Perception

People can be bad judges of risk. We carry with us a set of psychological disabilities that can make it next to impossible for us to visualize and communicate risks reliably.

The realization that we judge as badly as we do is relatively recent. In the 1970s, two psychologists, Kahneman and Tversky, began doing experiments on the ways in which people perceive and react to risks. Their results were strikingly counterintuitive and led to exciting generalizations. Cognitive psychologists, economists, sociologists and others took up the theme. They continue to discover quirks of human perception today.

To illustrate how badly we analyse, consider the information in Table 1.2. Each activity results in an equal additional risk of dying (one in a million) to the person who does them. The table is somewhat misleading because the numbers are averages drawn from large samples.

They may be interpreted by creating realistic scenarios. For example, assume that every 10 km a person rides on a bicycle is a random sample of all sets of 10 km ridden by themselves and all others, at least insofar as the chance of being hit by a car and killed is concerned. This allows us to apply a statistical definition of probability. The table

Table 1.2. *Activities that increase chance of death by about one in a million per year in the United States (after Wilson 1979, Stewart and Melchers 1997, NFPRF 2002). Here, we assume that the risks are simple causal relationships that accumulate additively with exposure*

Activities	Cause of death
Spending 1 h in a coal mine	Black lung disease
Spending 1 h as an agricultural worker	Accident
Spending $2\frac{1}{2}$ days as a firefighter	Accident
Travelling 15 km by bicycle	Accident
Travelling 500 km by car	Accident
Flying 1500 km by jet	Accident
Living 2 months in a brick building	Cancer from radiation
Working for 1 year in a nuclear power plant	Cancer from radiation
Living 50 years within 8 km of a nuclear reactor	Radiation (accident)
Living 20 years near a PVC plant	Cancer from vinyl chloride

The number of additional deaths per year for the example below are: a. 3840 (assuming 240 working days per year), b. 1920, c. 0.2, d. 110, e. 11. So, the 'right' order is c (least risky), e, d, b, a (most risky).

is an example of a standard form of communication about risks and, like most others, it gives no indication of the certainty of the estimates, nor does it evaluate the benefits of each activity (Kammen and Hassenzahl 1999).

Even though these statistics provide a seemingly unambiguous way of interpreting risk, they are notoriously difficult for people to interpret in a sensible way. Now, consider the following scenarios:

a. Riding a bicycle to and from a suburban home to work, a distance of 20 km each way, each working day for four years.
b. Working for one year in a coal mine (assuming a 40-h working week).
c. Living 4 km from a nuclear reactor for five years.
d. Working for one year as a firefighter.
e. Working for two months in a nuclear power plant.

We assume the risks accumulate additively with time and that there are no causes of death other than those listed. Each scenario involves a chance of dying. For this exercise, rank them from your most preferred to your least preferred option (that is, try to make these ranks reflect an ordering from the least risky to the most risky activity) without doing any arithmetic. An answer is given at the foot of Table 1.2.

People rarely get the order 'right', even if they have just seen the table. The risks in the example span five orders of magnitude. Most people overestimate the risk of living near or working in the nuclear reactor, and underestimate the risk of riding a bicycle. This observation is striking because interpretation may have direct consequences for individual chances of surviving and reproducing. People are poor judges despite presumably powerful selective forces to get it right.

In an influential and much-cited study, Fischhoff *et al.* (1982) described judgements of perceived risk by experts and lay people. It included Figure 1.3, which plotted judgements against independent technical estimates of annual fatalities for 25 technologies and activities. Each point represents the average responses of the participants. The dashed lines are the straight lines that best fit the points.

The experts' risk judgements are closer to measured annual fatality rates (indicted by the solid lines) than are lay judgements. But they are not as close as one might have expected. Both groups substantially underestimated the risks of high-probability events, and overestimated the risks of low-probability events. In some cases, the misjudgements were by two orders of magnitude for events that directly affect the lives of those making the judgements. We will return to the reliability of experts in Chapter 4.

Kahneman and Tversky (1979, 1984) found that people's perceptions are tuned to measure relative change in the magnitude of consequences rather than the absolute change. They also found that perceived utility of a risk depends on how it changes one's prospects, so that individuals will interpret risks differently, depending on how they affect their personal circumstances. In addition, they observed that a tiny risk is not necessarily a trivial one. Judgements about importance are substantially divorced from estimates of the magnitude of a risk, and depend on social context.

Suter (1993) provided a good example. He stated, 'Some exposures and effects are manifestly trivial' (p. 87). He went on to say that it is absurd, for example, to consider restricting a pollutant because it causes one copepod species to replace another. This statement horrifies aquatic invertebrate ecologists and outrages copepod specialists. They believe that copepod species are important for themselves, and that the ecological implications of replacement may be indicative of qualitative shifts in whole ecosystems. People will sometimes 'tolerate' a risk, implying a willingness to live with a risk to secure benefits. The risk should not be seen as negligible or something that might be ignored, but as under review and to be reduced further as soon as possible (Pidgeon *et al.* 1992).

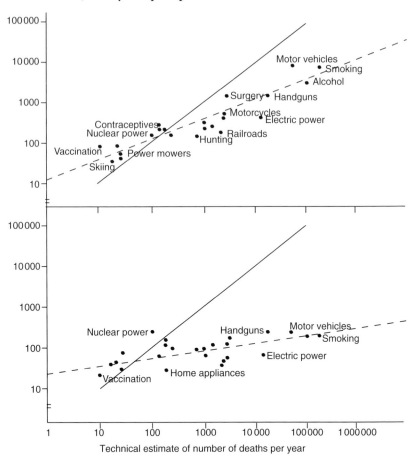

Figure 1.3. Averages of risk estimates made by 15 experts (top) and 40 lay people (bottom) plotted against the best technical estimates of annual fatalities for 25 technologies and activities. The experts included geographers, biochemists and hazardous materials regulators. The lay people were from the US League of Women Voters. The dashed lines are the straight lines that best fit the points. Both groups were biased. The experts' risk judgements were more closely associated with annual fatality rates than were lay judgements (from Slovic *et al.* 1979, Fischhoff *et al.* 1982). The correct assessments are given by the solid lines.

When scientists debate risks publicly, they sometimes try to influence risk perception by emphasizing (or trivializing) numbers. For example, Cauchi (2003) reported in a newspaper that scientists inserted a gene into the chloroplasts of tobacco plants and then looked for it in the plants' offspring. The technique of injecting modified genes away from a plant cell's nucleus was an attempt to eliminate the risk that the modified gene would find its way into related species by hybridization. In about 1 in 16 000 seedlings, the gene moved from the chloroplast to the nucleus where it was heritable.

One expert commented that the rate of 1 in 16 000 'sounds like a very large number'. Another said the rate was 'not going to be measurable' because only 1 seed in about 10 000 made it to a neighbouring nonmodified field, giving a total rate of 1 seed in 160 million. Both opinions are conveniently framed. 'Large' relative to what? We need to know, at the least, the number of seeds in the average field. And we need to know what it is we are trying to avoid. If any escape is unacceptable, 1 is a large number. If less than 1% hybridization is tolerable, 1 in 160 million might be small depending on the time frame involved. Such arguments can be difficult to evaluate in the absence of a complete evaluation of potential costs and benefits.

These examples illustrate the importance of context and framing. These issues are important in designing risk assessments, communicating the results and managing risks subsequently, the basis of discussions about risk management in Chapter 12.

1.4.1 Risk aversion and framing

People make inconsistent (but not necessarily incorrect) choices when a problem appears in different frames (Kahneman and Tversky 1984). The answers to questions about risk depend on the setting. Risk aversion occurs when people prefer to have a smaller reward with greater certainty, than a larger reward with less certainty. This may be so, even when, on average, a person would do better to choose the high–risk, high–reward scenario. Risk aversion makes sense whenever the cost or the failure to gain a reward is unacceptable. All decisions involve 'objective' facts and subjective views of the desirability of what is to be gained or lost by the decision. The preferences and circumstances of the individuals who experience the outcome determine its desirability. It is inherently a subjective quantity, influenced by context, and may change capriciously.

A framing effect occurs when a change in the presentation of a choice influences choice behaviour, even when the objective characteristics (outcomes or probabilities) are not changed (Kahneman and Tversky 1979, 1984). Risk aversion can be modified by the frame as well as by emotional motivation. For example, I put the following two sets of questions to a class of about 250 undergraduates each year for five years:

1. Is it acceptable for each species of Australia's endemic mammals to have a 50% chance of persisting for 50 years? What about a 90% chance? What about a 95% chance? Is it an acceptable level of risk for each species to have a 95% chance of persisting for 100 years?

Very few (<1%) raised their hands in answer to the first question. Between 25% and 50% of people raised their hands to the last question. Consider the alternative question, asked a little later:

2. Australia has the worst conservation record on Earth, when it comes to mammal extinctions. Of the roughly 70 species that have become extinct globally in the last 400 years, more than 20 have become extinct in Australia in the last 200 years. Who thinks this is an acceptable record?

The second question was prefaced by an implicit value judgement (the woefulness of Australia's conservation record). Few raised their hand to this question. Most were insensitive to the fact that Australia has 200 species of endemic mammals, of which around 20 are extinct. This translates (roughly) to a loss of 5% of the fauna every 100 years, a rate equal to that stipulated in the final form of question 1, which a substantial proportion of people thought was acceptable when framed differently.

Decisions such as those made by the students about acceptable levels of threat to species may appear to be irrational because the driving force is loss-aversion. The same thinking applies to choices on the stock market. Bernstein (1996) recognized the importance of this phenomenon when he stated, 'Few people feel the same way about risk every day of their lives. . . . Investors as a group also alter their views about risk, causing significant changes in how they value the future streams of earnings that they expect stocks and long-term bonds to provide' (p. 263).

Thaler (1991) provided an example in which the prospect of loss influenced decisions more strongly than the prospect of gain. He asked groups of people, 'How much would you pay to eliminate a one-in-a-thousand chance of immediate death?' and, 'How much would you have to be paid to accept an additional one-in-a-thousand chance of immediate death?' Median answers were $200 for the former question, and more than $50 000 for the latter question. Such observations led Bernstein (1996) to conclude that it is absolutely important for an investment adviser to know the personal circumstances, commitments and sensitivities of his or her client. Without it, the adviser is unable to give sensible advice.

The vagaries of money markets, mortgages and investments may seem divorced from environmental decision-making, but the analogy is close. There are potential costs and benefits to any decision. A decision by an individual not to develop a site because of potential environmental harm, for instance, is essentially the same as deciding to invest in one arena rather than another. Exactly how the costs and benefits should be computed and compared depends on the values, perceptions and circumstances of the individuals involved, especially for things that do not have monetary value.

1.5   The pathology of risk perception

Perceptual idiosyncrasies colour judgements about risky situations (Adams 1995). Cognitive psychologists and sociologists have made some useful generalizations, summarized below (drawn from Tversky and Kahneman 1974, 1982a,b, Kahneman and Tversky 1979, 1984, Fischhoff et al. 1982, Slovic et al. 1984, Morgan 1993, Plous 1993, Adams 1995, Fischhoff 1995, Freudenburg 1996, Freudenburg et al. 1996, Morgan et al. 1996). It is important to remember that not everyone reacts like this, just that the majority of people do. These attributes of risk psychology may be viewed collectively as a pathology, a set of identifiable symptoms that are characteristic of an underlying malaise.

1.5.1   Insensitivity to sample size

Perhaps the most debilitating psychological flaw, from the point of view of risk assessments, is insensitivity to sample size. Most people (including experienced scientists) draw inferences from data that can only be justified with much larger samples. Cognitive psychologists have termed this the belief in 'the law of small numbers', making an oblique, somewhat sarcastic, reference to the law of large numbers. In environmental science, the law of small numbers leads to:

- underestimation of risk by proponents,
- overestimation of risk by those faced with dealing with the consequences,
- research based on underpowered samples,
- undue confidence in early trends and apparent patterns, and
- undue confidence in the failure to detect impacts.

The causes and consequences of this failing will be explored in Chapters 4 and 11. Its existence is one of the primary reasons why formal, transparent and repeatable risk assessments are necessary.

1.5.2 Overconfidence

Unfortunately, unjustified optimism is a pervasive feature of risk assessments. Typically there is little relationship between confidence and accuracy, including eyewitness testimony in law, clinical diagnoses in medicine and answers to general knowledge questions by people without special training (see Gigerenzer 2002, Chapter 4).

Capen (1976) asked a group of professional geophysicists attending a conference to guess the number of beans in a jar, and to provide 90% confidence intervals for their estimates. One interpretation of this interval is that the scientists would have been willing to accept a 9 to 1 bet, winning $1 if they were right, and giving up $9 if they were wrong. Most people were optimistic about their ability to enclose the truth. Only 40% (14 out of 34) of the respondents included the true value in their interval.

This result is easy to replicate. Figure 1.4 repeats Capen's experiment for a group of 63 senior undergraduate students. Only 46% participants (29 out of 63) included the correct value.

The result in Figure 1.4 is important because it illustrates the unreliability and optimism typical of subjective estimation. In fact, the example underestimates the problem because, in general, people involved in risk assessments know far less about the physical and ecological properties of a system than they do of a jar of beans, and the example is free of the linguistic uncertainties that affect more realistic applications.

1.5.3 Judgement bias

People are overconfident in assessing the quality and reliability of their own judgements. An important source of overconfidence is a failure to appreciate the nature and tenuousness of assumptions. People consistently exaggerate what could have been anticipated.

1.5.4 Anchoring

Anchoring is the tendency to be influenced by initial estimates. When asked to guess a number or a property, people will be drawn to the guesses made by others, and will defer their judgements to people they believe have greater authority, even when asked to make an independent

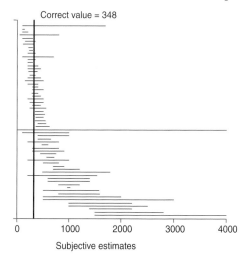

Figure 1.4. Subjective estimates for 90% 'confidence' intervals. Participants were asked to guess the number of beans in a jar, and to provide an interval around their guess such that they were 90% certain it would enclose the true value. This was taken to mean that they would accept a 9 to 1 bet that their interval enclosed the correct value.

judgement. This bias is closely related to *cognitive availability*, the tendency to judge the probability of an event by the ease with which examples are recalled. More vivid and memorable examples are overestimated. This implicit weighting gives them undue influence in subjective estimates.

1.5.5 Arbitrary risk tolerance

The perception of risk is linked tenuously to its magnitude and conse-quences. Other factors filter individual responses and mitigate tolerance, including:

- *Level of personal control.* When people feel they are in control of the situation, they will tolerate higher risks. People will tolerate higher risks of driving a car (when they are behind the steering wheel) than they will when riding in an airplane (when someone else steers). This is not the only factor determining acceptable risks for cars and planes, but it contributes.
- *Voluntary acceptance.* Related to the level of personal control, people will tolerate greater risks when they are given a choice, than they will when the risks are imposed.

- *Fear of the unknown*. People will not always tolerate greater risks just because they understand technical detail and causal mechanisms. Lack of experience reduces tolerance of risk. New technologies, in particular, are susceptible. Explanations of technical detail are rarely sufficient.
- *Uncertainty about the consequences*. If consequences are uncertain, people are less likely to tolerate a risk than if the consequences are known and relatively certain. Any new technology that includes uncertain consequences for subsequent generations is particularly difficult to sell.
- *Dreadfulness of the outcome*. Tolerance of risk is strongly motivated by how terrible the consequences appear. This is also known as the 'kill size' or the 'outrage factor', and is a primary concern of those trying to communicate risks to the public. People are particularly sensitive to large numbers of instantaneous deaths, impacts on 'innocent' people (especially children) and substantial involuntary impacts.
- *Equitability of distribution of the risk*. People are less likely to tolerate a risk that they bear alone or with a subset of society, than they are to tolerate one shared more broadly.
- *Visibility of the hazard*. Risk tolerance is influenced by the profile of potential hazards. Deaths in aeroplanes and nuclear power systems are much more newsworthy than deaths in cars and coal-fired power systems, although the latter are much more common. Media and other forms of information influence risk tolerance.

1.5.6 Race, religion, culture, gender

Many of the average responses to questions about risk by human populations can be explained by culture. Cultural differences contribute substantially to perceptions and acceptance, so that different social groups react differently when confronted by the same hazards (Rohrmann 1994, 1998, Slovic 1999). For example, gender and race are associated with apparent differences in attitudes to environmental risks and risk-taking behaviour in the United States (Kalof *et al.* 2002, Weber *et al.* 2002, Figure 1.5).

Similarly, on the whole, most Japanese believe they are worse than average drivers and need collision insurance, even though there is no law requiring it. In contrast, most American drivers believe they are better than average drivers and are required by law to have collision insurance. Hayakawa *et al.* (2000) explained the cross-national differences by

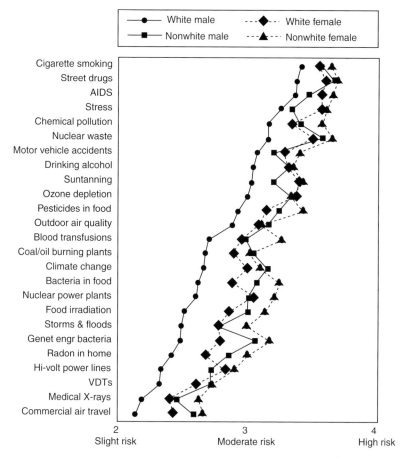

Figure 1.5. Mean risk-perception ratings for health hazards, differentiated by race and gender (after Slovic 1999; see also Kalof *et al.* 2002). The 'white-male effect' was caused by a subset of about 30% of the white male population who were relatively well educated, earned higher incomes and were politically conservative. Although they were identified *post hoc*, they were substantially and consistently more optimistic about health risks than most others in the population. The majority of the white male population (the other 70%) was not substantially different from the other groups.

different traffic environments: fatal accidents in Japan are much more likely to involve cars killing motorcyclists, bicyclists and pedestrians, whereas US fatalities usually involve two cars.

Wen-Qiang and Keller (1999) reported the attitudes of Americans and Chinese to risky situations. One involved a choice either to take a

1% chance of 100 people dying (and a 99% chance that no-one dies), or to take one individual out of 100 (say, by a lottery) and sacrifice them to ensure that the other 99 remain safe. The former circumstance was considered fairer by both Chinese and Americans, but most Chinese chose the latter course and most Americans chose the former. The Chinese population was more risk averse, seeking to avoid catastrophe. The authors speculated that the views reflected the greater importance attributed to collective actions in Chinese society.

Judgements are clouded by different value systems. For instance, Goklany (2001) argued that threats to human life and health should take precedence over threats to the environment, although there may be exceptions, depending on the nature, severity and extent of the threat. Others take a less anthropocentric view (see Brown 1996, MacLean 1996).

1.6 Discussion

The evidence is compelling that the psychological disabilities above are shared by us all, to varying degrees. We will return to a detailed examination of other cognitive attributes in Chapter 4, examining the ability of experts to contribute to risk assessments. In general, people are poor judges of risk. Risk analysts need to be aware that the people they work with and they, themselves, are susceptible to the vagaries of human perception. Cognitive biases are heightened by the politically charged and value-laden contexts of most risk assessments. The analyst should try to clarify information so that it is as free as possible of the subjective filters, preferences and values of those involved in its construction.

The extent of social and cognitive influences, taken together with the paucity of data in most circumstances, led Adams (1995) to conclude that '. . . risk may be viewed as culturally constructed, and context dependent. Risks can be changed, magnified, dramatised, or minimised, and so are open to social definition and construction. . . . Risks are culturally constructed because sufficient facts are unavailable. Cases of genuine uncertainty are far more common than are cases in which risk is quantifiable. We must proceed in the absence of agreed facts.' Arguments for the social construction of risk can be compelling. If they are right, technical views of risk assessment are misplaced. Credibility and trustworthiness may be more important than data.

Despite cognitive biases, people process new information systematically. They compare it to information they already have, and form approximations and rules of thumb to make decisions. Cognitive biases may be as much a problem of communication as of estimation (Trumbo and McComas 2003). We return to this theme in Chapters 4 and 12.

Risk assessments are invariably subject to distorting influences, perhaps more so than other types of scientific analysis, because of the public setting of many of the problems. Typically there is considerable pressure on risk analysts to produce reliable projections, to diffuse social tension. Unfortunately, analysts and the experts they employ are themselves susceptible to the same set of pressures. They cannot occupy the independent, objective ground that politicians and policy makers wish them to.

There are exceptions to the general rule of abysmally poor performance in assessing risks. The performances of professional bookmakers, weather forecasters and bridge players are substantially better than those of untrained or inexperienced people who attempt the same tasks (Morgan and Henrion 1990, Plous 1993). Yet we don't usually associate these groups with good predictions. They have a number of things in common. They practice. They make predictions on a routine basis and receive immediate unambiguous feedback on their performance. Consequently, they learn from their mistakes. Often, their judgements reflect on them personally when they get them wrong.

This book is devoted to making the rest of us behave more like weather forecasters, bookmakers and bridge players, but only as far as is reasonable. People who deal successfully with technical risks work in narrow domains with enviable replication and relatively little ambiguity and social pressure. The rest of us are doomed to working in circumstances that guarantee that we can never achieve the kind of reliability boasted by these groups. The book provides some help with the slippery issues that arise when statistical probabilities don't apply.

2 · *Kinds of uncertainty*

Uncertainty pervades the natural environment and obscures our view of it. To organize ideas about uncertainty, this book uses a taxonomy of uncertainty created by Regan *et al.* (2002a). At the highest level, it distinguishes between epistemic and linguistic uncertainty.

Epistemic uncertainty exists because of the limitations of measurement devices, insufficient data, extrapolations and interpolations, and variability over time or space. There is a fact, but we don't know it exactly. This is the domain of ordinary statistics and conventional scientific training.

Linguistic uncertainty, on the other hand, arises because natural language, including our scientific vocabulary, often is underspecific, ambiguous, vague, context dependent, or indeterminate. It is distinguished from epistemic uncertainty because it results from people using words differently or inexactly.

This chapter introduces the concepts of epistemic and linguistic uncertainty. It defines several subtypes within each broad category. It provides some examples of each and outlines the methods that may be best suited to treating them.

2.1 Epistemic uncertainty

Epistemic uncertainty reflects incomplete knowledge. It has several main types: measurement error, systematic error, natural variation, model uncertainty and subjective judgement. Each arises in different ways. A variety of well-known statistical methods are available to treat them (e.g. Sokal and Rohlf 1995).

2.1.1 Variability and incertitude

The terms 'variability' and 'incertitude' make a simple taxonomy of epistemic uncertainty worth describing because of its utility. Variability is naturally occurring, unpredictable change, differences in parameters attributable to 'true' heterogeneity or diversity in a population. Incertitude is lack of knowledge about parameters or models (including parameter and model uncertainty).

Incertitude usually can be reduced by collecting more and better data. Variability is better understood and more reliably estimated, but is not reduced by collecting additional data.

These words have been useful in routine applications of risk assessment (Hoffman and Hammonds 1994, Finkel 1995, US EPA 1997a,b, Kelly and Campbell 2000). While all parameters exhibit both characteristics to some extent, the distinction can help structure quantitative investigations of risk (Chapter 10) and influence priorities for collecting data (Hoffman and Hammonds 1994, Regan *et al.* 2003). A more detailed taxonomy of epistemic uncertainty follows.

2.1.2 Measurement error

Measuring equipment and observers are imperfect. Measurement error results in (apparently) random variation in a quantity. Repeated measurements vary about a mean. In the absence of other uncertainty, the relationship between the true quantity and the measured quantity depends on the number of measurements taken, the variation amongst them, the accuracy of the equipment and the skill of the observer. This type of uncertainty can be dealt with by applying statistical techniques to multiple measurements and reporting bounds such as confidence intervals. For example, various attempts have been made over time to estimate such fundamental constants as the speed of light and Avogadro's number. Theory tells us there is a fact, a single value for each constant. Any uncertainty derives from our inability to measure correctly.

It is clearly incorrect to believe that the confidence intervals associated with each set of measurements capture all uncertainty. For instance, the confidence intervals associated with reports of measurements of these physical constants rarely overlap the intervals associated with measurements made subsequently (Figures 2.1a,b). The values from one time to another include differences between experimental protocols, people and equipment as well as measurement error.

2.1.3 Systematic error

Systematic error occurs when measurements are biased. It is defined as the difference between the true value of a parameter and the value to which the mean of the measurements converges as sample sizes increase.

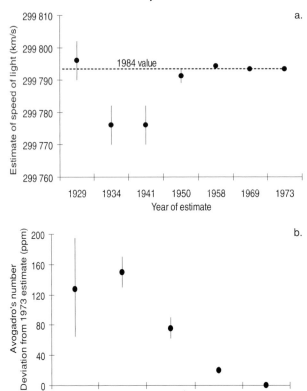

Figure 2.1a,b. a. Measurements of the speed of light together with reported uncertainties (confidence intervals) for the period 1929–73 (Henrion and Fischhoff 1986, in Morgan and Henrion 1990). b. Deviations of estimates of Avogadro's number from the 1973 value together with reported uncertainties (confidence intervals) for the period 1952–73 (Henrion and Fischhoff 1986, in Morgan and Henrion 1990). Avogadro's number is the number of atoms present in 12 g of the carbon-12 isotope (a mole of C_{12}).

Unlike measurement error, it is not (apparently) random. Measurements subject to systematic error alone do not vary about a true value. Systematic error can result from the deliberate judgement of a scientist to exclude (or include) data that ought not be excluded (or included). It can result from consistent, unintentional errors in calibrating equipment or recording measurements. For example, the large differences in measurements of physical variables among different experiments are the result of bias (Figures 2.1a,b).

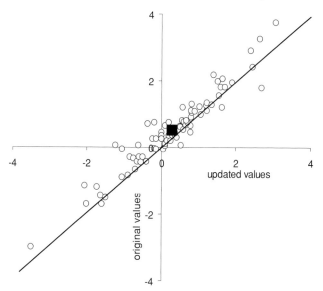

Figure 2.2. Bias in air pollution data resulting from a programming error (redrawn from Kaiser 2002). The vertical distance of the circles from the diagonal line shows how much the estimated death rate was wrong for each of 90 cities (not all 90 are shown here). The black square shows the location of the pooled estimates (the updated estimate was 0.27% per 10 $\mu g/m^3$ of PM_{10}, and the original was 0.41%).

The US EPA felt the consequences of bias when it was discovered in 2002 that studies linking deaths to very fine particulate matter (such as diesel exhaust) were biased because of a default setting in a computer program. When daily levels of soot rise, slightly more people die from heart and lung disease. In 1997, the EPA regulated by limiting permissible levels of the pollutants, in part based on the results of the statistical analysis. The correction resulted in a revision of the estimate from a 0.41% rise in mortality per 10 $\mu g/m^3$ of fine particles, to a 0.27% increase (Kaiser 2002; Figure 2.2). The standard could change as a result.

Some biases exist because observations are affected by theories (Chalmers 1999; termed the 'theory-ladenness of observation'). We observe, in effect, what our theory instructs us to observe. For example, a formal taxonomic description is a theory about a species. Museum or herbarium records of species often contribute to estimates of distributional ranges. If a taxon is not recognized as a species, frequently it is not collected, catalogued or recorded. If it is subsequently recognized as a good species because taxonomists re-evaluate it, estimates of its range will be biased because it has not been collected thoroughly or systematically. Burgman *et al.* (2000) observed this phenomenon when estimating range changes for *Acacia* species.

Sometimes, systematic error can be recognized and removed. A correction may be applied when the magnitude and direction of the bias are known, as in Figure 2.2. Such corrections underlie double-sampling methods (Gilbert 1987).

Systematic error, however, is notoriously difficult to recognize, except on theoretical grounds. The best strategies for dealing with it include diligent inspection of experimental procedures, comparison of estimates with scientific theory, independent studies, replication and careful attention to detail.

2.1.4 Natural variation

Natural variation is environmental change (with respect to time, space or other variables) that is difficult to predict. In ecology for instance, populations of plants and animals experience natural variation because individuals die while others are recruited into the population at rates that depend on food availability, weather conditions, predators, disease and so on. The true values of survival or birth rates change as a result of changes in independent (driving) variables. In ecotoxicology, the concentration of a chemical in a stream varies with temperature, particulate matter, flow rates and other variables, all of which fluctuate naturally.

The true values of parameters may be extraordinarily difficult to measure across the full range of temporal and spatial values (or other related variables). In some taxonomies, natural variation is considered to be irreducible. Many of the examples in this book are dominated by natural variation.

Genuine cases of inherent randomness are hard to find (see Regan *et al.* 2002a). Even random experiments like tossing coins and throwing dice are deterministic – it's just that we don't have enough information about the dynamic processes and initial conditions to make reliable estimates about the outcomes. For similar reasons, complex or chaotic systems such as ecosystems or weather patterns are not inherently random. Chaotic systems are entirely deterministic but are unpredictable unless the deterministic processes generating them and the relevant initial conditions are known (Sugihara *et al.* 1990).

2.1.5 Model uncertainty

Model uncertainty occurs because models are simple abstractions of reality. Models may be based on language, diagrams, flow charts, logic trees, mathematics or computer simulations, among others. They may be used as conceptual tools, to assist in the understanding of the structure of the system in question. They may be used to predict future events or to answer questions about a system.

Model uncertainty arises in two main ways. First, usually only variables and processes that are regarded as relevant and important for the purpose at hand are featured in the model. Texts that describe model building advise that models should be a compromise between the level of understanding of the system and the kinds of questions it is necessary to answer (e.g. Levins 1966).

Second, the choice of a way to represent observed processes involves further abstractions. For example, consider describing how populations change in time. Some individuals die, others reproduce, eat available food, encounter one another, disperse and so forth. Each of these activities affects population abundance and may be represented in a variety of mathematical forms. We may choose the logistic equation, based on a theory about intra-specific competition, to represent all important natural processes.

Similar kinds of uncertainty arise from curve fitting (including interpolation and extrapolation) and approximation. For instance, when a system of continuous differential equations is used to represent chemical diffusion, it is often necessary to employ a numerical algorithm to solve the equations. In such cases, a meta-model is constructed to make predictions and answer questions about the original model (see Regan *et al.* 2002a and references therein). The most appropriate mathematical, verbal, or diagrammatic representation of a process is a matter of opinion and understanding.

For example, several different curves would fit the two sets of data in Figure 2.3 equally well. It's rare to see more than one alternative explored.

Leon and Bonano (1998) predicted that model uncertainty will become a focus of legal challenges to risk assessments. Uncertainty associated with model selection is a difficult area because there are no accepted methods for treating it and there are no general guidelines for measuring the adequacy of a model for its intended use. Mostly, individual scientists use a given model because it is convenient or they are familiar with it.

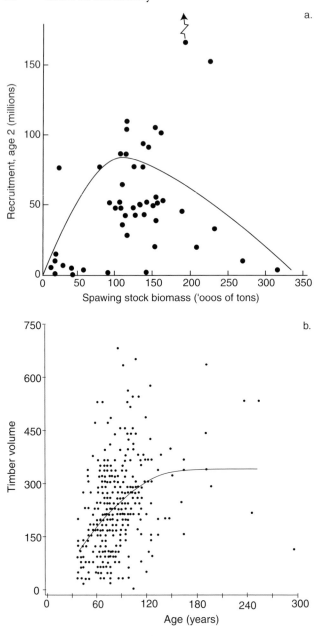

Figure 2.3a,b. Two examples of model uncertainty. a. Recruitment of two–year–old fish as a function of the biomass of adults. b. Volume (yield) versus age for plots in Canadian forests (from Dennis *et al.* 1985; see also Cushing 1995).

Model uncertainty is difficult to quantify and impossible to eliminate. It may be possible to find the 'best' available model for particular purposes (Morgan and Henrion 1990). It may be better to treat models as plausible alternatives, and to accumulate evidence over time that may winnow the selection (Hilborn and Mangel 1997).

The only way of determining how appropriate a model is for prediction is to validate it by comparing predictions with outcomes. It will be reliable if predictions are within an 'acceptable' margin of error. This is as much a social decision as a scientific one.

2.1.6 Subjective judgement

When there is a fact (for example, a true parameter or model structure), uncertainty due to subjective judgement results from the interpretation of data. Often, there are insufficient measurements to make reliable estimates. When there is an imperative to proceed with some course of action, the judgement of an expert may be used in place of or to interpret data. Often, the expert's judgement will be based on observations and experience, both of which constitute data. In all such assessments there is an element of uncertainty (see Chapter 4).

The standard way of dealing with this type of uncertainty is to assign a degree of belief in the form of a subjective probability. For instance, an expert might assign a probability of 0.9 to the event 'there will be an algal bloom in the lake this month'. This judgement might be based partly on frequency data. An expert's judgement will be uncertain but it should coincide with the results of data if they are available (see Chapter 1 and references in Regan *et al.* 2002a).

2.2 Linguistic uncertainty

Linguistic uncertainty arises because language is not exact. Vanackere (1999) gave the example 'it's raining', which may apply when it is pouring, or when a few drops sprinkle down. The need for precision depends on context. It is not practical to subdivide meaning to cope with all circumstances. We can't have a different word for all degrees of rain intensity. Language needs to be compact. Generality is necessary for communication.

Linguistic uncertainty can be classified into five distinct types: vagueness, context dependence, ambiguity, indeterminacy and underspecificity

(Regan *et al.* 2002a). Several arise in natural and scientific language, and can impact on environmental applications.

2.2.1 Vagueness

Vagueness arises because language permits borderline cases. Algae are always present in fresh water. At what point does an algal population become an 'algal bloom'? In practical terms, a bloom occurs when the density of algae is such that water appears green (or red), smells are emitted, oxygen is depleted, the water becomes toxic and fish die. The bloom is defined in terms of its consequences. A range of algal densities may lead to them, depending on light, temperature, flows, suspended sediments, phosphorus and nitrogen. The term is vague because it permits borderline cases, including slight and severe algal blooms. A spurt of algal growth that turns the water slightly green may not count as a bloom to a fishing fleet if it has no impact on fish. The vagueness of the term 'algal bloom' ensures that there is no straightforward answer to the question of how many algal blooms occur in a lake in a year.

Many other terms in routine use in science are vaguely defined. A 'tree crown' is a perfectly serviceable word that communicates a simple concept: the foliage clustered at the top of a tree. However, this is a poor operational definition because it does not precisely define a crown's boundary. Different people observing the same trees produce wildly different measurements of crown depth because they identify different points at which it begins and ends.

A common strategy for eliminating vagueness is to replace the intuitive meaning with a technical term defined by an arbitrary, sharp boundary. For example, the operational definition of an algal bloom in some jurisdictions is that the number of cells exceeds 5000 per ml (usually in a single surface sample). In another example of a sharp boundary, one of the IUCN's (1994, 2001) operational definitions of the term 'endangered species' is that there are fewer than 100 mature individuals.

One of the most serious problems with sharp boundaries is called the *Sorites Paradox* (see references in Regan *et al.* 2002a). If sharp boundaries are used to define classes in a continuum, very small changes delineate cases close to the boundaries. A taxon with 100 individuals is classified differently to a taxon with 99 individuals. This is at odds with the meaning of the original term 'endangered' in which a taxon would not receive different treatment on the basis of a difference of one individual. Similarly, if a 1-ml water sample contains 4999 cells, no action is taken, but if it contains 5001 cells, then weirs and dams may be opened, drinking water imported and domestic animals removed from the shores of the lake.

Vagueness permeates far too much of our language to hold any serious hope for its elimination. Consider the US Federal Register's definition of 'species viability':

A species consisting of self sustaining and interacting populations that are well distributed through the species' range. Self-sustaining populations are those that are sufficiently abundant and have sufficient diversity to display the array of life history strategies and forms to provide for their long-term persistence and adaptability over time. (*US Federal Register, undated*)

Many terms in this definition are vague, including 'self sustaining', 'well distributed' and 'sufficiently abundant'. Vagueness is widespread in environmental management and difficult to eliminate (note also that some terms in the definition above may also be subject to additional sources of linguistic uncertainty described below).

There are other, better ways to deal with vagueness than attempting to eliminate it, including fuzzy sets which use degrees of membership to deal with borderline cases (see Walley and DeCooman 2001, Regan *et al.* 2002a). Regan *et al.* (2000) suggested their use in classifying conservation status. For instance, a species that has declined by 70% may have partial membership (say 0.25 on a scale of 0 to 1) in the set of threatened species (Figure 2.4). Species with higher rates of decline should have a higher degree of membership.

2.2.2 Context dependence

Context dependence is uncertainty arising from a failure to specify the context in which a proposition is to be understood. For example, suppose an oil spill is said to be 'small'. Without specifying the context, the audience is left wondering whether the oil spill is small for an oil container ship or a dinghy, small for a port or the open ocean.

Note that 'small' is also vague but that vagueness and context dependence are quite separate issues. The vagueness persists after the context has been fixed. That is, even after we are told the context – 'small-for-a-cargo-vessel-in-a-port', say – there are still borderline cases. The way to deal with context dependence is to specify context unambiguously and correctly. While the solution is clear, this kind of uncertainty is pervasive.

For example, Cooke and Kraaikamp (2000) described a risk analysis for an accident at a rural train crossing. The analysts calculated the risk of

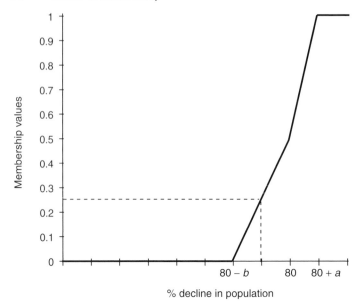

Figure 2.4. A fuzzy boundary for the classification of threatened species under IUCN (2001) protocols, based on the decline in the number of mature individuals (after Regan *et al.* 2000).

a signal failure anywhere in the country, rather than the correct context of a railway crossing in a wooded rural setting (where the chance of the event was much higher). They described the use of the wrong context as 'habitual sloppiness' and one of the 'familiar pitfalls in probabilistic reasoning'.

2.2.3 Ambiguity

Ambiguity is uncertainty arising from the fact that a word can have more than one meaning and it is not clear which meaning is intended. For example, the word 'cover' used routinely to describe vegetation can mean projective foliage cover (the proportion of the ground covered by a vertical projection of the aerial parts of plants, e.g. Kershaw 1964, p. 15), or crown cover (the area encompassed by a vertical projection of tree or shrub crown perimeters, e.g. Philip 1994, p. 132). The former definition excludes gaps within crowns from estimates of cover. The latter definition includes them, relying on the polygon formed by the outer edges of the crown.

Ambiguity is often confused with vagueness. However, the two types of uncertainty are quite distinct. The ambiguity in the word 'cover' does not give rise to borderline cases in the way 'endangered' does – there is nothing borderline between projective foliage cover and crown cover.

Shrader-Frechette (2001) documented the use of different terms to refer to nonindigenous species; she found examples of terms including 'alien', 'exotic', 'invasive', 'imported', 'weedy', 'introduced', 'nonnative', 'immigrant', 'colonizer' and 'naturalized'. In most cases the authors she reviewed defined the terms imprecisely or not at all. As a result, she pointed out, there is no common vocabulary in which to evaluate claims about invasions or the impacts of nonindigenous species. Careful definition would resolve at least part of the problem.

An important example is given by the term 'no observed effect concentration', the highest amount of a substance for which no significant effect was found (at $\alpha = 0.05$) in a statistical test between a treatment and a control (Chapter 7). The acronym is easily and frequently interpreted to mean a no *observable* effect level (Laskowski 1995). The lack of significance is taken to mean there is no effect, which is wrong (Chapter 11).

2.2.4 Underspecificity

Underspecificity occurs when there is unwanted generality. For example, a weather forecast may state that there is a 70% chance of rain. Possible interpretations are:

- rain during 70% of the day (so that if you emerge from a building at a random moment, you have a 70% chance of getting wet),
- rain over 70% of the area,
- 70% chance of at least some rain at least somewhere in the area,
- 70% chance of at least some rain at any point in the area,
- 70% chance of at least some rain at a particular point (the weather station).

Typically, weather forecasts assume the latter, although it is rarely stated and almost never understood by the people receiving the information. In most instances, the degree of specificity in the statement 'there is a 70% chance of rain' is sufficient to make a decision about how to dress, but it may be underspecific depending on a person's sensitivity to the outcome. Farmers, for instance, may prefer to have information reported using the second definition.

Underspecificity also arises where data could have been obtained but are no longer available. For example, in fauna and flora surveys it was once sufficient to provide imprecise locations such as 'inland Siberia' or 'north of Edmonton', or to provide no location information at all. Today, many locations are recorded with Global Positioning Systems accurate to a few metres. This distinction is important because opportunistic observations are used to assess temporal trends in species distributions. Underspecificity of location makes records unusable.

The best we can do is to provide the narrowest possible bounds on estimates given the data, and to make available all the information behind such statements. For example, a risk assessment of the ecological effects of a lobster fishery considered impacts on sea lions (IRC 2002; this example is explored in more depth in Chapter 6). Sea lion pups sometimes drown in lobster pots, attracted inside by the lobsters and unable to escape afterwards. The assessment concluded that an additional 3–4% mortality of sea lion pups was probably unimportant because sea lion pup mortalities between birth and five months of age 'vary naturally between 7% and 24%'.

This statement provided a range without a sample size, and therefore gave an incomplete picture of the magnitude of variation. Furthermore, it did not say if this variation was experienced over time within a single population or over several populations within a year. It may be some combination of the two. Lastly, it didn't say how old the pups were that died in lobster pots. With this information, it is impossible to assess the assertion that an additional 3–4% mortality is tolerable.

2.2.5 Indeterminacy

Uncertainty arises from indeterminacies in our theoretical terms. The problem is that the future usage of terms is not completely fixed by past usage. Some of our terms may not be ambiguous now, but they have the potential for ambiguity. This is sometimes called the open texture of language.

Uncertainty arising from this source is different from and more insidious than ambiguity. When we encounter a case of ambiguity we can always use other words. But in the case of theoretical indeterminacy, this can only be done after the fact when the new usage of the word comes into effect.

For example, the theoretical indeterminacy associated with the species concept emerges after taxonomic revisions. Taxonomic revisions of *Acacia browniana* in 1995 reclassified the taxon into *Acacia browniana*, *A. grisea*,

Table 2.1. *Sources of epistemic and linguistic uncertainty and their most appropriate general treatments (see Regan* et al. *2002a for a more complete treatment)*

Source of uncertainty	General treatments
Epistemic uncertainty	
Measurement error	Statistical techniques; intervals
Systematic error	Recognize and remove bias
Natural variation	Probability distributions; intervals
Inherent randomness	Probability distributions
Model uncertainty	Validation; revision of theory based on observation; model averaging; information-gap theory
Subjective judgement	Degrees of belief
Linguistic uncertainty	
Numerical vagueness	Sharp delineation; fuzzy sets
Non-numerical vagueness	Construct measures then treat as for numerical vagueness
Context dependence	Specify context
Ambiguity	Clarify meaning
Indeterminacy in theoretical terms	Make decisions about future usage
Underspecificity	Specify all available data

A. lateriticola, A. luteola, A. newbeyi and *A. subracemosa* (Burgman *et al.* 2000). Prior to 1995, taxonomists did not know that there was an ambiguity in the name *Acacia browniana* and would not have had the taxonomic concepts or the scientific vocabulary to treat the entities separately. This contributed to bias in range estimates.

Theoretical indeterminacy is dealt with by making decisions to anticipate the future usage of theoretical terms. It means that there must be tradeoffs; for example, between acquiring new collections and allocating resources to superficially redundant collections that may become important, if a taxon is revised. In general, anticipating future use is difficult because the future usage must be consistent with the past usage and it must be theoretically well motivated and fruitful.

2.3 Discussion

All risk assessments involve at least some elements of each of the sources of uncertainty outlined above. Table 2.1 summarizes the taxonomy. In any application, uncertainties from different sources compound, including uncertainties from epistemic and linguistic sources (e.g. Regan *et al.* 2001). The quality of a risk assessment is determined by the extent to which they are recognized and dealt with in a comprehensive, transparent and repeatable manner.

The categories presented here are not the only ones possible. Not all uncertainty can be neatly classified into one and only one of the categories defined by Regan *et al.* (2002a). But the framework is relatively complete and unambiguous. It identifies the elements of uncertainty that are encountered most routinely in environmental risk assessments.

Other taxonomies have been devised to provide different emphases. For instance, Morgan and Henrion (1990) noted linguistic uncertainty but did not distinguish between its different forms. Klir and Harmanec (1997) nominated vagueness and underspecificity (defined as above) and added conflict uncertainty to their taxonomy. Conflict arises when evidence is inconsistent and may be resolved by the weight of evidence (the legal model) or by collation of additional information. Ben-Haim (2001) used a slightly different interpretation of linguistic and epistemic uncertainty. Some taxonomies confuse epistemic and linguistic uncertainty. Others introduce redundant categories. Many in environmental science treat only epistemic uncertainty (e.g. Chesson 1978, Hilborn 1987, Shaffer 1987, Burgman *et al.* 1993, Shrader-Frechette 1996a).

No single method can treat all sources of uncertainty. Methods outlined in Table 2.1 to treat epistemic and linguistic uncertainty differ. For instance, none of the methods to deal with linguistic uncertainty are probabilistic, while many strategies for treating epistemic uncertainty are. Colyvan (2004) argued that probabilities are not suitable for some types of uncertainty. We will return to the question of how to deal with different kinds of uncertainty, and to applications that combine and propagate uncertainties, in later chapters.

Many methods for dealing with uncertainty are largely untried in environmental risk assessment. For instance, interval probabilities can be assigned to represent an expert's degree of belief where lower and upper bounds encompass the range of beliefs (Walley 1991, see Chapters 4 and 8). Treatments of subjective uncertainty may use interval arithmetic, imprecise probabilities, Dempster-Shafer belief functions and related tools (see Regan *et al.* 2002a), some of which are outlined in the chapters that follow.

Sources of epistemic uncertainty have attracted attention in environmental risk assessment mainly in relation to decision-making (e.g. Taylor

1995). In some cases, consideration of even a subset of the full spectrum of uncertainty has been considered to be debilitating. Beissinger and Westphal (1998) argued that there are insufficient data to parameterize models dealing with epistemic uncertainty for all but a handful of species. But there has been no comprehensive evaluation of the importance of the full spectrum of uncertainties for decision-making. Although linguistic uncertainty is common, it is often ignored and only epistemic uncertainty is considered. Clear understanding of the nature of uncertainty will assist analysts to use appropriate methods.

3 · Conventions and the risk management cycle

Sit in a room of engineers and listen as they plan a risk assessment, and you'll hear a view of how it should be done. Sit in a room of ecotoxicologists, epidemiologists or conservation biologists and you'll get quite a different view. So different, in fact, that you might think the four groups were discussing four different topics. The assumptions, methods for data collection, models, use of experts and so on would differ from group to group. Some would be wildly different, although it would be difficult to tell through the blankets of jargon.

Some terms are useful in setting up a broad, common context for risk assessment.

- A hazard is a situation that in particular circumstances could lead to harm (Royal Society 1983).
- Stressors are the elements of a system that precipitate an unwanted outcome.
- Environmental aspects are human activities, products or services (such as emissions, chemical handling and storage, road construction) that can interact with the environment (Zaunbrecher 1999).
- An environmental effect is any change to the environment, whether adverse or beneficial.

These definitions help to describe activities and their relationships to detrimental (and advantageous) environmental outcomes. Management goals embody social values and management aspirations. Stressors and environmental aspects have effects on (consequences for) valued ecological attributes, processes and services. The broad scope of environmental aspects makes them valuable in organizing ideas for assessing complex operations. Table 3.1 provides some examples of these terms for an assessment of an irrigated catchment.

In some engineering, toxicology and human health risk assessments, hazards are measured by accidents, deaths, illnesses or injuries. In most environmental assessments there are so many potential measurement

Table 3.1. *Examples of terms in the environmental effects hierarchy applied to a risk assessment of an irrigation system (after Hart et al. 2003)*

Term	Example
Management goal	Ecologically sustainable catchment
Hazard	Irrigation
Stressors	Nutrients, pesticides, flows
Effect/consequence	Algal bloom
Assessment endpoint	Phytoplankton community
Measurement endpoint	Algal abundance in surface water

endpoints that it is not possible to generate a complete list. In Table 3.1, the algal bloom is listed as a consequence. It may be more suitable to think about the algal bloom as a hazard, the probability of which is affected by irrigation practices. The consequences of the hazard may be fish kills, the deaths of farm animals or decreased revenue from tourism.

Often, in operational circumstances, it is easier to define hazards in terms of their consequences. If the hazard is an oil spill from a grounded ship, there are different kinds of ships, and different kinds and sizes of spills, even for a single type of ship. It may be easier to classify oil spills as small, moderate and large (taking care to define the kind of oil, the terms small, moderate and large, and the context of the spills), and then to estimate the likelihood of each consequence.

In many instances, there is no substantive reason for different disciplines to perform risk assessments differently. The conventions arose in response to the timing of the development of risk analyses, the kinds of problems that presented themselves and the social context in which solutions were sought.

The purpose of this chapter is to outline the features of different professional conventions (paradigms). It describes an environmental effects hierarchy proposed by Suter (1993). It proposes a broad, unifying framework, taking ideas from adaptive management and applying them to risk assessment. It echoes Walters (1986) who defined adaptive environmental assessment as a framework that states the objectives of management, reviews and assesses any current information, identifies knowledge gaps and uncertainties and constructs alternative management scenarios. The broad professional obligation of risk analysts to undertake 'complete' analyses may be achieved by adhering to the framework.

3.1 Risk assessments in different disciplines

Some disciplines such as the nuclear power industry began conducting risk assessments on data-rich issues with relatively well-understood processes, and were protected to some extent from public scrutiny. These assessments have a highly technical flavour and are driven by experts. Other risk assessments grew out of public concern over things such as the effects of contaminants on human health. The conventions in these domains involved a closer relationship with stakeholders and included attempts to influence public policy through advocacy.

3.1.1 Ecology

Fisheries biologists began to develop models in the early 1900s to estimate sustainable harvest. More recently, fisheries managers have used risk assessments together with adaptive management strategies (Holling 1978, Walters 1986). An emerging operational strategy uses detailed models of biological processes to challenge the performance of management options. Harwood and Stokes (2003) urged more general adoption of this system, pointing to the management of the southern gemfish fishery where management was based on a simple model, in preference to a detailed age-structured model. The simple model resulted in less variable catches and the data required to monitor the resource effectively were relatively cheap to collect (Punt and Smith 1999). We explore the utility and broad applicability of these ideas in Chapter 12.

Conservation biologists developed a preoccupation with estimating objective risk in the form of declines of population size. Conservation biology emerged in the early 1980s focused on risk assessments for threatened populations (see Beissinger 2002). Shaffer (1981) defined a 'minimum viable population' to be the smallest isolated population having a 99% chance of surviving 1000 years. This led to the development of population models to estimate extinction risk under the heading of 'population viability analysis' (Soulé 1987).

A pre-existing framework of population dynamic modelling was adapted for risk assessment that allowed analysts to make judgements about probabilities of population decline and loss (Burgman *et al.* 1993). These models evolved to include spatially explicit incidence models and frequency models, individual-based models, logic trees and related quantitative tools (see Beissinger and McCullough 2002, Reed *et al.* 2002).

Other tools that accommodate subjective parameter estimation and attitudes to risk (Akçakaya *et al.* 2000; see Ralls *et al.* 2002) have emerged to assist with risk ranking. Their use is in its infancy.

3.1.2 Engineering

Nuclear power safety provided strong motivation for risk assessment when the first plants were constructed in the 1950s. The people who developed the ideas for nuclear power were also responsible for developing numerical methods to support Monte Carlo simulations (Ulam 1976).

Engineering applications are dominated by quantitative risk assessments and technical philosophies that assume probabilistic interpretations of risk (Kumamoto and Henley 1996, Stewart and Melchers 1997). They use probability calculus and explicit models to estimate risk as the probability of an adverse event and its consequences over time (Kaplan 1997, Hayes 1997, 2002a).

Petroleum geologists embarked on quantitative risk analyses in the early 1970s (Capen 1976). They share the probabilistic view of risk held by the nuclear industry (e.g. Vose 1996) and we explore some of these methods in Chapters 8 and 10. Usually, risk assessments are conducted by specialists independently of stakeholders. Experts are viewed as unbiased interpreters of facts. Nuclear engineers and petroleum geologists support their empirical, frequentist view of probability with extensive 'failure rate' data bases and accident statistics.

Subjective risk ranking exercises described in detail in Chapter 6 are one of the most common forms of environmental risk assessments for engineering, mining and other industrial settings. Estimates of likelihoods and consequences are classified into one of a few categories. These generate a small number of risk classes that are used to guide decisions and set priorities for mitigation and remediation. This approach has been approved by institutions as diverse as the British Society of Actuaries, The Royal Society for Civil Engineering (ICE/FIA 1998) and Standards Australia (1999). It has its roots in methods developed by the aerospace industry in the USA during the 1960s. Current applications have a number of substantial flaws but, fortunately, some at least can be fixed (Chapter 6).

3.1.3 Ecotoxicology

Ecotoxicology took a different turn. Ecotoxicologists estimate the intensity, frequency, duration and extent of exposure of populations (human and others) to chemical, physical or biological agents (Suter 1995, Davies 1996). In ecological risk assessments, conceptual models are developed around the concepts of stressors (things that may cause harm) and receptors (elements of ecosystems at risk). As formulated by the US EPA, ecological risk assessment is a broad framework in which communication and management are treated as separate processes from analysing risks (Figure 3.1).

The ecotoxicology paradigm (Chapter 7) matured into the formal process represented in Figure 3.1 during the 1970s and 1980s. It includes standard laboratory tests, equations, default values and conventions for dealing with some uncertainties. For someone wanting to sell or use a chemical, for instance, compliance with regulations involves following a set of stylized procedures, simple conceptual models, laboratory results and arithmetic calculations. Assessments are supported by data bases of toxic substances and their effects on standard test organisms (Calow and Forbes 2003).

This paradigm is influential because it is deeply embedded in many government authorities around the world devoted to 'environmental protection' (e.g. US NRC 1983; see Kammen and Hassenzahl 1999). At least some of its mystique derives from its arcane, specialized jargon and from the heavy use of acronyms.

3.1.4 Public health

Physicians' judgements about individual patients are exercises in risk assessment that rely on mental models, sometimes supported by statistical evidence and formal decision tools. Epidemiologists focus on risk assessments at the level of populations.

In the USA, health risk assessments grew out of food, drug, pesticide and pollution legislation. In the United Kingdom, health risk assessments grew out of the need to set occupational exposure limits, mitigate floods and manage safety in the offshore oil industry (Pollard *et al.* 2002).

The public health risk paradigm shares a number of features with ecotoxicology, including a focus on exposure pathways, contaminants, mortality and morbidity. For example, like ecotoxicology, international benchmarks for food intake of pesticide residues are guided by standard equations and parameter values (Crossley 2000).

Public health risk assessments use different regulatory mechanisms (Byrd and Cothern 2000). In the USA, the FDA accomplishes regulation by publishing lists of 'allowed' substances and tolerances. Foods containing

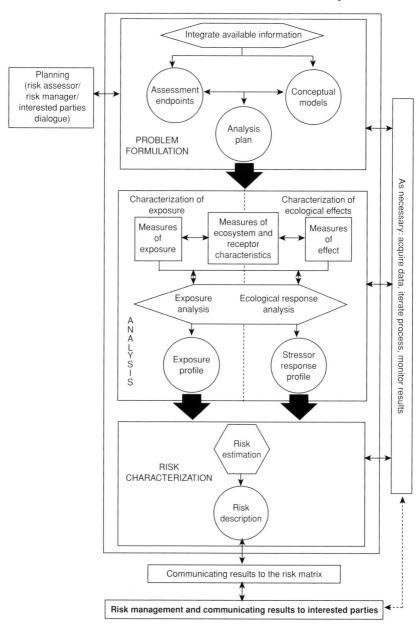

Figure 3.1. The US EPA's ecological risk assessment framework (from US EPA 1997a,b).

unlisted substances are adulterated by definition. This contrasts with the EPA, which publishes lists of maximum permitted exposures to substances based on ecotoxicological tests (Chapter 7).

Public health risk assessments also operate in a unique social context, typically involving a closer relationship with stakeholders than do other disciplines (Beer 2003). Assessments include problem definition, elicitation of risk factors, intervention and advocacy. Public health authorities in most countries are empowered to act to protect public health in ways that are not available in other domains of risk assessment (Byrd and Cothern 2000). In particular, human health risk assessment differs because its proponents are advocates for public health. They may intervene to influence public policy, something that engineers and ecotoxicologists rarely claim to attempt.

3.1.5 Economics

Economics was perhaps the first profession to embrace risk assessment. The stock market provides immediate feedback and unambiguous interpretation of outcome, an ideal template for the development of ideas about risk. Insurance actuaries began to develop ideas about risk in the 1600s and 1700s. Modern economics had the benefit of some of the best thinkers in this area, including Keynes (1921, 1936), Arrow (1971) and Thaler (1991). Today, analyses of acquisition and investment risk, exposure, hedging and risk-adjusted preferences are commonplace (e.g. Bell and Schleifer 1995).

Like engineers, economists generally subscribe to a frequency interpretation of probability when it comes to things like forecasting stock movements. But Bernstein (1996) noted that factors that temper the thinking of environmental risk analysts, such as context and nonmonetary values, are also important to business risk assessments. Keynes (1921) argued that some knowledge is not amenable to exact measurement or probabilistic description, that there are instances of uncertainty where there is no scientific basis on which to calculate a probability. Even when 'there is a relation between the evidence and the event being considered, . . . it is not necessarily measurable' (Keynes, 1921, p. 3). The value of what is to be gained or lost depends on the perspectives of the people the decision affects.

Perceptions, nonmonetary values and risk aversion cloud otherwise 'rational' economic judgements. As noted in Chapter 1, utilities are benefits discounted by the chance that they will occur. Utilities are inherently subjective. In addition, the parameters we need to make decisions are unavailable. Subjective judgements of uncertainty are commonplace and the tension between the different concepts of probability is always present.

3.1.6 Attributes of risk assessments

The list above is not complete. For instance, international quarantine services use risk assessments routinely. They assess, among other things, human health, animal and plant diseases and invasive species. They use an empirical, frequentist view of probability. Like many engineering, epidemiological and ecotoxicological applications, their assessments are supported by large data bases and conventional protocols (Morley 1993, Lonsdale and Smith 2001).

The evolution of conventions has a lasting effect on how risk analyses are conducted. Some include officially approved model assumptions and parameter values. The EPA in the USA has set mostly conservative defaults. In contrast, the US Nuclear Regulatory Commission defaults generally are close to the mean of parameter estimates (Bier 2004). These differences reflect different attitudes to risk adopted by different professions.

The approaches share a common belief in the epistemic nature of risk: there is a fact and the job of the risk analyst is to estimate it. An opposing philosophy is that risks are subjective and context dependent, and cannot be analysed formally (Jasanoff 1993, Adams 1995; Chapter 2). Adams (1995) argued that risk assessments always involve decisions about values and preferences, and are coloured by the personal experiences and prospects of the individuals conducting the assessments. This view objects to the artificial separation of risk analyst and manager/decision-maker present in frameworks such as Figure 3.1 (Kammen and Hassenzahl 1999). Instead of using technical analysis, risk assessments are conducted through stakeholder engagement, elicitation of preferences and values, and consensus building (e.g. Stirling 1999).

Adams may be right. Certainly the importance of psychology and context outlined in Chapter 1 provide strong support. The answers generated by quantitative risk analysts may be little more than smoke and mirrors, reflecting the personal prejudices and stakes of those conducting the analysis. It is likely that at least some of the problems alluded to by Adams will affect all risk analyses. The extent to which they are felt will depend on the nature of the problem, the amount and quality of data and understanding, the personal outcomes for those involved in the analysis, and the degree to which their predispositions can be made apparent.

Methods for dealing with qualitative and subjective uncertainties are introduced in Chapters 4, 5 and 6. For the majority of socially based risk assessments, risks are not usually amenable to probabilistic treatment. Risk sociologists apply a suite of different methods including multicriteria methods outlined in Chapter 12, and consensus building and stakeholder assessment outlined in Chapter 4.

3.2 A common context for environmental risk assessment

Suter (1993) created a system of thinking to help people to define environmental hazards and their consequences. He defined endpoints as an expression of the values that we want to protect. There are three broad kinds:

1. Management goals are statements that embody broad objectives, things such as clean water or a healthy ecosystem. They are defined in terms of goals that are both ambiguous and vague, but they carry with them a clear social mandate.
2. Assessment endpoints translate the management goals into a conceptual model that satisfies social objectives. Clean water may be water that can be consumed and bathed in by people. A healthy ecosystem may be one in which all ecological stages are represented, all natural ecological processes continue to operate, and populations of important plants and animals persist. But even assessment endpoints cannot be measured.
3. Measurement endpoints are things that we can actually measure. They are operational definitions of assessment endpoints that are, in turn, conceptual representations of management goals. Thus, measurement endpoints for fresh water may include counts of *Eschericia coli* and the concentration of salt. Measurement endpoints of a healthy ecosystem may be the abundance of several important species (threatened species or game species), and the prevalence of diseases and invasive species (e.g. Table 3.1).

3.2.1 Selecting endpoints

Selecting assessment and measurement endpoints in environmental systems is difficult because natural systems are complex and there are many potential candidates. Selection also needs to take heed of social context. Suter (1993) suggested that endpoints should be:

- biologically relevant,
- important to society,
- unambiguously defined,
- operationally feasible,
- predictable and measurable, and
- susceptible to the hazard(s).

If endpoints are to provide credible evidence that management goals have been satisfied, then they must also satisfy the expectations of policy makers, politicians and others with a stake in the use of the environment (Suter 1993).

The process of identifying measurement endpoints presupposes that a conceptual model of the system has been created and hazards have been identified. Conceptual model building and systematic hazard identification are explored in Chapter 5.

3.2.2 Targeting risk assessments: ecosystems and indicators

Ecosystem services are the processes through which natural ecosystems sustain human life. They include goods such as food, fuels and pharmaceuticals, and services such as the maintenance of biodiversity and waste assimilation (Daily 1997, 2000). In general, management goals have an ecosystem process or service in mind.

Some measurement endpoints sample ecosystem attributes (examples include algal blooms, river flow rates and the number of species in a mud core). Others sample the hazard itself (such as the toxic plume in groundwater) or a single environmental feature (such as the abundance of a threatened species). This book argues that monitoring is an immutable part of a complete risk assessment. Without it, it is impossible to revise ideas or improve estimates and decisions.

Risk assessments may be conducted, at least in theory, on populations, single species, multiple species, communities, ecological processes, or natural resources. Measures of impact may include changes in:

- genetic variability within and between populations,
- relative abundance of a stage,
- relative abundance of a species,
- numbers of species and their relative abundances,

- the abundances of functionally different kinds of organisms,
- species turnover from place to place in the landscape within a community,
- the value or magnitude of ecosystem services,
- species turnover among communities, and
- the number, size and spatial distribution of communities (Weaver 1995).

'Indicators' may be selected as measurement endpoints. Indicators are biological entities whose interactions with an ecosystem make them especially informative about communities and ecosystem processes. They should reflect changes in ecosystem dynamics at multiple spatial, temporal and organizational scales (Landres *et al.* 1988, Noss 1990).

Very often, species are chosen to act as indicators. It is usually assumed implicitly that these species act as 'umbrellas' in the sense that conserving them will result in the protection of numerous other species (Noss 1990). Such claims are rarely verified in specific applications. Single indicators are unlikely to reflect a full range of possible responses to ecological perturbations (Landres *et al.* 1988). Single attributes can be monitored and managed to the detriment of others. In particular, a single species focus may be detrimental to other species and ecosystem processes (Simberloff 1998). Indicators should come from a broad taxonomic spectrum and a broad range of spatial, temporal and organizational scales (Simberloff and Dayan 1991). Unfortunately, operational and budget constraints rarely allow such breadth.

The choice of measures should depend on the motivation for the risk assessment. More often it is determined by operational constraints. Impediments to comprehensive ecosystem risk assessments include (Suter 1993):

- expense and time,
- a lack of general standards for conducting ecosystem-level risk assessments,
- little experience in laboratories,
- problems with definition of endpoints (such as the list above),
- effects of variations in test conditions are uncontrolled,
- generally, replicates are few, variance among replicates is high, and replicates tend to diverge over time, and
- the complexity of most ecosystems defies parameterization within the time frames and with the budgets available in most circumstances.

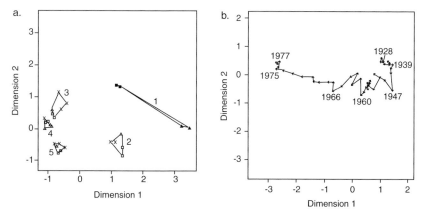

Figure 3.2a,b. Trajectories in the relative abundances of fauna shown in a reduced dimensional space using NMDS (from Philippi *et al.* 1998). a. Samples of zooplankton collected on five different days in mid 1986 from a cooling reservoir in the Savannah River, South Carolina. Lines enclose all points sampled on the same date. Site A (cross), Site B (square), Site C (triangle). b. Changes in the composition of land-bird communities over time in Stockholm, from 1928 to 1979.

A few attempts have been made to establish risk assessment protocols using assemblages of plants or animals. Baseline conditions are characterized by sampling biota either prior to impact or at a set of 'reference' sites. Changes in the biota are monitored over time and compared to the baseline or the reference sites. The changes can be visualized by plotting the multivariate data in a reduced space (using tools such as non–metric multidimensional scaling (NMDS, e.g. Figure 3.2)). However, it is difficult to judge how much of a deviation from the baseline condition is acceptable.

Figure 3.2a shows samples of zooplankton in a cooling reservoir in the Savannah River, South Carolina. Pairs of samples were taken from three sites five times in 1986. Site A was in the main channel of the reservoir, whereas sites B and C were in protected coves. The first sample, indicated by the solid shapes, was taken while the reactor was operating, heating the channel water. Water temperature in the coves remained about 15 °C cooler than the channel while the reactor ran, but was the same as the channel otherwise. Philippi *et al.* (1998) noted that pairs of replicates lay close together and samples taken from the same time formed clusters. Philippi *et al.* (1998) interpreted Figure 3.2b as showing three periods

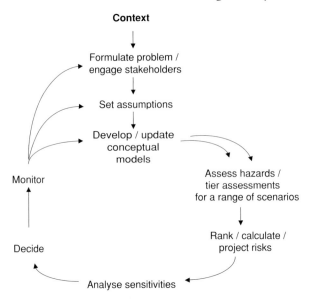

Figure 3.3. The risk management cycle.

of relatively stable species assemblages (1928–44, 1951–60, and 1975–79). NMDS may be used to visualize ecological change. Other methods, such as control charts and explicit ecosystem models, are discussed in later chapters.

3.3 The risk management cycle

A complete risk assessment involves learning. The environmental risk management cycle includes problem formulation, hazard identification, risk analysis, sensitivity analysis, decision-making, monitoring, communication and updating (Figure 3.3). Completion of the cycle ensures that ideas will be validated and revised or reinforced as knowledge accumulates, so that environmental decisions improve over time.

The details of the various steps in this process are explored in later chapters. The same general framework was outlined by Harwood (2000), who emphasized the need in environmental risk assessments to account for different kinds of uncertainty. Stakeholders generate scenarios. Monitoring data are used to update models and predictions. The feedback should be used to revise estimates of the relative costs of different kinds of decisions.

The iterative nature of risk assessment and decision-making emphasized in Figure 3.3 has a number of features in common with the US EPA's framework (Figure 3.1). Both outline a process that defines the scope of the problem and provides for a feedback between prediction and model building that relies on monitoring.

The risk management cycle differs from the US EPA's framework by emphasizing the importance of iteration, the critical roles played by monitoring and updating, and the importance of involving stakeholders closely in the risk analysis. Communication of results is not treated as a separate issue, but is part of the cycle itself. As we'll see later, the involvement of interested parties can and perhaps always should be continuous.

The remainder of this section sketches the main elements of the risk management cycle. Each is treated from various perspectives and in much greater detail in subsequent chapters.

The risk management cycle in Figure 3.3 is modified from adaptive management proposed and developed by Holling (1978) and Walters (1986). It differs from other schemes such as that of the US EPA by elevating the importance of monitoring and the involvement of stakeholders (see Beer 2003). These topics are treated in detail in Chapters 4 and 12.

3.3.1   Context: who pays and what do they want?

Context, the setting of the problem at hand, influences outcomes. The people who control context can guide the outcomes of risk assessments. To a large extent, the context determines the focus of the assessment, the kinds of endpoints used and the time frames considered. Context determines to whom the assessment is answerable, who is responsible for conducting it, who participates and who determines if the detail, solutions and plans for review are acceptable. The budget and deadlines determine the extent to which data can be collated and synthesized.

Control is usually in the hands of the people paying for the exercise. For instance, they may be able to influence the choice of experts. They may express preferences for the presence of specific stakeholders. They may have direct or indirect control over access and use of data, and so on. Many issues may be resolved, or at least understood by stakeholders, if goals and their links to human values are stated clearly.

Some resource companies conduct in-house risk assessments to support investments in new projects. This context influences the kinds of analyses deemed acceptable. In many countries, risk assessments are a legislated requirement preceding the development or market release of a new chemical. Regulations sometimes specify the details of procedures for conducting an assessment.

Nevertheless, the people paying for the exercise retain substantial control over context and problem formulation. This fundamental issue is rarely addressed explicitly in risk assessments. Instead, assessments are assumed to be a kind of impregnable, objective scientific process. We will see that the context and human frailties of the participants are such that this cannot be true. Analytical methods go some way towards fixing the problem if used to assist stakeholders to examine their ideas, but they do not eliminate it entirely.

3.3.2 Problem formulation

Problem formulation answers the questions: 'What is the scope of the problem?' and 'What do we need to know to make a decision?'. The first step in problem formulation establishes the scope of the problem, including the ecological, social, geographic, temporal and political limits. These things define the kinds of solutions we seek. They identify who and what is affected by decisions.

Defining scope is intimately related to hazard identification and the construction of the conceptual model. The problem boundaries determine what may be considered when answering the question: 'What can go wrong?'.

Problem formulation includes the creation of a set of alternative actions, a task that can only be achieved reliably by involving all those potentially affected by any decisions. Prior to specifying the details of conceptual models, the analyst and other participants should scope alternative management options, identifying as many as possible. Chapter 5 provides an example in which the identification of a hazard (salt fallout from a cooling tower) led the design team to create an alternative solution (a freshwater cooling tower) that was less costly than the original proposal and that eliminated the hazard entirely. Scenario analysis is a more formal approach to creating alternatives and it is outlined in Chapter 12.

When risk assessments fail, it is often because no-one thought of a particular possibility, event or process. Avoiding surprises depends on creating a complete list of hazards. Because of the importance of this step, Chapter 4 examines the range of people who may be considered stakeholders and Chapter 5 devotes some energy to outlining methods for defining and compiling hazards.

The second step in problem formulation is deciding on the form and type of risk assessment. Different approaches embody different assumptions and make different demands for data and expertise. Social and scientific context limit what kind of risk assessment is acceptable. Time and money devoted to the task affect the collection of data, the number of stakeholders and the kinds of analyses that can be attempted.

3.3.3 Conceptual modelling and hazard assessment

Conceptual models define components in a system. They identify input and output, flows, cycles, system boundaries, causal links and so on. They can be verbal models, diagrams, logical trees, or sets of mathematical equations. The level of detail is determined by context and problem formulation. The model should be sufficiently detailed to provide answers to relevant questions, and no more detailed. Hazard identification and assessment depend on the construction of an appropriately detailed model. For example, if interest lies in the fate of toxicants, then the analyst may decide to explore endpoints in different parts of an exposed mammalian body (Figure 3.4). This model attempts to encapsulate processes by which the contaminant enters, is absorbed into, distributed within and excreted from the body.

Hazards are possibilities, without probabilities. They are all those things that might happen, without saying how likely they are to happen. A hazard is often seen as an intrinsic potential to harm persons, property or the environment (Beer and Ziolkowski 1995, Potter 1996). The term is defined more usefully as a function of both intrinsic properties and circumstances (Royal Society 1983), a state that may result in an undesired outcome (Suter 1995).

Hazard identification involves compiling a list of hazards associated with a proposal or activity (Suter 1995, Potter 1996). Hazard assessment involves describing the potential consequences of each hazard and the nature of their undesirable effects (Beer and Ziolkowski 1995).

Sometimes, to solve a problem it is sufficient simply to document the kinds of events that would flow if a hazard eventuated. If the prospect of the event is unacceptable, the process designers may then take the trouble to re-engineer the system to eliminate the hazard entirely.

Risk assessment frequently is a tiered or phased process, moving from simple analyses and 'conservative' assumptions that overestimate risk, to site-specific and more detailed characterizations, and less conservative

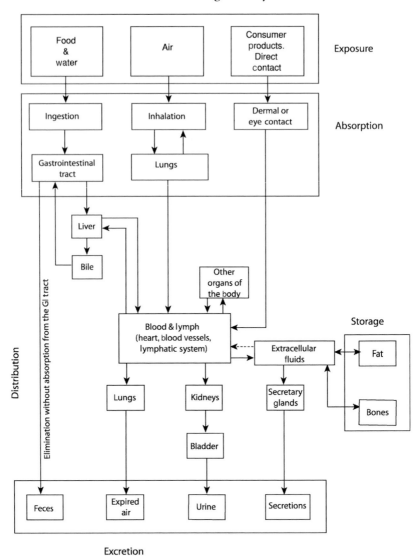

Figure 3.4. Conceptual model of components of a mammal that may be important in understanding the fate of a toxin, and that may assist the selection of measurement endpoints (from Klaassen 1996).

assumptions. The rationale is that if a hazard results in an acceptable risk, even if assumptions are pessimistic, it is safe to forgo further detailed consideration. The hazard assessment phase is linked to the risk analysis phase by assessing the appropriate tier for each hazard.

3.3.4 The risk analysis phase

The conversion of a hazard assessment to a risk assessment involves a probabilistic element: the likelihood (or relative likelihood) of a hazard having an effect. The paradigms for risk assessment outlined above have in common the processes of problem formulation, conceptual modelling and hazard identification and assessment. Differences emerge primarily in the ways in which risks are estimated.

As a result, there are important and sometimes arbitrary differences between the ways in which risk analyses are done. Some are driven by convention, others by the nature of the problem and the kinds of data available. Some reflect philosophical positions. Chapters 6–12 document and explore methods for risk estimation at length.

In some circumstances, particularly when dealing with relatively well-understood problems, it is worthwhile expressing the conceptual model in a formal mathematical framework. This involves rendering the concepts in mathematical notation, which guarantees internal logical consistency. Then, the analyst has to estimate parameters from data or expert judgement.

From the system of equations and parameters, the analyst may estimate frequencies and consequences of individual hazards. This approach offers the benefits of formal calibration and validation. It facilitates exploration of interactions between hazards. Furthermore, sensitivity analysis is a well-developed concept in this domain, and the analyses that support decision-making are relatively easy to conduct. However, they come at a cost of additional technical skills, time and resources.

These methods appear as though they require more data than other approaches. In fact they simply make apparent the absence of data and the necessity for simplifying assumptions. In other approaches, these uncertainties are submerged in assumptions.

3.3.5 Sensitivity and decision-making

Risk assessments depend on a set of assumptions and simplifications. Before a decision can be made, it's important to know how sensitive the predictions are to model assumptions. Sensitivity can be measured by manipulating the model and evaluating changes in predictions. This can be done for qualitative and conceptual models, as well as for detailed mathematical models.

If the assumptions make a substantial difference to expectations, then it is part of the risk assessor's job to evaluate whether the decision would change, given different, plausible choices. If nothing of importance changes as a result of the sensitivity analysis, then the decision may be considered robust to the uncertainties in model structure, parameters and/or the judgements of experts.

Risk assessments usually are performed to support an action or decision (act/don't act, allow/disallow a proposal). A model that focuses on the central tendency of the system provides an answer to what is most likely to happen. But it does not provide a feel for possibilities, for the kinds of things that might go wrong. Numerical models to assist decision-making in the presence of risk are outlined in Chapters 10 and 11.

3.3.6 Monitoring, updating and communicating

Once a hazard has been identified as posing an unacceptable risk, the next step in the risk management cycle involves eliminating or reducing the risk. Certainly, the analyst is responsible for ensuring that the risk is communicated clearly to those people affected by the risk.

Risk assessments unravel when monitoring is omitted. A primary weakness of risk assessments is that when the sensitivity analysis is finished, the task is viewed as complete. Monitoring should provide information that allows managers to react to changes in the system, to evaluate assumptions and uncertainties, and to modify models as knowledge grows. Methods for achieving these important goals are explored in Chapter 11.

3.4 Discussion

The risk management cycle emerged in fisheries management (e.g. Punt and Hilborn 1997). We will explore applications of risk management in Chapter 12 that involve stakeholders and scientists in analytical risk assessment.

Most environmental decisions are set in socially charged contexts. People stand to gain or lose substantially. Arguments are clouded by linguistic ambiguity, vagueness and underspecificity to which analysts themselves are susceptible. Prejudice gets in the way of constructive discussion. A transparent framework helps to relieve these impediments.

In my experience, many disagreements among stakeholders are resolved by seeing clearly what the other participants want, and why they want it. Risk assessments can have their greatest utility in meeting these challenges. They work because they are logically robust and relatively free from linguistic ambiguity. They are not necessarily any closer to the truth than purely subjective evaluations. But they have the potential, if properly managed, to communicate all the dimensions of an environmental problem to all participants. They may do so in a way that is internally consistent and transparent, serving the needs of communication, given appropriate skills in the analyst.

Risk analysts sometimes believe themselves to be technical specialists, closeted with the data and a computer, providing results to managers and other interested people. This book takes the view that analysts are translators of ideas and perceptions. Successful implementation, then, relies on a range of people to diffuse the imprint of the analyst and to counteract the prejudices of expert participants.

Risk assessment is just as important as a kind of social grease as it is an instrument of technical analysis. It may provide a focus for people who disagree to define what it is that they agree about and where they differ substantively and ethically. The process illuminates the thinking of people who advocate different solutions. Thus, risk analysts need substantive knowledge in the domain in which they work (earth science, engineering, ecology, chemistry and so on), analytical skills that are sufficient to select and use the right tool for the task at hand, and skills to facilitate the process and make technical detail accessible.

4 · Experts, stakeholders and elicitation

Environmental risk assessments almost always depend on expert judgement. In risk assessments, it is important to consider what the stakeholders and experts stand to gain or lose. It is also wise to consider what the person performing the analysis stands to gain or lose.

For example, there was a terrible accident in Thredbo, Australia in 1997 in which part of a hill subsided, demolishing two ski lodges and killing 18 people (Figure 4.1). Some of the people living in lodges adjacent to those that subsided assessed the risk and decided to remain. In doing so, they disagreed with an assessment by consulting geologists and engineers that concluded many lodges (including theirs) were at 'high risk' of slippage because of swamp deposits, high water tables and significant surface water flow. The report went on to say that sites at 'very high risk' showed evidence of soil creep and signs of past instability.

Consider the positions of the people for the period between the accident and the completion of engineering works. Lodge owners lose income unnecessarily if they decide the site is unsafe and close their lodges, and it turns out to be safe. The consulting engineers damage their professional reputation and incur substantial costs if they conclude the site is safe, and later it subsides. For both parties, wrong assessments result in considerable personal costs. The consulting engineers cannot benefit substantially from an assessment that concludes the site is safe. The lodge owner may. That is, lodge owners and consulting engineers will have applied different utilities when they made decisions.

In the period following the accident, few people would accept an invitation to stay in the adjacent lodges. In the climate of shock following the accident, the benefits (of a holiday) were not worth the perceived risk (of dying). The tone of the newspaper's headlines suggests that the newspaper accepted the experts' arguments.

The process of the landslip is deterministic, but our knowledge is such that we cannot adequately describe it. The remedy is to cast the problem as a probabilistic one. It turned out that people who stayed on made

Thredbo lodges at risk

· **New landslide fear**
· **Call for urgent repairs**

By ADRIAN ROLLINS and GERVASE GREEN

Parts of Thredbo alpine village remain vulnerable to devastating landslides, with at least 30 lodges at high or very high risk, an engineers' study has found.

Five months after a landslide swept away the Carinya and Bimbadeen lodges, killing 18 people, the study concluded that extensive engineering works were needed around the village to eliminate the possibility of another disaster.

Figure 4.1. Excerpt from a front-page newspaper report on the risks of landslip faced by mountain resort lodges, recounting details of an engineering report (*The Age*, Melbourne, 31 December 1997).

the right prediction. In hindsight, the remaining lodges were perfectly safe during the period following the event. This is not an argument to disregard the advice of engineers. It is a starting point to examine the reliability of expert judgement.

Crawford–Brown (1999) defined five categories of scientific evidence for risk assessments:

1. *Direct empirical evidence*: direct experimental observation of cause and effect, probability or frequency.
2. *Extrapolation*: observations outside the range at hand.
3. *Correlation*: statistical associations between measures.
4. *Theory-based inference*: relationships and causal mechanisms inferred from understanding of physical or ecological principles.
5. *Expert judgement*.

The direct empirical evidence, extrapolations and theories for an assessment are almost never available. In their place, we turn to experts. Experts can use frequency concepts of probability to specify point estimates, intervals or statistical distributions without the use of formal mathematical tools or measurements. Experts can also offer subjective beliefs.

Their judgements can be used in a range of qualitative or quantitative analysis.

Kerr (1996) reported an expert group elicitation procedure to estimate radioactive waste risks. The question was: 'How likely is a volcanic eruption to rupture the repository at Yucca Mountain, Nevada, which is designed to keep thousands of tonnes of high-grade radioactive waste undisturbed for 10 000 years?' The process included:

- selecting 10 experts,
- conducting workshops and field trips to provide background and context,
- face-to-face individual interviews to elicit location, frequencies and uncertainties of eruptions,
- inclusion of estimates in multiple alternative models of volcanic processes, and
- using the resulting values to predict the probability that a magma conduit would cut through the repository.

The result was an estimate of 1/10 000, ranging from 5/10 000 to 5/1 000 000. Kerr (1996, p. 913) claimed the result '... cut through the miasma of scientific discord about volcanoes around Yucca Mountain ...'. This optimism was not shared by Shrader-Frechette (1996a,b). We shall see below that there are solid grounds for her scepticism.

Morgan and Henrion (1990) expressed the opinion that it is almost always possible to make a quantitative risk assessment, but at 'some point' the effort isn't worth the output. Hora (1993) advised that risk assessments require formal methods for acquiring reliable expert opinion. But how reliable are experts? Which experts should you use? How many? At what point does expert opinion become unreliable? Should you include or exclude experts who have a personal stake in the outcome of a risk assessment? Should you use experts who are retained by or who act as advocates for a stakeholder?

There are several steps in working with experts, including:

1. Defining necessary knowledge and skills for the problem at hand (deciding who might be considered an expert).
2. Selecting the experts.
3. Eliciting information.
4. Evaluating the reliability of the information.
5. Combining information from different experts.
6. Using the information in estimation, calculations or decisions.

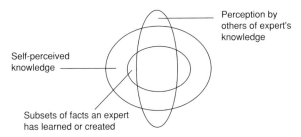

Figure 4.2. Expert knowledge and the region of overconfidence, which lies between the subset of facts known by the expert and the subset they believe they know (modified from Ayyub 2001).

Rarely is the reasoning or the methodology behind any of these steps made transparent. The aim of this chapter is to outline options for using expert judgement in environmental risk assessment, and to explore some of the questions outlined above.

4.1 Who's an expert?

Often, the property of being–an–expert is taken to be self-evident. The person or team in setting the context of a problem (Chapter 3) 'defines a set of issues and selects a set of respondents who are experts on the issues' (Cooke 1991, p. 13). At the risk of placing myself in the midst of a tautology about expert judgement on expert judgement, it is important to explore the question.

An expert may be defined as someone who has knowledge of the issue at an appropriate level of detail and who is capable of communicating their knowledge (Meyer and Booker 1990). Technical experience and training are sometimes called substantive expertise. Normative expertise is the ability to communicate. It involves knowledge of, for instance, the statistical principles and jargon of a field.

This definition is superficially appealing but it is difficult to put into operation. It can be difficult to distinguish between a competent expert and an incompetent one, and to have a clear picture of what an 'expert' knows.

Experts themselves have difficulty knowing the limits of their own expertise. It is especially difficult to judge the likelihood of events for which there is no precedent, or for which experts have no direct experience. Most experts have a region of overconfidence, a domain between the subset of facts they have learned (Figure 4.2), and the subset they think they know. It varies between experts and depends on the elicitation process. Freudenburg (1999, p. 108) called it 'insufficient humility about what we do not know'.

Experts are most dangerous when the region of self-perceived but unjustified belief about knowledge overlaps the region of perception by others. An expert thinks they know something, other people think they know it and, in fact, they don't. The self-serving nature of the provision of expert judgement on a consulting basis further complicates things. Krinitzsky (1993) studied the use of expert opinion in assessments of earthquake risks and concluded experts may be '... fee-hungry knaves, time servers, dodderers in their dotage.... Yet, these and all sorts of other characters can pass inspections, especially when their most serious deficiencies are submerged in tepid douches of banality'. The remainder of this section explores legal and philosophical definitions of an expert that shed light on these difficulties.

4.1.1 Legal definitions

Lawson (1900) wrote the foundation rules for expert and opinion evidence for the US legal system. In these rules, opinion is not admissible in evidence except, '... on questions of science ... persons instructed therein by study or experience may give their opinions. Such persons are called experts' (p. 1–2). This definition was reiterated by the US Federal Rules of Evidence 702 (1992; see Imwinkelried 1993) that stated a witness may qualify as an expert by, '... knowledge, skill, experience, training, or education'.

Expert opinion is seen as necessary to inform the court as completely as possible about facts that might be otherwise unattainable because they are future probabilities, contingencies or facts, 'not within positive knowledge' (Lawson 1900, p. 236). In Australian Federal courts, opinion evidence is admissible if it assists, '... the trier of fact in understanding the testimony, or determining a fact in issue' (ALRC 1985, pp. 739–40). Scientific validity is established through falsifiability, peer review, acknowledged error rates, general acceptance of ideas and valid methods (Preston 2003).

Science is presumed to act as a check on bias or prejudice. Scientists reinforce this view of themselves. For instance, Cooke (1991) argued that if individuals follow the scientific method scrupulously, then they may arrive at results that have a claim on rationality and impartiality. The same thinking permeates current views of expert judgements in environmental risk assessments. We will see below that the assumption may not be justified. Despite some legal and philosophical views of science as an impartial analytical tool, the context of environmental science demands a different set of strategies.

It is interesting to note that Lawson's view was that expert judgement is more akin to belief than empirical frequency (one of the two kinds of probability outlined in Chapter 3, and not the conventional one of formal statistics). This view was especially clear when he stated (Lawson 1900, p. 236), 'The witness swears as to the present conviction of his own mind as to an actual fact'.

Courts have the luxury of being able to question experts and to reject them. The qualifications of an expert may be tested by the opinions of other experts. In adversarial systems, the accuracy of testimony may be tested under cross-examination. The substantive knowledge of an expert may be tested by hypothetical questions. Opinions may be 'impeached' by proof that, on a former occasion, an expert expressed a different opinion. We will see below that the ability to question experts is an essential part of the use of expert judgement. Often, this element is lacking in environmental risk assessments.

4.1.2 Courts as gatekeepers

In most jurisdictions, courts act as gatekeepers, deciding if experts and their testimony are admissible. In some countries such as France and Germany, courts maintain lists of experts from which they may choose someone to assist them (Chapman 1995). Judges have discretion and may ignore the official lists. In Germany, experts are publicly appointed to assist courts with specific issues (Reynolds and Rinderknecht 1993).

Until 1992, expert testimony in the USA required the expert's view to be broadly accepted by the scientific community. New Federal Rules of Evidence and subsequent decisions have broadened the definition to include opinions '... of a type reasonably relied upon by experts in the particular field' (Imwinkelried 1993). The critical elements are scientific reliability (accounting for error rates, procedural care, predictive reliability) and grounding in scientific principles, relevant methodology and reasoning (Kirsch 1995). Peer review and acceptance are no longer completely necessary, though they certainly help (Walton 1997). Legal standards of scientific credibility vary between disciplines. For instance, in medical matters within legal contexts, epidemiologists show consistent empirical associations between a substance and an effect, demonstrate dose–response relationships in laboratory tests, and specify plausible exposure pathways and physiological and other causal links (Kirsch 1995).

The responsibility for deciding admissibility has always been a question for the court (Lawson 1900). But more recent definitions place a greater onus on the judge to discriminate; they assume judges are competent to do so. ALRC (2000) noted that where a substantial disagreement arises, a judge may have no criteria by which to evaluate the opinions.

Difficulties judges face in evaluating conflicting expert evidence may result from the specialized and technical nature of the evidence. Australian judge Justice Kirby remarked that '. . . the experience of lawyers, and their education is such as to make the detailed examination of the language of science and technology uncongenial or even impossible' (1987, in Stewart 1993, p. 13).

4.1.3   Advocacy, adversaries and authority

Scientists are advocates of a scientific view and spend their time trying to convince others of their position, even if they are unaware of it. In adversarial legal systems, potential expert witnesses are selected overwhelmingly for their credentials and for the strength of their support for the lawyer's viewpoint (Shuman *et al.* 1993, in Freckelton 1995). Lawyers search for appropriate attributes in an expert and develop strategies to maximize their chance of winning a case. A close association between a lawyer and an expert may orientate the expert's opinion to provide greatest benefit for the person who retains them. Success often depends on the plausibility or self-confidence of the expert, rather than the expert's professional competence (ALRC 2000).

The difficulties of dealing with advocates have led to other models, such as witness conferencing (Wolfgang 2002). This involves the confrontation of two teams of experts, allowing simultaneous, joint hearing of expert evidence. This system claims to be efficient, clarifying facts and technical issues and eliciting expert judgements more reliably than other systems (to some extent, an antidote for over-confidence; see below).

In some jurisdictions including Australia, Britain, France and Germany, an expert is primarily responsible to the court, not to the person who retains them (Chapman 1995, Lord Woolf 1996, CPR 1999). Experts are obliged to be independent, as far as possible, of the context of the legal proceedings. A court may direct experts to confer and develop a consensus position or it may appoint a specialist to assist the court.

In the latter case, if the expert's opinion is not questioned, the information may be believed simply because of the status of the person providing it. The assumption is that the scientist can act as an impartial observer, relatively free from bias and guided by impartial scientific method. If opinions are unassailable, the expert may assume a false mantle of objective certainty (Walton 1997).

This view of science contrasts starkly with the view of scientists as advocates. If we see scientists as advocates, valid questions from any source should be considered. That is, it should not only be experts (or well-informed lawyers) who can put critical questions to an expert. Anyone with a stake in the outcome should be able to question an expert's opinion.

While some bias is inevitable in adversarial systems because experts are paid and instructed by one party, the system encourages critical questioning of expert's evidence. The ability of courts and opposing lawyers to question credentials and substantive knowledge counteracts biases, but it has limits.

The dialogue between the expert and the users of expert advice is critically important. It underlines the importance of normative skills in experts. In risk assessment, the dialogue is mediated by the risk analyst. It implies that the expert is accessible and open to critical questions from the analyst, other experts, stakeholders and other users of the advice. A hallmark of an assessment that attempts an honest evaluation is that it exposes experts to unfettered, critical evaluation.

4.1.4 Philosophical definitions

Hart (1986) identified three attributes that characterize an expert:

1. *Effectiveness*: they use knowledge to solve problems with an acceptable rate of success.
2. *Efficiency*: they solve problems quickly.
3. *Awareness of limitations*: they are willing to say when they cannot solve a problem.

Awareness may be generally weak among experts, as we will see below. Johnson and Blair (1983) and Walton (1997) argued that appeals to scientific experts are reasonable if their authority is open to challenge. This is a feature of some legal contexts. Suppression of critical questioning occurs when arguments are coercive or confused, when the context or presentation daunts the questioner or precludes questions.

Walton (1997) and Hart (1986) proposed that if an appeal is made to an expert (E) in a domain (D) to make an assertion (A) within their domain of expertise, the appeal is reasonable if it passes a number of critical questions, including:

1. Is E credible as an expert?
2. Is E an expert in the domain that A is in?
3. Is D a domain in which authoritative knowledge exists?
4. What did E assert that implies A?
5. Is E personally reliable?
6. Does E have a special interest in A being accepted?
7. Is A consistent with the assertions of other experts?
8. On what evidence is E's assertion based?

These questions provide a basis for evaluating the worth of an expert's opinion a priori. Other methods outlined below are useful for evaluating an expert's opinion once it has been provided.

4.2 Who should be selected?

How should a risk analyst decide which, and how many, experts to interrogate? How should the analysts themselves be selected? We assume there is a pool of available people, appropriately trained, with sufficient experience and adequate normative skills. But the pool is sometimes small or nonexistent. Often, it is composed of people with different opinions and values.

Meyer and Booker (1990) stated that in face-to-face elicitation, four is too few to get reliable consensus; more than nine is too many to manage effectively. These are general rules of thumb and there is no known way of providing more specific advice. There are, however, many examples of expert groups with fewer than four and more than nine participants.

Group assessments of uncertainty have some advantages over individual assessments. In general, if we want to estimate a fact, if participants act rationally (from the perspective of classical probabilities) and are provided with information that reflects utilities and expectations in an unbiased way, group (consensus) judgements can be expected to be better than individual ones. However, group judgements can fail, attenuating individual biases. Groups have difficulty with small probabilities and their assessments may be biased if the backgrounds of participants are inadequate or unrepresentative (Krinitzsky 1993, Bottom *et al.* 2002).

Literature reviews are a useful way to define the expert pool. Meyer and Booker (1990) recommended asking experts to suggest other experts who suggest others, and so on (termed 'snowballing', Bernard 1988). They also suggested stratifying the experts by sectors such as industry, university and government.

Cognitive psychology theory predicts that conflict between experts may be minimized if the people involved in the estimation process share common values, experiences, professional norms, context, cultural background, and stand to gain or lose in the same way from outcomes. Selection of experts with a narrow set of social attributes will tend to underestimate uncertainty in the subjective elements of the risk assessment. Choosing mature males, for instance, may generate assessments that are more optimistic than those resulting from a more diverse group (Chapter 1).

Strategic appointments may be used to produce deliberate biases. For example, the group leader or analyst may select the majority of participants who advocate one scientific or technical position. Anonymous opinions exacerbate the problem because experts cannot be held personally accountable (Krinitzsky 1993).

Stratification of the pool of expert(s) is a critical step in the selection process. Some dimensions include motivation (stakeholder membership; see below), technical background, gender and cultural background.

4.2.1   Examples of expert selection

A report on the storage of high-level nuclear wastes by Kerr (1996; also see Kammen and Hassenzahl 1999) highlighted some of the issues that arise in using experts. A proposed facility at Yucca Mountain in Nevada would need to store radioactive material for 10 000 years. To evaluate the question of how likely it is that magma will intrude into the facility sometime in the next 10 000 years, the risk analysts went to experts.

The project leader chose 10 expert earth scientists on the basis of expertise, institutional affiliations and normative qualities (including their ability to communicate, interpersonal skills, flexibility and impartiality). These normative attributes were not defined or reported, but instead were interpreted subjectively by the group leader. Such unstructured and unstratified expert selection processes are the norm and are more likely to result in biased judgements than stratified selection processes.

A joint FAO/WHO (2001) Expert Consultation used some stratification. The group was formed to examine the potential for microbiological pathogens to infect food and to prepare information for exposure assessment and hazard characterization. The opportunity to participate was advertised. A complete list of 'qualified' experts was posted on web sites. Selection criteria included:

- technical expertise,
- professional recognition (panels, editorial boards, conferences),
- publications, and
- the ability to participate in group discussions and draft clear reports.

Experts were selected from the published list by a panel made up of 'eminent' experts (invited scientists) and representatives from the organizations involved. In making their decision, the panel considered diversity of scientific background, geographic region, gender and economic development, an improvement on the strategy used by the nuclear waste panel. This step is important because, as we will see below, the composition of the expert group affects the outcome of a risk assessment. The selection process should take into account the ways in which different expert opinions will be aggregated, carried through assessments and subsequently revised.

4.2.2 *Post hoc* calibration

It is always possible that an unreliable expert is selected. People will provide an opinion, even if their information base is scant, simply because they are asked. It is difficult to detect unreliable opinion unless opinions are tested against independent information.

A better solution is to use explicit mechanisms to exclude or adjust opinion, developed before the expert advice is received. If not, the result will be contaminated by the prejudices and context of the analyst or the group that has editorial control over the final report.

For example, Stewart and Melchers (1997) described an example in which the United States Nuclear Regulatory Commission asked 12 experts to estimate the average annual frequency with which pipes of a particular kind rupture. Their estimates ranged from 5×10^{-6} to 1×10^{-10}. The government report concluded that the failure rate was 1×10^{-10} and provided bounds. Eight of the 12 experts provided estimates of failure rates higher than the upper '95th' limit of the final recommendation (Figure 4.3). These results suggest *post hoc* calibration in which in-house experts adjudicated on the expert advice. There was no indication that the in-house judgements were based on better data.

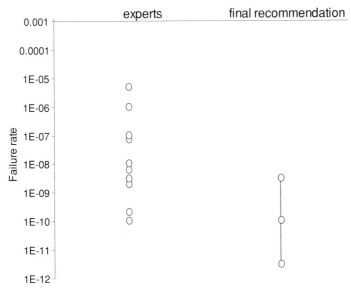

Figure 4.3. Match between expert judgement and final recommendations for pipe failure rates in a nuclear facility. The lines joining points under the final recommendation link the '5th' and '95th' quantiles (after US Nuclear Regulatory Commission 1975, in Stewart and Melchers 1997, p. 87).

The final recommendation had much narrower bounds than the spectrum of expert advice, especially considering the log scale of the failure rates, and the fact that experts were not asked for bounds. It was much more optimistic about pipe failure. A sceptical mind may ask, for example, could the cost of more robust pipes required as a result of a more pessimistic estimate have weighed on the minds of those making the final assessment? It may be that the bias in the external expert judgement was adjusted with independent, empirical data. We will explore means for calibrating expert opinion in the sections below.

4.3 Eliciting conceptual models

Often, experts have strong and divergent opinions about cause and effect relationships. Structural uncertainty reflects different ideas about how systems function. It is difficult to deal with because there are few formal methods for elicitation or representation.

How do you go about sorting through different possible models? One of the main difficulties is that there are many. If you have two variables, X

Table 4.1. *The number of different possible cause and effect relationships (the number of graphs) for N variables (from Shipley 2000)*

N	Number of graphs
2	4
3	64
4	4 096
5	1 048 576
6	1 073 741 824

and Y, and there are no other processes or variables involved, then there are four possible causal relationships between them:

X causes Y,
Y causes X,
X and Y cause each other, and
X and Y are independent (Shipley 2000).

In general, there are $N!/(4^{2!(N-2)!})$ combinations of N variables (Table 4.1).

The search for the right structure is often intuitive. Some differences of opinion about causal relationships may be resolved by data. Empirical evidence should be used to examine plausible alternative models and assumptions critically. An example in Chapter 8 shows a dichotomy of ideas about whether blooms of poisonous algae kill fish, or dead fish cause blooms of poisonous algae.

Elicitation of causal models is a subjective process, usually based around verbal elicitation. The analyst asks questions about causes, effects, actions that will prevent effects, variables that moderate or enable effects, synergies and interference between variables, and so on. Often, elicitation draws on conceptual models constructed previously. The analyst should try to confirm opinions with relevant theory (see Korb and Nicholson 2003).

Assessments are especially error prone when they require experts to judge outcomes of complex models (Freudenburg 1992). Disaggregation (decomposition) may help. It involves reducing a complex and unfamiliar problem into a set of underlying, simpler and more familiar processes and structures. The belief is that it will result in more reliable estimates. Each element in the system should become tangible and easy to envisage (Vose 1996, Bier *et al.* 1999).

Visual representations such as Figure 3.4 are useful. In particular, they make it easy for the analyst to move from a conceptual model to influence diagrams, causal networks and more detailed mathematical models (Chapters 8–11).

If alternative models make a difference, the analyst is obliged to decide among them, or to weight them and perhaps to combine the results. The analyst should retain alternatives until there are grounds to discount or dispense with them. Bayesian methods provide a formal way for weighting alternative models (Hilborn and Mangel 1997). Some approaches to 'model averaging' are described briefly in Chapter 10.

4.4 Eliciting uncertain parameters

The reliability of expert estimates of parameters depends, to some extent, on how the experts are interrogated. Methods for elicitation may include mail surveys, face-to-face individual interviews, traditional meetings, structured group meetings aimed at achieving consensus, and structured group meetings that combine consensus with numerical aggregation. Meyer and Booker (1990), Morgan and Henrion (1990), Cooke (1991), Vose (1996), Hoffman and Kaplan (1999), Ayyub (2001) and Walley and DeCooman (2001) provide practical advice on how to elicit and analyse expert information.

Elicitation is usually preceded by distribution of background information. Conceptual models (Chapter 5) outline the context and assumptions of the problem at hand. Alternative conceptual models encapsulate competing ideas about how a system works or how best to manage a situation. Part of the elicitation process involves the expert(s) and stakeholders in the development and revision of conceptual models.

Expert assessments are made substantially more reliable if the analyst takes care to eliminate as many sources of linguistic uncertainty as possible, prior to the commencement of the elicitation process. This may involve testing elicitation procedures against known variables (Cooke 1991; see Section 4.10).

If the analyst considers carefully what is needed to answer a question, it may provide opportunities to disaggregate individual parameters. Instead of estimating an entire distribution, it may be sufficient to estimate the probability of a discrete event. For example, ocean storm damage is often represented as a function of the maximum wave height (Bier *et al.* 1999).

Similarly, a probability may be defined by the tail of a distribution. For example, the chance of failure of a mine tailings pond may be viewed as the tail probability of the distribution of flood heights (Bier *et al.* 1999). The way in which a problem is structured and presented can have a powerful effect on the results of an elicitation exercise.

4.4.1 Verbal representations

Words may be used to approximate probabilities. They provide a convenient and easily understood vehicle for elicitation (Lichtenstein and Newman 1967). Verbal summaries of risk estimates are commonplace in environmental impact assessments. A useful approach is to rank relative risks using a set of verbal cues. Vose (1996) proposed a list:

Almost certain,
Very likely,
Highly likely,
Reasonably likely,
Fairly likely,
Even chance,
Fairly unlikely,
Reasonably unlikely,
Highly unlikely,
Very unlikely, and
Almost impossible.

The phrases are ranked from highest to lowest probability. The order in the list reflects relative probability. The expert is asked to choose a phrase that equates best with each hazard. Once a word or phrase has been selected for each hazard from the list, the expert is asked to relate as many of the events as possible to observed frequencies or to other representations of probability.

'Kent' scales sometimes are used to translate the linguistic interpretations of uncertainty into quantitative values (e.g. Table 4.2). In practice, it is virtually impossible to distinguish between terms such as 'very unlikely' and 'highly unlikely'. It is often effective to use relatively few (five to seven) categories.

Watson (1998) reported the successful use of a Kent scale together with logic trees (Chapter 8) to assess the probability of success of new

Table 4.2. a. *A Kent scale used by the US Defense Intelligence Agency in 1980 to relate linguistic terms to probabilities (after Cooke 1991)*

Expression	Synonyms	Rank	Percent probability
Near certainty	Virtually certain, highly likely	5	91–100
Probable	Likely, chances are good, we believe	4	61–90
Even chance	About even	3	41–60
Improbable	Probably not, unlikely	2	11–40
Near impossibility	Almost impossible, a slight chance, highly doubtful	1	1–10

Table 4.2. b. *A Kent scale used to evaluate geological risk of petroleum exploration prospects (after Watson 1998)*

Expression	Synonyms	Rank	Percent probability
Proven	True	8	98–100
Virtually certain	Convinced	7	90–98
Highly probable	Strongly believe, highly likely	6	75–90
Likely	Probably true, chances are good	5	60–75
Even chance	Slightly better, slightly less than even	4	40–60
Probably not true	Unlikely, chances are poor	3	20–40
Possible but very doubtful	A slight chance, very unlikely	2	2–20
Proven untrue	Impossible	1	0–2

oil drilling ventures. The same kind of strategy can be used to elicit information about the tails of distributions. Create scenarios for particular events that represent the tails, and rank them. Then link as many as possible to observations of frequency.

Ayyub (2001) listed more than 40 words that describe different degrees of probability. The terms have a natural order but are ambiguous and vague (for instance, what would the correct ordering be for the words faintly possible, quite unlikely, rare, quite improbable and low risk?).

Linguistic probabilities may be modelled numerically with 'fuzzy' methods (see Chapter 2). This avenue exists but has rarely been applied in ecological risk assessment (see Burgman *et al.* 2001, Regan *et al.* 2001 for examples, and Ayyub 2001 and Klir and Wierman 1998

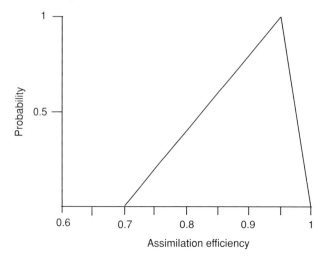

Figure 4.4. A triangular distribution representing uncertainty in knowledge about the efficiency with which insects assimilate PCBs, part of a food web model developed by Regan *et al.* (2002b) to estimate the exposure of mink (*Mustela vison*) to toxic chemicals.

for methods). Walley and DeCooman (2001) suggested associating intervals with linguistic interpretations. For instance, 'probably true' might be modelled as an interval [0.5,1] (see Chapter 9).

Windschitl (2002) argued that verbal representations of probability are more easily understood than numerical values. Numerical values are likely to elicit more deliberative, rule-based thinking. Thus, imprecise, verbal representations may be most useful when the analyst wants to elicit intuitive perceptions of an event's likelihood associated with underlying decision-making tendencies. Numerical answers may be more useful for assessing beliefs about objective probabilities.

4.4.2 Which distribution?

When experts estimate objective probabilities or parameters, they choose a best estimate and a model for uncertainty. Usually this means specifying bounds, a particular statistical distribution, or a set of distributions.

The triangular distribution is popular because it suits the elicitation process. People are reasonably happy to guess things such as the 'most likely' value or their 'best guess' (e.g. Figure 4.4). Usually, people can be coerced, goaded or teased into providing the maximum and minimum 'plausible' values for a parameter, knowing that more extreme outcomes are possible, if unlikely. The triangular distribution has the obvious disadvantage that it is rigidly linear between the three defining points. It seems an unnatural representation.

Morgan and Henrion (1990) reported instances in which experts believed the available knowledge was so poor that they did not have an adequate basis for a judgement. Unfortunately, all probabilistic methods require a distribution. It is tempting to use a uniform distribution when a person is unsure about the form of uncertainty, although it also seems an unnatural model.

While most people are comfortable with a best guess and extremes, they are generally much less happy about guessing means, modes, medians, variances, coefficients of variation, or skewness. Vose (1996) recommended the beta distribution to assist with elicitation. Like the triangular distribution, it is defined by three parameters (lower, best, upper) but has a more 'natural' shape.

We will return to the meaning of the bounds and how they may be elicited in Chapter 9. We will outline the details of some alternative distributions in Chapter 10.

4.4.3 Eliciting distributions and tails

Risk assessments usually are concerned with the tails of a distribution, focusing on extremes rather than the central tendency. This creates special problems because it is much harder to elicit reliable information about the tails, simply because people have less experience with them.

Sensible elicitation strategies extract as much information as possible about the tails, striving to find a distribution that reflects extreme events as faithfully as possible. In many respects, this is a more important process than judgements about the mean or median. Eliciting a realistic tail depends on the ability of the analyst to describe scenarios that result from correlated events, common failure modes, or the confluence of unusual occurrences from independent sources (Vose 1996).

When the mean and variance of a distribution are known with reasonable certainty, extreme value theory (Gumbel 1958) can be used to estimate the probabilities of tails, even when the exact functional form of the distribution is unknown. This approach may be unreliable if the physical mechanisms that generate events of different magnitudes are different (Bier *et al.* 1999).

Morgan and Henrion (1990) recommended two series of questions to elicit points on a distribution. They described the 'fixed value' method in which the assessor asks experts to judge the probability that the variable lies within a specified interval. The answers approximate points on a probability density function (a 'pdf').

The second approach is the 'fixed probability' method in which values of the variable that bound specified quantiles were elicited. A typical question is, 'Give a value of x such that you think the unknown quantity has a 25% chance of being less than x'. Answers to these questions approximate points on a cumulative density function (a 'cdf'). The assessor concentrates on medians, quartiles and extremes (such as the 1% and 99% limits).

Morgan and Henrion's (1990) experience was that fixed value methods that approximate pdfs are usually better calibrated, even though cdfs may be easier to interpret by nontechnical people. Most other analysts agree (e.g. Vose 1996) and my personal experience is the same. cdfs are harder because all cdfs look the same. Distributions are more easily visualized as pdfs.

The most direct method for eliciting a distribution is to ask the expert to draw it, usually as a pdf (Morgan and Henrion 1990, Vose 1996). The choices can be constrained by prior knowledge of the type of distribution the variable comes from.

Frequency formats are relatively robust to the inconsistencies of human perception (Gigerenzer 2002). For example, Borsuk *et al.* (2001a) elicited expert information about the time between mixing events in an estuary. Rather than asking for the probability of different intervals, they used a frequency context and asked, 'If you were to observe 100 vertical mixing events, how many do you think would be x days apart?'.

Morgan and Henrion (1990) suggested asking first for extreme values for an uncertain quantity. That is, the analyst elicits the maximum and minimum 'credible' values, before asking for the best estimate, to help overcome overly narrow bounds. Then, they suggested the assessor ask the respondent to think about scenarios that produce values outside the extremes.

Eliciting distributions is difficult enough. Eliciting dependencies reliably is more difficult. In general, people have poor intuition unless dependencies are strong (i.e. close to -1 or $+1$; Morgan and Henrion 1990). Cooke and Kraan (2000) developed a method that depends on a series of questions about conditional probabilities among pairs of variables of interest. Mechanistic understanding of dependencies may also be useful. A great deal remains to be done before eliciting dependencies is reliable.

4.4.4 How hard should the analyst try?

The imperatives for finding an answer, any answer, can be powerful or overwhelming. How does the quality of expert information deteriorate and how hard should the analyst try once an expert becomes reticent?

If an analyst fails to secure a judgement about a statistical distribution from an expert, then it may be that the analyst lacks the necessary elicitation skills. But experts sometimes make judgements just to be helpful and to retain the semblance of expert respectability (because the experts have been brow-beaten into providing an answer, or won't admit they can't). If the analyst tries too hard, the answers may be too unreliable to be useful.

Expert fatigue may corrupt expert knowledge or make it unavailable. Many conceptual models are complex. They may involve numerous functions and tens or even hundreds of parameters. For ecological models in particular, there may be more than one plausible opinion, requiring several alternative models or sub-models. Time is limited, the process is demanding and people tire of estimation.

The balancing act for the person undertaking the elicitation involves keeping the experts within the domain of their knowledge and putting aside sufficient time for the elicitation process. The region of overconfidence, between the subset of facts experts have learned (Figure 4.2) and the subset they think they know, varies between experts and depends on the elicitation process.

Elicitation is particularly error-prone when the topic involves low probability events such as catastrophic failure of a system, extreme weather, coincidences of independent events and so on. Sometimes, it may be possible to compare the situation with other low-probability events that are better defined. Alternatively, it may be possible to disaggregate the rare event into a sequence of more likely events that are easier to estimate, the combination of which generates the outcome in question (Bier *et al.* 1999). In these circumstances, logic trees (Chapter 8) or failure modes and effects analysis (Chapter 5) can be particularly useful.

It is a great shame, but there is no substantial empirical information and no theory to answer the question, 'How hard should the analyst try?'. Almost certainly, differences in the personalities of experts and their sensitivities to errors hinder any potential generalizations. In any case, tolerance to inaccuracy will depend on the local context.

One solution is to measure the performance of experts routinely, calibrating their reliability, weighing their contributions according to their usefulness and disregarding their contributions when they become too inexact to be useful (Cooke 1991; see below). For example, scoring rules measure the difference between actual outcomes and assessed probabilities. A scoring rule is 'proper' if it rewards assessors for giving their true opinion (Morgan and Henrion 1990, Cooke 1991). The 'reward' is a score, fed back to the expert. The efficacy of such rewards is questionable, especially when experts advocate a value-laden position. We will outline methods to weight expert opinion below, after first examining the kinds of things that make expert judgements unreliable.

In general, there is a powerful and unacknowledged tension between the analyst's desire to elicit a value, and the expert's reticence to do so. It is easy for the analyst to stray into the expert's region of overconfidence. The only way to be sure you have not overstepped the mark is to monitor and validate predictions.

4.5 Expert frailties

Experts are susceptible to the same range of cognitive biases as nonexperts (Chapter 1). Training can alleviate some of these problems. Routinely reliable estimates have only ever been demonstrated in people who make frequent, repeated, easily verified, unambiguous predictions so that they learn from feedback. Because of the importance of expert opinion in risk assessment, the following sections examine the fallacies to which experts are susceptible, and review the track records of experts in making judgements.

4.5.1 Format

The scale of measurement and the structure of questions influence judgements. For example, Slovic *et al.* (2000) reported a study in which the two scales in Figure 4.5 were presented to two groups of experts. Forensic psychologists and psychiatrists were shown case summaries of patients with mental disorders (Figure 4.6). The same questions were put to the two groups, about the probability that a person convicted of a particular crime would re-offend after release. Response scales that emphasized the lower end of the probability scale led clinicians and lay people to judge patients as posing lower risks of committing harmful acts.

The judgements could be influenced, but not necessarily improved, if data were reported in frequency format (Slovic *et al.* 2000). Gigerenzer

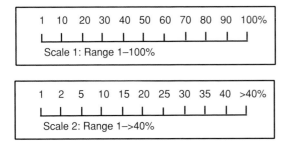

Figure 4.5. Two scales used to guide judgements about probabilities that violent criminals will re-offend (after Slovic *et al.* 2000).

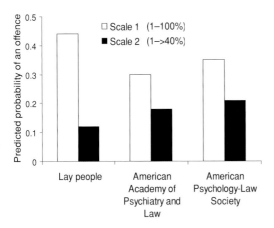

Figure 4.6. The influence of choice of scale of expert estimates of probability. The members of the American Psychology-Law Society were given a tutorial on probability theory prior to their assessment (after Slovic *et al.* 2000 and Gigerenzer 2002). In all cases, the sample sizes were several hundred individuals.

(2002) speculated that when numbers are presented, people tend to picture real events, whereas they remain detached when presented with probabilities.

This result reflects some of the earliest experiments on framing effects. Tversky and Kahneman (1981) described an experiment in which people chose a programme to combat a new disease expected to kill 600 people. The question was posed as follows:

If programme A is adopted, 200 people will be saved.
If programme B is adopted, there is a 1/3 chance that 600 will be saved and a 2/3 chance that no-one will be saved.

Seventy-two percent of people preferred A to B.

They put the question differently to a separate set of people, as follows.

If programme C is adopted, 400 people will die.
If programme D is adopted, there is a 1/3 chance that no-one will die, and a 2/3 chance that 600 people will die.

Twenty-two percent of people preferred C to D.

Of course, A and C are the same programme, as are B and D. The preferences of most people were influenced by how the question was framed. Tversky and Kahneman (1981) found the same responses among undergraduate students, university faculty and practising physicians. It didn't help to be an expert.

These results are disquieting because experts are easily influenced by a frame of reference or the way a proposal is worded. The experts in Figure 4.6 represent psychiatrists and psychologists who routinely make decisions that affect the lives of others in profound ways. They were operating within the accepted limits of their professional expertise.

4.5.2 Availability

Biases arise because of the relative ease with which different kinds of information are remembered by experts. The reliability of an estimate is determined by sample size – in effect, the expert's experience base. Recollections are influenced by the similarity of past experiences to the problem at hand. Catastrophic, newsworthy and recent events are more likely to be remembered (Meyer and Booker 1990). Stark and unusual events are more easily remembered than routine and diffuse events (Vose 1996).

4.5.3 Overconfidence

A few months before the meltdown of the Chernobyl nuclear reactor in Russia in 1986 and the release of substantial amounts of deadly radiation, the Ukranian Minister of Power estimated the risk of a meltdown to be 1 in 10 000 years. Before the space shuttle Challenger exploded on its 25th mission, NASA's official estimate of the risk of catastrophic failure was 1 in 100 000 launches (Plous 1993). Both estimates were wildly overconfident. The space shuttles have a failure rate much closer to 1:50, about the historical failure rate for solid fuel rockets (see Tufte 1997 for a discussion of the way evidence was used in the decision to launch the Challenger).

The tendency towards unjustified optimism is a pervasive feature of risk assessments. Typically there is little relationship between confidence and accuracy, including eyewitness testimony in law, clinical diagnoses in medicine and answers to general knowledge questions by people without special training.

Figure 1.4 shows the tendency towards overconfidence in a trivial assessment of the number of beans in a jar. It is perhaps more disquieting that expert judgements suffer from the same malaise. Vose (1996, pp. 156–8) found in a series of informal observations in a range of settings that experts provide intervals that contain the truth about 40% of the time, when asked for intervals that will contain the truth about 90% of the time.

For example, Baran (2000) described a 40-ha patch of sclerophyll forest and asked a group of professional ecologists at an international meeting to estimate the number of 0.1-ha quadrats that would be necessary to sample various percentages of the plant species. She also asked them to provide 90% credible bounds on their estimates. Field ecologists routinely estimate the number of quadrats necessary to adequately sample the plants in an area. The type of vegetation and the spatial scale were familiar to all of the ecologists. The context is one in which the analyst may have expected a reliable answer. Baran (2000) sampled the area intensively to validate their answers.

Unfortunately, estimates of central tendency and bounds were unreliable. The ecologists' estimates of the number of quadrats necessary to sample 75% of the plant species were about right but only 10 of the intervals estimated by the 28 respondents enclosed the correct value. They were substantially overconfident. Only 2 of 22 intervals captured the correct answer when estimating the number required to sample 95% of the plants (Figure 4.7). The median response substantially underestimated the number of samples.

Plant ecologists are insensitive to sample size (Chapter 1) and there is no simple feedback between estimation and outcome in routine plant surveys. As a result, the experts are poor judges of the effort required for reliable surveys.

Nordhaus (1994, in Kammen and Hassenzahl 1999) asked a group of researchers in global climate change to estimate the effect of a 3°C change by 2090 on gross world product. The participants included people researching scientific, economic, political, ecological and engineering aspects of climate change (Figure 4.8).

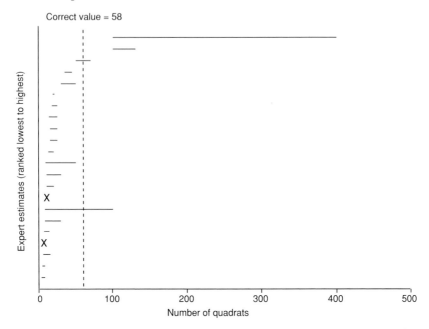

Figure 4.7. Expert estimates of the number of 0.1-ha quadrats necessary to sample 95% of the vascular plant species in a 40-ha patch of dry sclerophyll forest containing about 120 species, and their 90% credible bounds (after Baran 2000). The analyses assume plants were randomly distributed in the 40-ha patch, with respect to the sample quadrats.

These estimates share many properties with Figure 4.7 and Figure 1.4. In all cases there is a tail of high values. Higher values are associated with broader intervals. Several opinions are so divergent that intervals do not include the best estimates of other people. Other intervals fail to overlap at all.

Some of the results suggested motivational biases (see below). Estimates by economists were generally 20–30 times lower than those of natural scientists, although the sample size was small.

The problem of overconfidence is exacerbated by the fact that individual experts may be habitually optimistic or pessimistic. Cooke (1991) noticed that experts tend to be one or the other. For example, when experts judged the probabilities of an electric shock, most selected values were consistently too high or too low. Experts were consistently biased much more frequently than could be expected by chance (Table 4.3).

Table 4.3. *Assessments by 12 experts of the probabilities of 10 events (a–j) leading to an electric shock from a lawnmower (events were that the person was grounded, that power was supplied and so on; P means the person made a pessimistic estimate, and O means their estimate was optimistic)*

Experts	Events										
	a	b	c	d	e	f	g	h	i	j	Proportion pessimistic
1	P	P	P	P	P	P	P	P	P	P	1
2	P	P	P	P	P	P	P	P	P	P	1
3	P	P	O	P	P	P	P	P	P	P	0.9
4	P	P	O	P	P	P	P	P	P	P	0.9
5	P	O	P	O	P	O	O	P	O	P	0.5
6	O	O	O	O	O	O	O	O	O	O	0
7	P	P	P	P	P	P	P	P	P	P	1
8	P	P	P	P	P	P	P	P	P	P	1
9	O	O	O	O	O	O	O	O	O	O	0
10	O	O	O	O	O	O	O	O	O	O	0
11	P	P	O	O	P	P	P	P	P	P	0.8
12	P	P	O	P	O	P	O	P	P	P	0.7

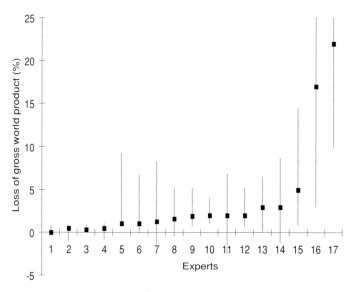

Figure 4.8. Best estimates and bounds for the loss of gross world product resulting from a doubling of atmospheric CO_2, resulting in an increase in global mean temperatures by 3 °C by 2090 (after Nordhaus 1994, in Kammen and Hassenzahl 1999, Figure 4.3). The intervals are 80% confidence intervals (the 10th and 90th percentiles). Each number on the x axis represents a different expert.

This may be due to anchoring (selecting one value and ordering others relative to it) or an inherent tendency towards optimism or pessimism.

Other factors may contribute to bounds that are too narrow. Frequently, experts are unwilling to consider extremes. In addition, the specification of relatively wide bounds is sometimes seen as an admission of a lack of knowledge. The implication may be that the person specifying the widest bounds has the least knowledge, and their reputation may suffer as a consequence (Vose 1996).

It is difficult to know how such a phenomenon will translate to other disciplines and other circumstances. Considering that Cooke's (1991) example was from a situation that was relatively well understood, and in which there were few differences among participants, it seems likely that the tendency for people to be consistently biased will be present, perhaps more strongly, in ecological risk assessments.

The surprise index is the frequency with which a true value falls outside the confidence limits of a judgement. Figures 1.4 and 4.7 give some instances. If experts are asked to provide 1st and 99th quantiles, for instance, surprises should occur roughly 2% of the time. Overconfidence can be measured by the frequency of surprises greater than the expected rate. The summary statistics from a large number of trials conducted on an array of different topics with different (mostly nonexpert) people (Figure 4.9) suggests that overconfidence is the norm, in both experts and nonexperts.

Morgan and Henrion (1990) concluded that all methods for elicitation show a strong tendency towards overconfidence, generating frequent surprises. Cooke (1991, p. 40) concluded that expert opinions in probabilistic risk assessments, '. . . have exhibited extreme spreads, have shown clustering, and have led to results with low reproducibility and poor calibration'. If experts were unbiased, then combining experts would converge on the truth and the best expert would be the one with the smallest variance. Because of clustering, convergence is not guaranteed. All experts may be biased.

4.5.4 Motivational bias

Well-intentioned people can provide biased assessments because it benefits them to do so. It can lead to overconfidence in predictions or other deviations from accurate and reliable assessments.

People in corporations acting as champions for a project are likely to emphasize potential benefits and understate potential costs. This is not due to malicious or cavalier attitudes in the proponents, but is a

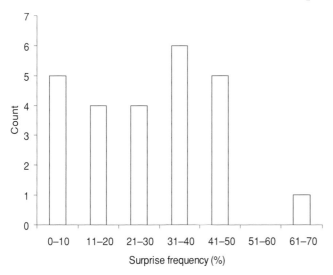

Figure 4.9. The frequency of surprises. The histogram gives the number of studies for which the percentage of judgements did not enclose the truth, after specifying that the range should include 98% of the distribution of uncertainty. Thus, the expected surprise frequency for all judgements was 2%. The data include 27 assessments from 8 separate studies, some preceded by training to elicit reliable bounds, and all involving in excess of 100 subjects per test (data from Lichenstein *et al.* 1982, in Cooke 1991; see also Morgan and Henrion 1990).

consequence of enthusiasm and ambition, both of which breed optimism. Technical experts may underestimate potential risks because their discipline and their own research and career prospects are tied to the development of the technology, even if indirectly. People may wish to appear knowledgeable and thereby underestimate uncertainty. Analysts themselves may introduce a bias by misinterpreting the information provided by experts, or by using it in models that have assumptions and functions that the expert would not agree with.

This creates particular problems for the assessment of new technology risks. The few people with substantive knowledge are also likely to possess a motivational bias.

Gigerenzer (2002) pointed out that roughly 44 000–98 000 people are killed in the USA each year by preventable medical errors, more than are killed by motor vehicles or acquired immune deficiency syndrome (AIDS). Safety systems such as those in commercial aviation that would be in the interests of the patient have not been set up in hospitals. Gigerenzer explained this anomaly by arguing that aviation safety is in the immediate interests of the pilot and crew, and the consequences of failure are highly visible and unambiguous. Doctors and patients, on the other hand, sometimes have different, or even opposing, goals.

One solution to these problems may be to ensure that analysts and experts do not gain or lose, personally, from the outcome of a risk assessment. Alternatively, experts may be required to carry a personal cost from errors that matches the costs experienced by other stakeholders. The analyst may decide to include a range of experts with a stake in the outcome, representing alternative scientific positions, or representing different stakeholder groups.

Morgan and Henrion (1990) suggested a systematic search for motivational biases. Meyer and Booker (1990) recommended testing expert understanding against known standard problems, making experts aware of potential sources of bias, providing them with training to improve their normative skills, and monitoring sources of bias during the elicitation process. Cooke (1991) provided some numerical methods for achieving these goals. These are outlined in Section 4.10. Motivational bias is related to social and philosophical context, as outlined in Section 4.5.7.

4.5.5   Lack of independence

Most technical disciplines have conventional (shared) wisdom. If the technical base is narrow, or if people are selected by inter-personal relationships, they may represent a single school of thought or be influenced by the same data (Bier 2002). If group interactions are not well managed, expert consensus may be influenced disproportionately by a single vocal or influential expert. Lastly, a set of experts may share the same motivational bias deriving from their membership of a professional domain, so that they stand to gain or lose in similar ways from the outcomes of decisions.

If experts are assumed to be independent when they are not, overlap between their estimates will be taken as stronger evidence than is justified. Even if there are small correlations between expert estimates, there may be little additional value in consulting more than four or five experts. Bier (2002) pointed out that infinitely many experts with correlations of 0.25 are equivalent to only four independent experts.

4.5.6   The conjunction fallacy

People find conditional probabilities hard to interpret correctly. Tversky and Kahneman (1982a) provided an example of Bill, an intelligent, unimaginative, compulsive and rather lifeless individual. They asked people to rank a number of possible attributes of Bill, from most to least likely. They included:

1. Bill is an architect.
2. Bill is an accountant.
3. Bill plays jazz for a hobby.
4. Bill is an accountant who plays jazz for a hobby.
5. Bill surfs for a hobby.

More than 80% of people, including statistically literate graduate students, judged (4) to be more likely than (2) or (3), even though (4) is a combination of (2) and (3). Most people ranked (4) before (5), even when told that many more people surf for a hobby than play jazz or are accountants.

Gigerenzer (2002) made the point that the way in which conditional information is worded can influence how experts interpret evidence. There are two ways of presenting it, as conditional probabilities and as raw frequencies.

First consider conditional probabilities. The probability that a woman of age 40 has breast cancer is about 1%. If she has breast cancer, the chance that she tests positive in a mammogram is 90%. If she does not have breast cancer, the chance that she tests positive is about 9%. What are the chances that a woman who tests positive actually has breast cancer? It is very difficult for experts or anyone else to give the right answer (9%) when information is presented like this (Figure 4.10).

Gigerenzer then re-stated the problem using raw frequencies. Think of 100 women of age 40. One has breast cancer. She will probably test positive. Of the 99 who do not have breast cancer, 9 will also test positive. How many of those who test positive actually have breast cancer? The answer, 1 in 10, is now much easier to see. The interpretation of the positive test depends on the reference group, made plain by the frequency information.

It is even easier to see as a tree with raw frequencies (Figure 4.11). Gigerenzer's point was that, while it is possible to work things out correctly if information is presented as raw frequencies, people rarely use them. The consequence is that experts make many avoidable mistakes. An example of a convention using raw frequencies is given in Chapter 12.

The conceptual fallacy that underlies the misinterpretation of conditional probabilities is allied with 'illusory certainty' (Gigerenzer 2002, p. 14), the belief that scientific tests such as mammograms are infallible or highly reliable. Lawyers suffer from the same flaw. It is termed the

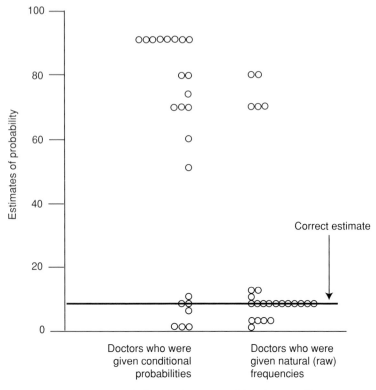

Figure 4.10. Estimates made by 48 physicians of the chance of breast cancer given a positive screening mammogram. Half were given the information as conditional probabilities and half were given it as raw frequencies. In the latter case, the estimates clustered more closely around the correct value of 9% (from Gigerenzer 2002).

'prosecutor's fallacy' in which the probability of an event (p(event)) is confused with the probability that someone is guilty, given the event (p(guilty | event)) (see Goldring 2003).

For example, juries are asked to decide whether blood at a crime scene belongs to a particular person. An expert says in court that one person in a million would have a DNA profile matching the crime scene evidence. The defendant's profile is a match. Unless the court is careful, the jury could be left with the impression that there is a million to one chance the evidence was from the defendant. However, if the evidence was found in the USA where there are about 300 million people, 300 other people are candidates. The DNA evidence, in isolation, should not be conclusive.

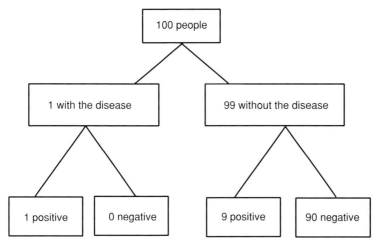

Figure 4.11. The meaning of a positive test. Out of 100 women, 10 will test positive and 1 will have the disease (after Gigerenzer 2002, p. 125).

The reference class problem is a related but separate issue. Courts sometimes consider a person to be more likely to be guilty, simply because they belong to a social group, a so-called reference class, within which such crimes are prevalent. Colyvan *et al.* (2003) made the point that people belong to infinitely many reference classes and care must be taken in defining them and drawing inferences.

4.5.7 Cultural, political and philosophical context

Experts are susceptible to the normal range of human emotions and values. In addition, context may impose an inescapable bias on the pool of experts.

Campbell (2002) used in-depth interviews to examine the attitudes of experts interested in the conservation and management of marine turtles. They were asked specifically about the sustainable, consumptive use of turtles and their eggs. Campbell's results revealed four 'positions' on use, all of which were defendable on scientific grounds (Table 4.4).

The positions were distinguished by how they dealt with uncertainty. Proponents of sustainable use were influenced by international conventions, local sociocultural norms, the potential to learn about the system by monitoring outcomes and the conservation benefits of improved economic status. Opponents of sustainable use were influenced by a lack of faith in free market economics, and moral and philosophical arguments

Table 4.4. *Positions on use and institutional affiliation of 36 marine turtle experts (after Campbell 2002)*

Positions on uncertainty	NGO	University	Government	Total
Consumption is supported, learn through use, use existing information	3	5	2	10
Support limited, consumptive use	2	2	0	4
Uncertainty dictates caution, consumptive use is not supported	5	7	7	19
Consumptive use is unacceptable, uncertainty is not an issue	0	2	1	3

against the use of wildlife. A total of 10 preferred adaptive management and learning through use whereas 3 experts were implacably opposed to commercial use, irrespective of uncertainty.

Campbell (2002) noted that experts saw opposing views as influenced by 'emotions', claiming dispassionate scientific objectivity for their own views, irrespective of their positions. Experts couched their arguments in scientific terms, downplaying the roles of other values. The experts attributed emotional involvement to people without scientific training, as though training protects people against emotional investment in their subject material. Few propositions are so plainly self-deluded.

4.6 Are expert judgements any good?

The acid test should be: 'Are expert judgments reliable?'. The impressive list of frailties above might lead you to conclude that there isn't much chance expert judgements will be reliable enough to use.

4.6.1 Performance measures

If the truth is known, expert judgements can be plotted against reality. These are termed calibration curves. For probabilistic judgements, answers partitioned into subsets with similar probabilities are plotted against the true proportion in each category. Well-calibrated experts lie close to the diagonal. Consistent bias results in a displacement of the median. Overconfident assessors overestimate low-probability events and underestimate high-probability events (Morgan and Henrion 1990).

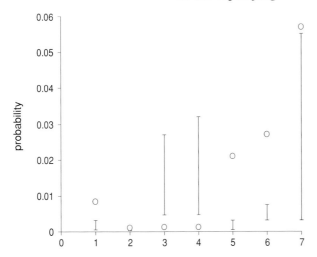

Figure 4.12. Ranges of expert judgements for probabilities of failure of nuclear reactor systems compared with actual outcomes (circles) at Oak Ridge in the USA (after Cooke 1991). The failures were (1–4, pressurized water reactor, 5–7, boiling water reactor) 1: small loss of coolant accident; 2: auxiliary feedwater system failure; 3: high-pressure injection failure; 4: long-term core cooling failure; 5: small loss of coolant accident; 6: automatic depressurization system demand; 7: high-pressure coolant injection.

A Brier score summarizes one of the elements of calibration between judgement and outcome. It is the weighted average of the mean squared differences between the proportion correct in each category and the probability associated with the category (Plous 1993).

Uncertainty often is quantified with an interval that we are x% certain encloses the true value. The surprise index is the percentage of true outcomes that lie outside expert confidence regions (e.g. Figure 4.9).

4.6.2   Performance measured

One of the main reasons expert judgements are used is that estimates are needed in circumstances in which it is difficult or impossible to acquire direct data. There are a few direct comparisons of expert estimates with actual probabilistic outcomes. The majority come from engineering. In Figure 4.12, for instance, the ranges of expert estimates from a study of reactor safety were compared to operating experience for seven different incidents. None of the expert ranges captured the actual outcomes.

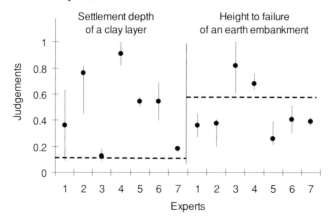

Figure 4.13. Opinions of geotechnical experts on two standard problems. The correct (measured) value for settlement depth was 1.5 cm and for height to failure was 4.9 m. The *x*-axis for both was rescaled so the maximum value was 1. Correct values are shown as dashed horizontal lines. The intervals show 'minimum' and 'maximum' values reported by the experts (after Hynes and Vanmarcke 1975 in Krinitzsky 1993).

In an independent study, Krinitzsky (1993) reviewed the accuracy of experts' earthquake probability assessments. Some experts provided ranges that did not make sense and appeared to lack fundamental knowledge. Despite their stature as experts, some participants did not have the requisite background to give competent judgements. Nevertheless, they were willing to do so.

Krinitzsky (1993; see also Fischhoff *et al.* 1982) also reported the abilities of geotechnical experts to assess the parameters in two standard problems. The results are not heartening (Figure 4.13). Seven internationally known geotechnical experts were asked to predict the height of fill at which an embankment would fail. In all cases, the true value fell outside their confidence intervals.

The geotechnical experts were provided with the data necessary to make calculations. They used a variety of methods. The answers given for the settlement depth problem show that averaging the experts' opinions will not converge on the truth. All were biased in the same direction. In both cases, the experts were overconfident.

Experts are not particularly good at judging the mean lifetime of components in an engineering system or the range of variation that might be expected for the mean lifetime in a large number of components (Figure 4.14).

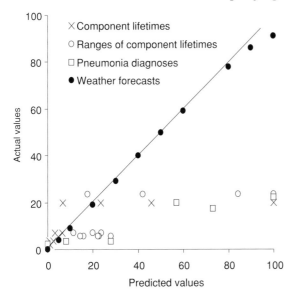

Figure 4.14. Calibration curves: expert predictions plotted against actual outcomes. Crosses are estimates by engineers of the mean lifetime of components in nuclear power systems, versus measured lifetimes. Open circles are estimates of the ranges for the mean lifetimes of the same components, versus measured ranges. Ranges are expressed as the maximum divided by the minimum. The components included pumps, valves and heat exchange units (after Mosleh 1987, in Cooke 1991). The squares are army doctors' subjective probabilities that patients have pneumonia, versus more reliable diagnoses based on radiography (after Christensen-Szalanski and Bushyhead 1981). Solid circles are meteorologists' predictions for the probability of rain on the following day, against the observed relative frequencies (after Murphy and Winkler in Plous 1993). The diagonal line provides the line of correct estimation for all sets of observations. Values are scaled so that the maximum value in each set is 100.

Capen (1976), a petroleum engineer, anticipated these results when he said 'Every test we have performed points in the same direction...The average smart, competent engineer is going to have a tough time coming up with reasonable probabilities...'.

Experts underestimated the lifetimes of long-lived components and overestimated the lifetime of short-lived components. The tendency to underestimate ranges seems more pronounced than the tendency to underestimate lifetimes, across the full range of values. Army doctors' subjective probabilities that patients have pneumonia were a poor predictor of the outcomes of more reliable diagnoses based on radiography. Performances of experts in business, energy consumption planning and military intelligence are likewise mixed and unimpressive (Morgan and Henrion 1990).

For instance, Van der Heijden (1996) documented consistent overestimates in projections for peak summer energy demand by utility planners in the Netherlands through the 1970s and 1980s. There was a similar bias in energy demand predictions for North America where the difference between use estimated in the mid 1970s and the amount actually used a decade later was equivalent to the output from 300 large nuclear power plants (Cooke 1991).

When judging the abilities of experts, it is important to be aware of the potential for normative and substantive weaknesses. The figures above may suffer from both. The information people might use to reach a decision may not be the same as the information they provide when asked a question.

Windschitl (2002) gave the analogy of asking basketball players the distance between the three-point line and a point on the floor directly beneath the hoop, in centimetres. Their responses may be unreliable because they do not have a good concept of a centimetre or because they miscalculate. They may be optimistic in judging 90% confidence intervals for their estimates. Question framing, context, the availability of recent information, the scale offered for the answer and opportunities for anchoring might affect the answers. These factors impair their normative skills (their ability to communicate an answer). Even so, they are likely to have a good appreciation of the distance. They lack the normative skills, even though they have sound substantive knowledge of the distance. Part of the inability of experts to judge parameters reliably may be due to poor communication skills.

To alleviate some of these problems, Krinitzsky (1993) suggested using direct data wherever possible. When experts are necessary, he emphasized the importance of justifying panel composition independently and of using people with well-developed skills in both communication and technical detail. Because experts don't know their own limitations, judgements that are demonstrably incoherent should be disregarded. Lastly, he advised that analysts should keep questions simple and make views transparent by attributing judgements to individuals.

4.6.3 Are expert judgements better than lay judgements?

There is some evidence that experts do better than untrained people, within their domain of expertise. In Figure 1.3, experts did better at

Table 4.5. *Rank correlations between data on mortality (number of deaths per 100 000) and lethality (number of deaths per 100 000, given a disease has been diagnosed) for 31 lethal events, and judgements of the same parameters made by insurance underwriters (experts) and students (novices). The subjects ranked the 31 events using direct estimates of mortality and lethality, and then ranked them by comparing all pairs of events (after Wright* et al. *2002)*

	Underwriters	Students
Direct marginal (mortality)	0.66	0.73
Direct conditional (lethality)	0.66	0.53
Paired marginal (mortality)	0.42	0.24
Paired conditional (lethality)	0.42	0.25

estimating the probabilities of dying from a range of day-to-day activities such as driving a car, getting a vaccination, or smoking, although they still underestimated the true values substantially. Figure 4.6 suggests that trained people may be somewhat less prone to bias caused by the reference frame when making judgements about the likelihood of violence.

Underwriters make routine assessments of risk for insurance companies. Wright *et al.* (2002) repeated the experiment of Slovic *et al.* (2000) more rigorously. They used insurance underwriters (experts) and students (novices) to judge marginal mortality (deaths per 100 000) and conditional lethality (deaths per 100 000 given the disease is diagnosed) for a set of 31 potentially lethal events. The people in the study estimated mortality and lethality directly, and then by making pairwise comparisons between events.

Students and experts were reasonably good at ordering the absolute values for mortality and lethality (Table 4.5), with rank correlations between estimated ranks and 'true' ranks ranging from 0.53 to 0.73 for the students, and 0.66 for the underwriters. Experts were better at ranking the pairwise comparisons of events. The differences in performance between students and experts were small and their biases were similar.

The broad conclusion seems to be that experts do better than untrained people, but not substantially better. Wright *et al.* (2002) concluded that if objective models and reliable feedback are unavailable, it may not be possible for experts to improve their performance above the performance of an untrained person.

Judgements by some professionals may be unlearnable, especially if they involve novel or infrequent events. Demands for unlearnable (and untested) judgements, together with unquenchable expert optimism, create a spectre of widespread, unreliable expert opinion wearing a mantle of scientific reliability. It may be especially prevalent in environmental domains where novel events are common and objective models, monitoring and feedback are rare.

4.7 When experts disagree

Environmental issues often are divisive, even in a technical realm. Krinitzsky (1993) argued that strong personalities influence outcomes. Participants advocate positions, views are anchored, change is resisted, people hold covert opinions that are not explained and there is pressure to conform. Despite these frailties, some analysts persist in the belief that group elicitation of risks is largely detached from subjective influences, even for circumstances as value-laden, uncertain and politically charged as assessing radioactive waste risks (e.g. Kerr 1996).

Salmon in the Pacific north west of the USA are economically, socially and culturally valuable. Ruckelshaus et al. (2002) described how management concentrates on reproductively isolated salmon populations (termed 'evolutionarily significant units') usually associated with a single catchment or geographic area. Widespread and substantial declines in wild salmon populations have been attributed to habitat degradation, dams, harvesting, fish hatcheries, el Niño events, predation and invasion of exotic organisms.

Experts disagree about the causes of decline. Data are often unavailable or equivocal. For instance, some salmon populations have recovered following reductions of harvest. Others have not. Recovery teams are composed of people with technical backgrounds but who represent a range of stakeholders (they are 'expert stakeholders', see below). Ruckelshaus et al. (2002, p. 691) noted, 'Major technical disagreements stemming from philosophical differences that seem to run as deep as religious beliefs are commonplace in such technical teams'.

Consistency of opinions among experts may be interpreted as a measure of reliability. Alternatively, differences may reflect honest, valid differences of opinion. The way in which differences are handled by the analyst should reflect a coherent philosophy about the nature of the uncertainties and the ways in which they affect decisions.

a.   b.   c.

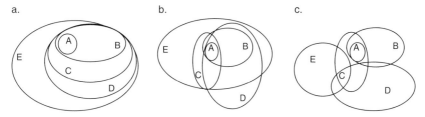

Figure 4.15. Representation of different kinds of disagreements among five experts (labelled A-E); a. consonant evidence, b. consistent evidence, c. arbitrary evidence (from Sentz and Ferson 2002).

Sentz and Ferson (2002) classified evidence from multiple sources. *Consonant evidence* can be represented by nested subsets (Figure 4.15a). For example, the information provided by experts 9, 10, and 11 in Figure 4.8 is consonant. The content of the smallest set is contained within the next largest, and so on.

Consistent evidence means at least one element is common to all of the subsets (Figure 4.15b). In Figure 4.8, experts 5–16 are consistent because they overlap to some extent.

Arbitrary evidence describes the situation where there is no element common to all subsets, even though some subsets have elements in common (Figure 4.15c). The opinions provided by experts 14–17 in Figure 4.8 are arbitrary because 14 and 17 do not overlap.

Disjoint evidence implies that any two subsets have no elements in common with any other subset. The opinions provided by experts 3, 10 and 17 in Figure 4.8 are disjoint.

The easiest way to avoid conflict is to ask just one person. The larger the set of experts, the greater will be the possibility for arbitrary or disjoint opinions. Stratifying the sample of experts to include a range of demographic and social attributes will further increase the chances of disagreement, and reduce the chances of bias.

Consulting multiple experts may be viewed as being equivalent to increasing experimental sample size (Clemen and Winkler 1999). Multiple experts with different backgrounds increase the extent of knowledge and experience contributing to an assessment. This view carries a hidden assumption that their opinions will converge on the truth. This assumption may be reasonable for some simple, empirical facts for which vague definitions, personal values and linguistic ambiguities do not arise strongly, and in circumstances in which experts have no motivational or cognitive biases, but even this is not guaranteed (Figure 4.13).

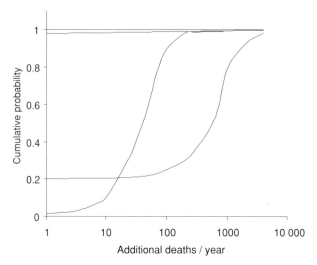

Figure 4.16. Assessments by four health experts, in the form of cumulative probability distributions, of the health impact in additional deaths from exposure to sulfate air pollution from a coal-fired power plant in Pittsburgh, Pennsylvania, USA. The curves give the probability that the total number of annual deaths will be less than the value specified (after Morgan and Henrion 1990).

Technically intensive disputes in particular rely heavily on expert judgement. Experts' knowledge, conceptual models and experiences differ, leading to different technical assessments. For example, Morgan and Henrion (1990) reported the results of a study in which four health experts were asked to assess the probabilities of additional deaths resulting from exposure to a toxin produced by a coal-fired power plant (Figure 4.16).

All participants believed that there would probably be fewer than about 5000 additional deaths annually in the surrounding population. Two experts were relatively optimistic, believing that it was likely that less than one additional death would occur annually. The experts' estimates of the probability of zero additional deaths per year ranged from 0 to 1.

Much of the work on risk assessment in decision theory, economics and statistics has focused on the development of normative theories of rational consensus: expert judgement takes the form of degrees of belief about facts. To estimate an uncertain quantity such as the parameter in a model, an analyst may combine the distributions provided by more than one expert. Alternatively, the group may wish to combine individual distributions to arrive at a collective decision or recommendation. In either case, there is an underlying assumption that there is a fact and the job of the experts is to do their best to estimate it.

There are two basic forms of aggregation: behavioural and numerical aggregation. The following sections outline their forms and uses.

4.8 Behavioural aggregation

4.8.1 Delphi and its descendants

The Delphi technique was invented with consensus in mind. Developed by the RAND Corporation in a joint US Air Force/Douglas Aircraft initiative, it began in 1948 (Cooke 1991, Ayyub 2001). There are different forms of the technique for technology forecasting and policy development but essentially it consists of the following steps:

- problem formulation and development of questionnaires,
- selection of experts,
- provision of background information and context to experts,
- elicitation of point estimates (performed independently; the experts do not see one another),
- aggregation of results with medians and interquartile ranges which are distributed to participants,
- review of combined results by experts and revision of answers, and
- iteration of feedback until consensus is achieved.

Answers and the thinking behind them are distributed among the experts without revealing who was responsible for each judgement. If experts persist in deviating substantially from the group consensus, they are asked to justify their position.

The method has been criticized for limiting interactions between experts. It does not deal adequately with uncertainty, it encourages uniformity and discourages dissent. Applications tend to overlook important interactions and misinterpret the joint probabilities of coincident events. The mechanisms for resolving differences tend to fuel expert overconfidence (see Cooke 1991, Ayyub 2001). Some of these criticisms have been addressed by changes in the protocol (Box 4.1).

Box 4.1 · *A modified Delphi technique for expert elicitation.*

Vose (1996) developed an approach aimed at having participants leave with a common perception of risks and uncertainties. It consists of the following steps:

1. Gather relevant information and circulate it prior to the meeting.
2. Bring experts together and lead a brainstorming session (Chapter 5).
3. Encourage discussion of uncertainty in each variable, including logical relationships and correlations.
4. Discuss scenarios that would produce extreme events.
5. Take minutes and circulate them afterwards.
6. Following the brainstorming session, conduct individual interviews to determine opinions about system structures and uncertainties in variables.
7. Reconvene the experts to discuss important differences of opinion.
8. Present residual differences and combine them in ways that retain the information (see below).

This approach gives experts a chance to consider the opinions and knowledge of others and does a better job of retaining uncertainties and exploring dependencies between system components.

4.8.2 Closure

Several methods seek consensus among experts. Some of these are appropriate for groups of people who represent interests of stakeholders (see Section 4.12 below). Most involve experts first expressing their views and then presenting them to others. Facilitators assist experts to revise their assessments as linguistic hurdles are overcome and information is disseminated (Clemen and Winkler 1999).

Valverde (2001) suggested a framework to achieve consensus expert opinion that has its foundations in an approach developed by Kaplan (1992). It involves diagnosing sources of disagreement and finding ways to resolve them. Valverde used the typology for arguments invented by Toulmin (1958) that recognizes five basic elements:

- *Claim*: an assertion or proposition, usually the end result of an argument, but neither necessarily certain or true.

- *Data*: information content of an argument, including physical evidence, observations and experimental values.
- *Warrant*: the causal laws, inference mechanisms, models or rules that link the data to the claim.
- *Backing*: background assumptions or foundations that support the warrant, including axioms, theory, and formal principles.
- *Rebuttal*: the conditions under which and the reasons why the claim does not necessarily hold, and which may apply to the warrant or the claim.

In the first step, experts advance a series of claims. They are subject to rebuttal by other experts. Disagreements are about fact, theories or principle (different scientific traditions or differences of a metaphysical or ethical nature). Disagreements have their origins in disagreements about data (the easiest to resolve), warrants or backing (the most difficult to resolve).

In the second step, the analyst diagnoses and treats sources of expert disagreement. Valverde (2001) expanded the typology of argument, requiring experts to make explicit numerical statements to represent degrees of belief. He created a second taxonomy to diagnose sources of disagreement:

- *Semantic disagreement*: experts misunderstand the meanings of words.
- *Preference disagreement*: experts have different preferences for methods and standards used to evaluate claims.
- *Alternative disagreement*: usually management options are offered that serve to frame a problem, but experts differ in their views about the admissible set of policy options.
- *Information disagreement*: experts differ in their views on measurements, the validity of experiments, the methods used to obtain observations, or the rules of inference.
- *Epistemic disagreement*: experts adhere to different scientific theories, professional conventions or ethical positions.

The taxonomy is applied sequentially, treating the most likely underlying cause of disagreement last. If, for instance, preference disagreement is thought to underlie a particular issue, the diagnostic procedure seeks first to identify and resolve semantic disagreements. If disagreement persists, the facilitator should introduce alternative information or explore the differences in understanding. The step is complete when the source of disagreement is understood.

The third step in Valverde's process is to seek 'closure' or 'resolution'. Closure may be achieved through (Engelhardt and Caplan 1986):

- *Sound argument*: a 'correct' position is identified and opposing views are seen to be incorrect.
- *Consensus*: the experts agree that a particular position is 'best'.
- *Negotiation*: an arranged resolution is reached that is acceptable to the participating experts and that is 'fair' rather than correct.
- *Natural death*: the conflict declines gradually and is resolved by ignoring it, usually because it turns out to be unimportant.
- *Procedure*: formal rules end sustained argument.

Closure assumes that expert judgements provide a rational interpretation of evidence and that 'political' influences can be divorced from the scientific process.

Recent advances in cognitive psychology outlined above make it plain that experts are not completely rational and that debates and expert disagreements are not isolated. In addition, procedures such as voting and preference voting create circumstances in which the order of choices offered to a committee can determine the outcome (Arrow 1950). In general, closure is not guaranteed and it is not always sensible to seek it.

4.8.3 Resolution

The alternative to closure is to seek 'resolution', which may include eliminating, mitigating or accepting differences of expert opinion. Valverde identified means for seeking resolution, some of which relate to approaches for closure. They include:

- *Co-optation*: experts acknowledge that the conflict is 'resolvable', and sound argument, consensus or negotiation may bring closure.
- *Supremacy*: expert disagreements are tested to determine the 'correct' position; experts agree on the grounds upon which 'supremacy' will be based, and what evidence would cause them to alter their position.
- *Replacement*: a new paradigm is introduced that integrates the best aspects of the competing viewpoints, in which the experts agree with the synthesis, leading to consensus or negotiation closure.

Any disagreement cannot necessarily be resolved. It may not be desirable in all circumstances because it may mask important, legitimate differences of opinion. Group judgement may lead to overconfidence and polarization of opinions in which groups adopt a position that is more extreme than that of any individual member (Clemen and Winkler 1999). The decision-oriented nature of risk assessments means that often an action will be taken, irrespective of the extent of disagreement. Political action does not depend on complete agreement on all issues.

There is no empirical or theoretical evidence to support one behavioural alternative over another in all circumstances. Group consensus has been used to estimate statistical distributions. For example, the International Whaling Commission used group consensus to develop priors for a Bayesian assessment of the recovery of Bowhead whales (Punt and Hilborn 1997). While it is always useful to eliminate arbitrary elements of disagreement (such as ambiguity and underspecificity), in the spirit of honest risk assessments, differences of opinion about scientific detail and ethical issues can be made transparent, without compromising the ability of a group to reach a decision. Methods for carrying uncertainty through a risk assessment are explored in the following chapters.

4.9 Numerical aggregation

Numerical aggregation uses quantitative strategies, rather than behavioural ones, to arrive at a combined estimate. If the information is probabilistic, then the tools of formal statistics are appropriate. In particular, Bayesian analysis provides a mechanism for combining knowledge from subjective sources with current information to produce a revised estimate of a parameter. Approaches such as evidence theory (see Sentz and Ferson 2002, Ferson *et al.* 2003) have been developed to cope with non-probabilistic uncertainty (see Chapter 2). When probabilities are themselves uncertain, the analyst may use bounds on probabilities (Walley 1991). Avenues for dealing with linguistic uncertainty and imprecise probabilities are outlined in Chapters 2 and 9.

Numerical aggregation involves tradeoffs between pieces of expert evidence. Sentz and Ferson (2002) described some extreme possibilities. Conjunctive pooling (A∩B) retains only those opinions that are common to all experts. Disjunctive pooling (A∪B) retains the full breadth of opinion offered by all experts. Tradeoffs involve something in between. For instance, disjunctive pooling may be used once the experts' opinions have been weighted for their reliability. The following sections outline some options.

4.9.1 Dempster's rule

Assuming there are no disjoint subsets, it may seem a good idea to use only the information about which everyone agrees. Dempster's rule applies the conjunctive AND to combine evidence. More formally, the probability density of the combined evidence is limited to the interval over which experts agree (see Klir and Wierman 1998).

The rule completely ignores areas of disagreement. Sentz and Ferson (2002) related a counterintuitive example from Zadeh (1986). A doctor believes one of her patients exhibiting neurological symptoms has viral meningitis with probability 0.99, or a brain tumour with a probability of 0.01. A second doctor believes the patient suffers from concussion with probability 0.99, but will also allow the possibility of a brain tumour, with probability 0.01. Dempster's rule would combine these by ignoring the conflicting opinions between viral meningitis and concussion, and conclude the patient has a brain tumour, with probability 1, a result that both doctors would consider to be unlikely.

Where there is substantial overlap between numerous experts, Dempster's rule may be used to trim the outliers and provide a central focus for further analysis. Dempster's rule is a generalization of Bayes' theorem (Sentz and Ferson 2002).

4.9.2 Bayes' theorem

Decision-making is deciding the 'best' course of action. We want to use, and therefore to combine, information from all available sources in a way that is repeatable and that gives due weight to the credibility of the evidence.

When using subjective estimates of a parameter, information may be combined with other evidence (updated) with Bayes' theorem. It asks, 'What is the probability that a proposition is true, given the data?'. It requires the experts to specify their subjective belief in a distribution, prior to the analysis, even if no data are available. This step would not be upsetting to anyone with experience in expert elicitation and empirical model building. But it can get under the skin of someone who trusts only direct measurement.

Estimates of the prior probabilities (or probability distributions) represent the probability that the data would be observed, if the various hypotheses were true. These values are combined with Bayes' theorem, giving *posterior probabilities*, the updated degrees of belief that the hypotheses are true. More formally, given two hypotheses, H_1 and H_2, and

Table 4.6. *Frequency of presence of a toxic chemical and a diatom species supposedly useful in indicating the presence of the chemical in 1000 independent samples of lake water (from Murtaugh 1996)*

Sample data	Contaminant present	Contaminant absent	Marginal totals
Diatom absent	6	800	806
Diatom present	54	140	194
Marginal totals	60	940	1000

some data, D, the probability that H_1 is true given the data, is:

$$p(H_1 \mid D) = \frac{p(H_1) \times p(D \mid H_1)}{p(H_1) \times p\langle D \mid H_1 \rangle + p(H_2) \times p\langle D \mid H_2 \rangle},$$

where $p(D)$ is the probability of observing the data given all possible hypotheses, $p(H_1)$ is the 'prior' probability of H_1, $p(H_2)$ is the 'prior' probability of H_2 and $p(D \mid H_1)$ is the probability of observing the data when H_1 is true.

Murtaugh (1996; see also Anderson 1998b, Wade 2000) provided the following example of data for the presence of a toxic chemical in a lake, and the presence of an indicator of the toxicant, a diatom species (Table 4.6). In the future, if the diatom species is again observed in a single sample taken during a standardized survey of a particular lake, what is the probability that the contaminant is actually present? The solution to this problem is seen most easily if the data are presented as raw frequencies in a tree (Figure 4.17).

Of the 806 observations where the diatom was absent, the proportion without the contaminant was $800/806 = 0.99$. Of the 194 observations where the diatom was present, the proportion with the contaminant was $54/194 = 0.28$. The data suggest the presence of the diatom is a poor indicator of the presence of the contaminant. They suggest that the absence of the diatom is a good indicator of the absence of the contaminant. The same analysis is presented as a numerical example of Bayes' theorem in Box 4.2.

Morris (1977) outlined the process of using Bayes' theorem to aggregate information from a set of experts. It can be used to aggregate point information or probability distributions. Clemen and Winkler (1999) described four approaches to combining discrete probabilities under Bayes' theorem, one of which is a linear weight function like the equation in

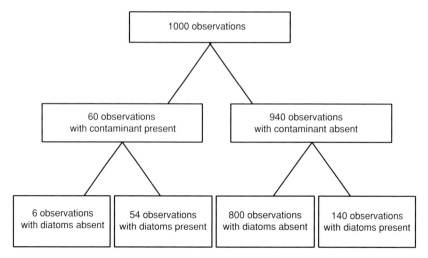

Figure 4.17. The observations in Table 4.6 rearranged in a logic tree (see Chapter 8) to give a frequency interpretation of Bayes' theorem.

Box 4.2 · *Numerical interpretation of Bayes' theorem for diatoms and a contaminant (see Murtaugh 1996)*

H_1: the contaminant is present in the lake.
H_2: the contaminant is absent from the lake.

The data are: the diatom was observed/not observed (a binomial problem). The first step is to estimate the necessary inputs for the Bayesian analysis:

$p(H_1)$, the 'prior' probability of H_1
   = probability that the contaminant is, in fact, present
   = 60/1000 = 0.06.

$p(H_2)$, the 'prior' probability of H_2
   = probability that the contaminant is, in fact, absent
   = 940/1000 = 0.94.

$p(D|H_1)$, the probability of observing the data, when H_1 is true
   = probability of observing diatoms when the contaminant is present
   = 54/60 = 0.90.

$p(D|H_2)$, the probability of observing the data, when H_2 is true
   = probability of observing diatoms when the contaminant is absent
   = 140/940 = 0.149.

Bayes' theorem (for two hypotheses) $p(H_1|D)$ results in an estimate for the probability that H_1 is true, given the data (Chapter 6). Using the Equation on p. 109, the probability that the contaminant is present, given the observation of diatoms in a standard survey, is:

$$p\langle H_1 |D\rangle = \frac{0.06 \times 0.9}{0.06 \times 0.9 + 0.94 \times 0.149} = \frac{0.054}{0.194} = 0.28.$$

The Bayesian analysis suggests that the presence of the diatom is not a particularly good indicator of the presence of the contaminant.
How about if the diatom species is not observed? What is the probability that the contaminant is absent?

$p(H_2|D)$, the probability that H_2 is true, given the data
= probability that the contaminant is **absent**, given **no** observation of diatoms in a standard survey.

$$p\langle H_2 |D\rangle = \frac{p(H_2) \times p\langle D|H_2\rangle}{p(H_1) \times p\langle D|H_1\rangle + p(H_2) \times p\langle D|H_2\rangle}$$

$$p\langle H_2 |D\rangle = \frac{0.94 \times 0.851}{0.06 \times 0.1 + 0.94 \times 0.851} = \frac{0.8}{0.806} = 0.99.$$

As above, the Bayesian analysis suggests that the absence of the diatom is a good indicator of the absence of the contaminant.
The posterior odds ratio measures the relative confidence in each of two competing explanations:

$$\text{odds} = \frac{p(H_1|D)}{p(H_2|D)} = \frac{0.28}{0.99} = 0.28 : 1.$$

That is, the contaminant is roughly four times more likely to be present when diatoms are present than it is when diatoms are absent.

Section 4.9.3 below, assuming expert information is composed of independent trials.

A 'robust' Bayesian analysis would implement all of these (and others) in a sensitivity analysis and would interpret the full set of potential posterior distributions. It can be difficult to account for differences in the precision and bias of individual assessments, and correlations between experts. We will return to robust Bayesian methods in Chapter 11, but for the moment, it is sufficient to be aware that these problems are common to all methods of aggregation. Bayesian methods make the difficulties apparent.

Posterior distributions are a compromise among expert judgements. In an example analogous to the diagnosis problem described above by Sentz and Ferson (2002), Clemen and Winkler (1999) described a situation in which two experts give estimates for a parameter:

$$g_1 = 2,$$
$$g_2 = 10.$$

Both experts estimate the variance to be 1. The Bayesian posterior will be $g = 6$ and $s = 0.5$, values that neither expert would agree with. A bimodal distribution would represent their opinions better. It is necessary to be careful to assess how aggregation works before deciding on the most appropriate method.

Because prior probabilities for Bayesian analysis may be estimated subjectively, they are subject to the same suite of cognitive illusions that affect all subjective science. In routine applications, risk analysts who use Bayesian estimation find that overestimation of single-event probabilities is quite common. Mutually exclusive events should sum to 1, but often people provide estimates that together exceed 1. Conjunction fallacies, such as $p(A+B) = p(A) + p(B)$ occur when people confound probability with plausibility, tendency or confidence (Chapter 1 and Section 4.5). These issues are largely unresolved.

4.9.3 Mixing and averaging

Probabilites associated with belief may be combined as weighted linear combinations of opinions (Cooke 1991, Vose 1996, Valverde 2001):

$$p_{1...n}(A) = \frac{1}{n} \sum w_i \, p_i(A),$$

where w_i are non-negative weights, p_i is the subjective belief of expert i associated with the value of A and n is the number of experts. This is sometimes termed mixing or averaging.

Cooke (1991) generalized this expression to include a range of options from arithmetic averaging to logarithmic weights based on the products of the probabilities given by the experts,

$$p(A) = k \prod_{i=1}^{n} p_i(A)^{w_i}.$$

This is useful when input estimates are probabilities. k is a normalization value to ensure probabilities add to 1. The formula can be used with cumulative probabilities and p-bounds (Chapter 10, Ferson et al. 2003). However, there is no underlying general theory that can justify the selection of the weights or the function that combines them (Clemen and Winkler 1999) beyond the empirical strategies developed by Cooke (1991, see Section 4.11 below).

Sentz and Ferson (2002) continued the example of two doctors (1 and 2) and a patient with neurological symptoms. As above, the doctor believes her patient has viral meningitis with probability 0.99, or a brain tumour with a probability of 0.01. A second doctor believes the patient suffers from concussion with probability 0.99, but will also allow the possibility of a brain tumour, with probability 0.01. Mixing their evidence and assigning equal weights to the reliability of the two experts results in:

P_{12}(meningitis) $= \frac{1}{2}(0.99) + \frac{1}{2}(0) = 0.445$
P_{12}(brain tumour) $= \frac{1}{2}(0.01) + \frac{1}{2}(0.01) = 0.01.$
P_{12}(concussion) $= \frac{1}{2}(0) + \frac{1}{2}(0.99) = 0.445.$

Mixing retained the full suite of uncertainty in the original advice, and gave a sensible distribution of frequencies of different values within the interval of possible values.

Simple numerical averages of individual judgements seem to perform slightly better than group judgements by behavioural consensus when the focus is on unambiguous, value-free and sharply defined parameter estimates (Gigone and Hastie 1997, Clemen and Winkler 1999). More complicated formulae are possible, involving, for example, updating of weights following discussion among experts or the acquisition of new data (Cooke 1991), or based on the performance of an expert against a set of known values (Goossens and Cooke 2001).

There is unlikely to be a single rule or set of weights appropriate for all circumstances. Ferson et al. (2003) outlined a range of alternatives. The choice of rules for combining evidence is an important aspect of a risk assessment. It should be explicit and its consequences explored through sensitivity analyses.

4.10 Combined techniques

Cooke (1991) pointed out that subjective estimates of probabilities and utilities may only be meaningful for individuals. He argued that they cannot be defined for a group. If so, the solution when experts disagree is not to try to maximize group utility, but to develop a position on which a sufficient number of experts agree. A few methods have been developed that combine elements of numerical aggregation and behavioural consensus.

4.10.1 Consensus convergence

Behavioural consensus techniques work when participants agree to negotiate to resolve conflict. Lehrer and Wagner (1981) combined opinions based on measures of respect provided by members of a group for the opinions of other members.

Suppose there are n experts who each assign probabilities (or degrees of belief) p_1^0, p_2^0, ..., p_n^0 to a set proposition. The first step involves each expert, i, assigning a weighting, w_{ij}, to the other experts', j, opinions, where

$$\sum_{j=1}^{n} w_{ij} = 1.$$

The higher the weight, w_{ij}, the greater the respect expert i has for the opinion of expert j. Typically, these weights are assigned before the other experts' opinions are known. Experts are obliged to assign a positive weight to at least one other member of the group apart from themselves.

Initial probabilities are updated to incorporate the opinions of the other members of the group according to the weights assigned to them. A revised, weighted average is formed for expert i's probability assignment as:

$$p_i^1 = w_{i1} p_1^0 + w_{i2} p_2^0 + \cdots + w_{in} p_n^0, \quad i = 1, \ldots, n.$$

The group may then view the weights and agree among themselves on a consensus position. If consensus is not reached on the first iteration, the process may be repeated with the same weights (if the experts have not changed their opinions about other members) or with new weights (if new information is available or people change their opinions). Matrix algebra may be used to find consensus positions for the group (Box 4.3).

4.10.2 The 'Procedures guide' for structured expert judgement

Cooke (1991; see also Cooke and Goossens 2000, Goossens and Cooke 2001) developed a protocol for elicitation in risk assessment called the

---

Box 4.3 · *An algebraic solution for group consensus (Lehrer 1997)*

If experts keep the same weights, then the second round of aggregation will give a probability for expert i:

$$p_i^2 = w_{i1} p_1^1 + w_{i2} p_2^1 + \cdots + w_{in} p_n^1, \quad i = 1, \ldots, n.$$

When all experts are considered simultaneously the consensus model becomes:

$$W = \begin{bmatrix} w_{11} & w_{12} & \cdots & w_{1n} \\ w_{21} & w_{22} & \cdots & w_{2n} \\ \cdots & \cdots & \cdots & \cdots \\ w_{n1} & w_{n2} & \cdots & w_{nn} \end{bmatrix}, \quad P = \begin{bmatrix} p_1^1 \\ p_2^1 \\ \cdots \\ p_n^1 \end{bmatrix}$$

where W is the table of constant weights, w_{12} is the weight assigned by expert 2 to the opinion of expert 1, and P is the column of initial probability assignments for each of the n members in the group. Probabilities from the first round of aggregation are given by the matrix multiplication WP. When the same weights are maintained for m rounds, probabilities are calculated by $W^m P$. When m is large, the probability assignments converge on a consensus probability (p_c, i.e. the probability such that $p_c = p_1{}^c = p_2{}^c = \cdots = p_n{}^c$).

Convergence may be blocked, for instance, by one expert assigning weights $= 0$ to all other members. Constraints may be applied to the matrix of weights to guarantee convergence in a variety of circumstances (Lehrer 1997). In its simplest form, convergence is guaranteed when weights are constant and positive.

'Procedures guide' (Box 4.4). It is intended for 'practical scientific and engineering contexts' (Goossens and Cooke 2001) in which uncertainty is removed through observation, and where decision-making is supported by quantitative models with large uncertainties.

Expert judgement is, for them, another form of scientific data. They assume that some unique, real value exists but we are uncertain what the value is, so uncertainty may be described by a subjective probability distribution with values in a continuous range.

Their method embodies performance measures of expert assessments based on assessments of variables whose values are made known to experts after they have completed their assessments. The goal is to combine

Box 4.4 · *The procedures guide for expert judgement (after Cooke 1991)*

The procedure has 15 steps:

1. *Definition of case structure*: this is achieved by creating a document that specifies all the issues to be considered during the expert judgement exercise. It provides information on where the results of the exercise will be used, and outlines the physical phenomena and models for which expert assessment is required.
2. *Identification of target variables*: a procedure is used to select variables for expert elicitation, to limit them to a manageable number. Variables are included if they are important and if historical data are insufficient or unavailable.
3. *Identification of query variables*: if target variables are not appropriate for direct elicitation, surrogate 'query' variables are constructed that ask for observable quantities, using questions formulated in a manner consistent with the experts' knowledge base.
4. *Identification of performance variables*: performance (seed) variables are supported with experimental evidence that is unknown to the experts, but known to the analyst, usually from within or closely associated with the domain of the enquiry at hand.
5. *Identification of experts*: as large a list as possible is compiled of people whose past or present field contains the subject in question and who are regarded by others as being knowledgeable about the subject.
6. *Selection of experts*: a subset is selected by a committee based on reputation, experimental experience, publication record, familiarity with uncertainty concepts, diversity of background, awards, balance of views, interest in the project and availability.
7. *Definition of elicitation format*: a document is created that gives the questions, provides explanations and the format for the assessments.
8. *Dry run exercise*: two experienced people review the case structure and elicitation format documents, commenting on ambiguities and completeness.
9. *Expert training*: experts are trained to provide judgements of uncertain variables in terms of quantiles for cumulative distributions, anchoring their judgements to familiar landmarks such as the 5th, 50th and 95th quantiles.
10. *Expert elicitation session*: each expert is interviewed individually by an analyst experienced in probability together with a substantive

expert with relevant experience. Typically, they are asked to provide subjective judgements for the query variables as quantiles of cumulative distributions.

11. *Combination of expert assessments*: estimates are combined to give a single probability distribution for each variable. Experts may be weighted equally, or by assigning weights reflecting performance on seed questions.

12. *Robustness and discrepancy analysis*: robustness may be calculated for experts by removing their opinions from the data set, one at a time, and recalculating the combined functions. Large information loss suggests that the results may not be replicated if another study was done with different experts. A similar evaluation may be conducted for seed variables. Discrepancy analysis identifies the items on which the experts differ most.

13. *Feedback*: each expert is provided their assessment, an informativeness score derived from the robustness analysis, weights given to their opinion, and passages from the final report in which their name is used.

14. *Post-processing analysis*: aggregated results may be adjusted to give appropriate distributions for the required input parameters.

15. *Documentation*: this involves the production of the formal report.

experts' judgements into a 'rational consensus' that is transparent, fair, neutral and subject to empirical quality control.

To elicit expert opinion about the relative ranks of unknown attributes (as in the subjective risk assessment of consequences and likelihoods, Chapter 6), the attributes may be presented pairwise to the experts. For each pair, the experts are asked to indicate their preference. Each attribute is evaluated $n - 1$ times. Within-expert consistency is measured by the frequency of inconsistent triads (e.g. if after specifying A>B and B>C, they specify A<C). Between-expert agreement is measured by the frequency with which pairs of assessors agree, compared to random assignments. Between-expert concordance is measured by the similarity in ranks assigned to attributes.

Cooke (1991) measured the 'information' contained in an expert's distribution with the Shannon entropy index, where

$$H(p) = -\sum_{i=1}^{M} p_i \ln p_i$$

and p_i is the probability assigned to integer (interval or class) i. $H(p)$ measures the spread and the evenness of the expert's judgements. The quantity is maximized at $\ln(M)$ when all $p_i = 1/M$ and it is minimized if any $p_i = 1$. The wider the range of judgements and the more even the assignment of probabilities to each possibility, the greater will be the entropy. The expert whose judgements give the lowest entropy is preferred (all other things being equal).

Calibration was defined as the likelihood that the expert's probabilities correspond with a set of repeated experimental results; namely, the probability that the difference between the expert's judgement and the observed values in the seed distribution have arisen by chance. Cooke (1991) measured the degree to which experts are 'calibrated' by the probability of observing by chance the differences between the expert distribution and the true distribution, where

$$2nC(s,\, p) = 2n \sum_{i=1}^{M} s_i \ln \frac{p_i}{s_i}$$

in which s_i is the true (observed) frequency and n is the sample size. The quantity $2nC$ is χ^2-distributed with $M - 1$ degrees of freedom.

To derive uncertainty distributions for model parameters, opinions are pooled with the linear weight function (Section 4.9.3 above). Weights are derived from calibration and information performance measured on seed variables. Global weights may be calculated for each expert from the product of their calibration and information scores.

Goossens *et al.* (1998) reported the performance of experts against seed values for tasks ranging from crane failure and space debris risk to groundwater and water pollution problems. Two environmentally relevant examples are shown in Table 4.7.

Experts may be given a weight of zero if their calibration score falls below a threshold. In the groundwater transport example in Table 4.7, 'domain' seed variables were used. These have the same physical dimensions as the variables of interest, measured from past studies at the same site or from similar circumstances. The water pollution example used 'adjacent' variables. They have different dimensions from the variables of interest but were drawn from the experts' relevant knowledge base.

Only gross differences in calibration scores (factors of 2 or 3) were regarded as important. Information scores cannot be compared across studies.

Table 4.7. *Calibration and information for two sets of expert judgements of parameters (frequencies of events) (after Goossens et al. 1998)*

Case	Groundwater transport	Water pollution
Number of experts	7	11
Number of variables	48	21
Number of seeds	10	11
Calibration		
Performance weights	0.70	0.35
Equal weights	0.05	0.35
Best expert	0.40	0.16
Information		
Performance weights	3.01	1.87
Equal weights	3.16	1.75
Best expert	3.97	2.76

Most studies identify habitually overconfident experts. They are people whose information content is high (the spread of judgements is small) but calibration is low (they are far from the truth) compared to other experts. Typically, performance-based weights give more informative estimates than equal weights. The best expert is only rarely better than the performance-weighted group.

Cooke (1991) did not seek closure or resolution. His philosophy differs from that of Valverde (2001). Expert judgement is assumed intrinsically not to be consensual. Different experts are expected to have different, legitimate expectations (Cooke and Kraaikamp 2000). Instead, the method relies on accountability (transparency), empirical quality controls and methods that encourage honesty by giving experts their maximum scores when they state their true beliefs (Cooke and Goossens 2000).

There are other, related ways of attacking the problem of elicitation. Kaplan (1992) for instance, outlined a method that uses some of the structured approach of Cooke (1991) together with some elements of consensus and negotiation described by Valverde (2001). It is similar to the structured expert judgement procedure because it assumes a real valued parameter that experts try to estimate in a structured elicitation process. It is similar to the framework developed by Valverde (2001) in insisting that experts document and communicate the evidence they bring to any argument, and that they should be prepared to negotiate to a group consensus.

4.10.3 Eliciting judgements for biosecurity

Horst *et al.* (1998) used experts to estimate the chances that new diseases would arrive in the Netherlands on imported foods. They were concerned about viral diseases including foot-and-mouth disease, swine fever, swine vesicular disease, Newcastle disease and avian influenza.

Horst *et al.* (1998) used 43 disease control experts. Social strata included people with policy, research and field expertise. Participants answered questions about only one of the diseases. Elicitation used a single workshop (one evening). Participants used self-explanatory computer systems designed to '*minimize interaction of the participants with each other and with the workshop facilitators*' (p. 255). After the workshop, individuals were presented with group results and their own estimates and were asked if they wanted to revise their estimates.

Three-point estimation (minimum, maximum, best estimate) was used to obtain information about high-risk periods during which viruses were most likely to spread. The experts were provided with a list of hazards, including importation of livestock, importation of animal products, tourists, wildlife movements and so on. They were asked to rank their relative importance in contributing to disease introduction. They also estimated the expected number of outbreaks of various diseases in different European countries. The analysts used median results to aggregate expert opinions.

The protocols employed in Horst *et al.*'s (1998) analysis are a mixture of the methods outlined above. They did not make use of seed variables with known quantities, and therefore did not weight the contributions of different experts. Protocols such as this could be informed by statistical data bases on disease prevalence maintained by organizations such as the Office Internationale des Epizooties. There were no indications that Horst *et al.*'s (1998) results would be validated subsequently. The breadth of uncertainty and areas of disagreement among experts did not contribute in any transparent way to their results or discussion. While there can be no single method appropriate for all circumstances, some of the methods outlined above have the potential to improve elicitation and aggregation protocols in environmental risk assessments generally.

4.11 Using expert opinion

Overconfidence can be substantially improved by asking people to consider the reasons why they may be wrong (Morgan and Henrion 1990).

Performance is also enhanced when experts possess appropriate models and are trained to translate subjective assessments into numerical estimates. Using a consistent format allows experts to familiarize themselves with a consistent means of communication, providing them with the necessary normative skills.

The compelling conclusion is that the people who are asked to provide expert judgement of uncertain quantities must have the substantive knowledge to provide a reliable answer, and must also be trained to provide unbiased estimates. Training may take the form of estimates of relatively frequent and infrequent events, and immediate feedback to provide experience in estimating the central tendencies and tails of distributions. Guidelines may be tailored for specific circumstances. For example, Andelman *et al.* (2001) outlined options to elicit expert judgement to assess the viability of threatened species under the US Forest Management Act. The results of Wright *et al.* (2002) suggest strongly that experts require training and ongoing feedback to be competent.

Bier *et al.* (1999) noted that people have opinions about the output of risk assessments as well as the inputs. It is common to have people revise their estimates of parameters once they have seen the consequences of their beliefs. It is difficult for people to integrate parameters through complex functions intuitively. When the operations are done explicitly, the results may be viewed as impossible or unlikely, forcing a revision of the parameters or of the functions that link them.

While some analysts resist the revision of judgements based on knowledge of the result, I believe this is one of the fundamental benefits of formal risk assessments. Revision embodies the notions of learning and model updating that close the loop for environmental risk management.

4.12   Who's a stakeholder?

Proponents of risk assessment who appeal to them as a vehicle for 'sound science' take a traditionally narrow view of risk. Technical risks analysts, usually without training in sociology, often ignore the issue of public participation or treat it superficially with a 'public comment' phase.

The usual interpretation of the term 'stakeholder' in law is someone who has custody of the possessions of other people; for instance, when the possessions are held in trust or their possession is contested. In business, stakeholders usually are shareholders, the people to whom most corporations owe primary legal responsibility.

Dewey (1927) defined 'public participation' as deliberation on the issues by those affected by a decision. Interest group politics were a 'separate issue'. Recently, definitions of stakeholders in business have broadened to include shareholders, employees, suppliers, customers and the communities in which the firm operates (Walker and Marr 2001; see Freeman 1984).

More generally, stakeholders may be virtually anyone who might have an interest in an issue, and may include public interest groups, protest groups, government agencies, trade associations, competitors, unions, employees, customers and shareowners (Jennings 1999). In a social context, a stakeholder usually is an individual or a representative of a group affected by or affecting the issues in question (Glicken 2000). Some definitions include disenfranchised social groups and nonhuman entities such as threatened species of plants and animals.

Rights and obligations under stakeholder theory depend upon the understanding of the term 'stakeholder'. Whenever people decide to co-operate in seeking benefits that entail risk, there exists the possibility of free-riding. People may receive the benefits without bearing risk or making a contribution. Fairness among the participants in a cooperative scheme is achieved by allocating benefits in proportion to risk and contribution (Jennings 1999). But such views must be tempered by the opportunity to participate, and should consider a broad spectrum of costs and involuntary or indirect risks.

Donaldson and Preston (1995) created a definition of a stakeholder based on a concept of property rights. To them, all contributors to a firm's (or government's or undertaking's) success or those who carry some of the burden of risk associated with its activities have a legitimate claim to its success. They argued that anyone having a legitimate 'stake' should have recognized stakeholder status and be treated as an 'owner'. They distinguished the role of a stakeholder from that of an agent acting on behalf of stakeholders and that of a steward who holds things in trust, without any personal stake in the outcome (Preston 1998).

Windsor (1998) noted counter-movements favouring stronger stock-holders' rights and corporate-governance standards, in which economic considerations predominate. For instance, Jennings (1999) argued that nothing precludes a company from doing more than the law requires or less than the law allows but, she argued, those standards should not be imposed by 'roving bands of stakeholders' who have not invested their resources and who do not share the risk of failure of the enterprise.

When public resources are disposed of or social equity and risk tolerance are debated, there are no general rules that ensure that all relevant stakeholders are adequately represented. Stakeholder maps can assist to ensure coverage is relatively complete. Some common elements of most maps include (Glicken 2000):

- government organizations,
- scientists,
- residents,
- traditional land owners,
- nongovernmental organizations (NGOs),
- industry representatives,
- unions, and
- consumer groups.

The character and the weight given to each group will change as issues change. Geography, organizational structures, local politics and history of the issue contribute to the construction of stakeholder maps. Social landscapes may be surveyed. The scope of the survey should depend on what is wanted from the process.

Freudenburg (1999) outlined sources of information to support understanding of the sociopolitical settings of risk analyses. They included existing data (archival and agency information), first-hand data collection (stratified random surveys and targeted interviews) and methods to identify gaps and oversights (sensitivity analyses, interdisciplinary double-checks and public involvement techniques).

Models for stakeholder involvement include (Glicken 2000):

- *Paternalistic*: governments or regulators invite participation under strictly controlled conditions such as public comment phases.
- *Consensus*: self-designated representatives from all affected groups attempt to find consensus positions, with the aid of a chair or an independent facilitator.
- *Conflict*: resolution is by arbitration or litigation.

I believe the best solution is to involve those affected by the outcomes of risk assessments closely and continuously in the risk assessment process, making a marriage of the technical and social dimensions of risk. Nontechnical information should enter at the planning stage, and nontechnical participants should contribute to all the elements of the risk assessment.

Plans for stakeholder involvement usually need to anticipate resistance from technical experts who discount the opinions of nontechnical participants (Freudenburg 1999). Interest groups should be involved in initial scoping and hazard analysis. A suite of social values relevant to the issues of interest should be elicited from stakeholders. The elicitation process may include survey data, behavioural data, demographic statistics, facilitated meetings involving people who have been actively recruited, open public forums, or structured interviews of a sample of the population. These values may be converted into assessment endpoints, used to structure model formulation, data collection and analysis (Glicken 2000).

4.12.1   Stakeholder experts and expert stakeholders

Almost all experts have personal, value-laden opinions about the outcomes of environmental decisions (e.g. Campbell 2002). It is naïve to think that scientists are anything but advocates of scientific positions. They also advocate personal value systems.

In many circumstances, experts are members of interest groups. The stakeholder and the expert are the same person. In other circumstances, experts are retained by an interest group. If so, it is likely that their personal values will resonate favourably with the people who employ them.

When the analyst is a translator, as was proposed in Chapter 3, then 'experts' may include all those with a stake in the outcome. Including all views in construction of a model has the advantage that the result expresses what must happen if the assumptions and causal relationships specified by the stakeholders are true. Showing people the consequences of what they believe to be true can go a great deal of the way towards resolving conflicts and achieving consensus. If the analyst takes a realistic view and allows the participation of people with transparently divergent opinions, then the strategies for selection, elicitation and aggregation need to be explicit.

4.13   Discussion

Numerical approaches are useful mainly for circumstances in which experts estimate a parametric quantity that is, in theory, knowable, and in circumstances in which linguistic ambiguities and vagueness play a small role.

They are less appropriate when the risk analyst must contribute expert judgement to policy formation. It is unclear how they would function when the expert has a passionate commitment to the outcome of an assessment. Campbell's (2002) survey of marine turtle experts underlines the importance of the selection of experts, a topic on which so little is written.

Most formal risk assessment protocols adopt the view that risks are amenable to decomposition and objective measurement, and that subjectivity is the domain of risk communication (see Baker 1996). The degree to which an expert tends to be protective of the environment is strongly influenced by his or her values, and by what the expert stands to gain or lose, personally, by the outcome of a false-positive or false-negative judgement. When an analyst makes a judgement about an 'objective' element of a risk assessment, they provide a statement about confidence in available information, overlain by morals, values and beliefs of the analyst themselves, as well as those from whom information has been elicited. Baker (1996) suggested that risk decisions with subjective components can best be protected from damaging bias through consensus-building activities.

Sharpe (1996), in contrast, argued that acknowledging the values about what constitutes harm, who or what is harmed, and comparative risks and benefits, is a representation of the reality of human concerns. Differences should be retained and communicated. To avoid bias, it may be necessary to know what forces may influence a decision and what stake participants have in the outcome.

Scientific opinion involves advocacy of a position (a hypothesis) in which the advocate tries to convince others of the correctness of their position. Advocacy groups take advantage of this uncertainty to select experts whose position is sympathetic to a social position. They try to use the expert's authority to have views fixed in government, courts and other institutions.

Science reinforces obedience to authority, called the fallacy of the appeal to authority, by encouraging uncritical acceptance. Those who dispute scientific opinion may be dismissed as people who do not think rationally. Walton (1997) pointed out that scientific argument may be rigid in the sense that it is beyond refutation by nonexperts. Fischer (2000) and O'Brien (2000) argued that technical methods confer on the expert the appearance that they transcend partisan interests. Furthermore, the assessments become too complex to be challenged. When scientific opinion becomes irrefutable, it generates what Walton called a culture of technical control. The solution to decisions about new technologies, for instance, is seen to be through public education so that the correct solution embodied in scientific knowledge will be transferred to correct policy. There is a naked assertion in this situation that the identity of the proponent warrants acceptance of the proposal (Walton, 1997, p. 12).

Expert opinions have become widespread and accepted as scientific support tools in conservation and environmental management. Reliance on expert opinion relaxes the pressure to collect data (see Ruckelshaus *et al.* 2002). It is often difficult to distinguish direct measurements from expert opinion in data bases and official reports. This makes it hard to challenge expert opinion.

The appeal to expert opinion should be a fallible but legitimate strategy (Walton 1997). It is legitimate only if it can be challenged by anyone. It is broadly accepted that because experts often are advocates of a theory, and they therefore have a stake in the outcome, methods are required that balance different opinions. The methods outlined above attempt to achieve this goal. We will return to this issue in Chapter 12.

5 · Conceptual models and hazard assessment

Hazard assessment attempts to answer the question. 'What can go wrong?'.

A hazard is a situation that in a particular circumstance could lead to harm (Royal Society 1983). Hazard identification is the process of creating an exhaustive list of hazards. It involves documenting all events with unwanted outcomes that may result from natural circumstances, a proposal or human activity. Hazard assessment estimates the consequences of those hazards, if they were to occur. It relies on understanding cause and effect.

This chapter outlines the role of conceptual models in summarizing ideas and providing a platform for identifying alternative management options. It describes several hazard identification methods developed and applied mostly in engineering contexts, but which have much broader utility. It evaluates their strengths and weaknesses and provides some examples of environmental applications.

5.1 Conceptual models

All steps in a risk assessment, including hazard identification and assessment, depend on a decent conceptual model. Models are abstractions. They represent how we think the world works. We build models to answer specific questions, to assist us in making decisions.

As was outlined in Chapter 3, the purpose of a model determines its structure and limits. The kind of model and its complexity are a compromise between the questions and available time, expertise and knowledge. It is the responsibility of the person who builds a model to communicate the full range of uncertainties and assumptions.

A diagram is the simplest form of a conceptual model. Figure 5.1 shows the thinking of a group of hydrogeologists about the source and fate of toxic chemicals on a disused factory site in Italy. The diagram

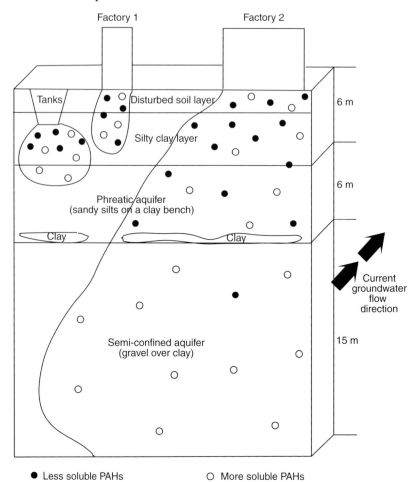

Figure 5.1. Conceptual model of the hydrology of a disused industrial site in Italy and the movement of a plume of polycyclic aromatic hydrocarbons (PAHs) from factories into groundwater and to adjacent sites (redrawn from Carlon *et al.* 2001). The model was used to develop ideas for building quantitative models to estimate risks of off-site contamination.

communicates several issues: the spatial scale of the problem, the level of detail (at least, the detail captured at this stage of problem formulation), conceptual compartments (factories, tanks, soil layers, groundwater layers and groundwater movement), soil composition in the layers, discontinuities such as clay caps and the boundary of the permanent aquifer, and at least two classes of chemicals. Thus, Figure 5.1 represents a conceptual

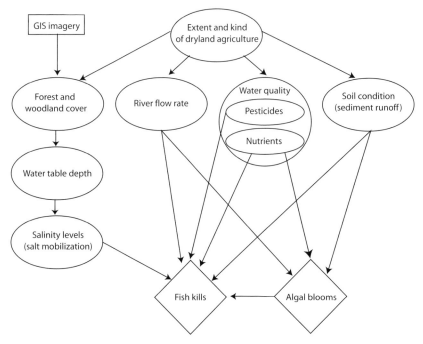

Figure 5.2. Influence diagram showing conceptual relationships among system components in a freshwater catchment (after Hart *et al.* in prep.). This diagram could be used as a starting point in an FMEA analysis (see Section 5.2).

model at a different organizational level than that represented by the fate of toxicants within a vertebrate body (Figure 3.4).

Influence diagrams represent the kinds of pictures in Figure 5.1 in a slightly different form, although there may be a one-to-one correspondence between them. They are a visual representation of the functional components and dependencies of a system. Shapes (ellipses, rectangles) represent variables, data and parameters. Arrows link the elements, specifying causal relations and dependencies (e.g. Figure 5.2).

The hazards in Figure 5.2 include salinity increases, changes in river flow, and deterioration in water quality and soil condition. Each contributes to at least one of the consequences that management is trying to avoid (the two diamonds). Data on changes in vegetation cover are available (shown by the rectangle). Major direct influences among system components are represented by arrows. Note, for example, that there are no compartments within the river. This captures an assumption that the chemical and biological composition within the river is more or less uniform. The figure assists in identifying causal relationships. It defines processes and pathways by which materials and energy flow through the system.

The model in Figure 5.2 postulates a number of relationships. As land is cleared of vegetation, transpiration rates decline. The water table rises, mobilizing salt from the soil profile as it moves. Agriculture can affect river flow rates by modifying direct runoff and by capturing water in dams. Water quality depends largely on agricultural practices and is defined in terms of pesticides and herbicides. Soil condition is a function of stocking densities and stock access to water courses.

The analyst works with experts, stakeholders and background information. The conceptual model may be developed iteratively until all participants in the risk assessment are happy that it captures adequately the way the system works. The model may then help to identify hazards or it may become a template for developing a mathematical model.

5.2   Hazard identification and assessment

Even with good conceptual models, it is easy to overlook something. A single hazard can lead to multiple adverse effects. Several hazards can have the same effect.

Consider identifying the hazards associated with burying high-grade nuclear waste underground (see Section 4.2). The facility needs to last 10 000 years, the time required for the wastes to degrade to the point where they are safe. What things would you consider? Earthquakes and volcanism may rupture tanks. The storage containers may corrode or leak, and material may be carried off site by groundwater. Are there any others?

In one such exercise, assessors simply overlooked the possibility that people over the next 10 000 years are likely to drill into the Earth's crust looking for oil, gas, minerals, fossils, water and so on. Some holes are drilled obliquely so even if the surface is protected, a drill may strike a long-term nuclear waste storage facility. It would generate a catastrophic release, perhaps even into the atmosphere.

It is often worthwhile separating proposals into phases. Things such as mines and storage facilities involve:

- exploration or site evaluation,
- benefit-cost (investment) analysis and facility design,
- construction,

- commissioning,
- operations and
- decommissioning.

Usually, different hazards need to be considered during construction (human health and ecological impacts of construction), commissioning (risks of release of radioactive materials during transport, for example) and operations (earthquake risk). Industrial production activities involve different phases, such as:

- research and design,
- feasibility trials and development,
- process design and purchasing,
- production,
- sales, training, use and customer service, and
- decommissioning and disposal.

The process may be iterative, including performance monitoring and redesign. The hazards (human health, economic, social and ecological) associated with each phase may be best treated separately.

A good hazard identification and assessment phase makes use of as many tools as possible, in an attempt to form as complete a list as possible. If you have been through this process once, you'll know how easy it is to overlook things, and how reckless it seems to perform this part of a risk assessment without using such tools.

In risk management systems, hazard identification continues throughout the life of a product or project. Hazards encountered by users, for instance, may not have been anticipated by developers or project managers. New hazards should be added to the register as they arise. Corrective management actions should then seek to reduce risks.

This chapter explores several approaches including hazards and operability analysis (HAZOP), failure modes and effects analysis (FMEA), and hierarchical holographic modelling (HHM) employed in engineering applications. Other useful tools such as fault and event trees will be covered in Chapter 8, although they can also be useful in conceptual modelling and in identifying and assessing hazards.

5.2.1 Checklists and brainstorming

Checklists and unstructured brainstorming are among the most common methods for assembling a list of hazards. Checklists are simple to construct and easy to use. They provide a record of the experience of people who have worked with the system. Brainstorming has the advantages of bringing new perspectives and of identifying causal relationships between system components and hazards.

Checklists and brainstorming usually identify most of the hazards that lie within the operating experience of the people involved. Their disadvantages are that they do not encourage the participants to extend their thinking to new possibilities. They may lead to the false impression that all potential hazards have been considered, particularly when existing lists are long and cumbersome (Hayes 2002a). They tend to overlook hazards when new technologies are introduced or new stresses are imposed on a system.

Lists and unstructured brainstorming are particularly susceptible to linguistic ambiguities and vague definitions. Words and phrases used to describe hazards may be misinterpreted by participants and by people who later use these lists to rank or quantify risks. They can be used most effectively in tandem with conceptual models and some of the other strategies outlined below. Because they represent a repository of collective experience and wisdom, they can be used after application of a more inductive strategy as an additional check on completeness.

5.2.2 Structured brainstorming

The Delphi and related techniques are outlined in detail in Chapter 4. While they may be used for a variety of purposes including elicitation of parameters, they are often useful in putting together a relatively complete list of hazards, and scoping alternative conceptual models. When used for conceptual modelling and hazard identification, some of the steps may include:

- problem formulation and development of questionnaires,
- selection of experts,
- provision of background information, definitions and context to experts,
- elicitation of conceptual models and lists of hazards (often performed by participants independently),
- aggregation of results,
- review of results by experts and revision of answers, and
- aggregation of results, or iteration of feedback until consensus is achieved.

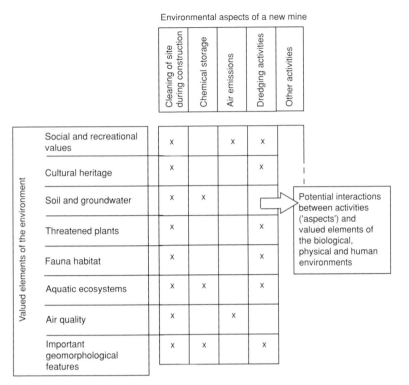

Figure 5.3. Part of a hazard matrix for an environmental impact assessment of a new mine (after Zaunbrecher 1999).

The drawbacks are that these methods can sometimes encourage uniformity. Some approaches do not give participants much opportunity to learn from one another. Variations on the theme detailed in Chapter 4 have useful applications in this arena, especially when many experts and stakeholders are involved.

5.2.3   Hazard matrix

Hazards may be characterized as a matrix of interactions between activities and components of the environment that may be affected by the actions (e.g. Figure 5.3). Hazard matrices are particularly helpful in identifying hazards that have multiple effects, and in identifying hazards associated with different operational components of a project. Construction depends usually on checklists and brainstorming.

The links between aspects and environmental components represented by the crosses in Figure 5.3 are a kind of conceptual model of environmental interactions, although the exact nature of the relationships is not specified. The matrix improves the probability that no interactions are overlooked, generating a more comprehensive list of hazards than brainstorming alone.

The Burrup Peninsula in Western Australia is home to a suite of unique marine and terrestrial biological communities. The broader region of Dampier Archipelago was listed as having 'extreme conservation value' by the Western Australian Department of Environmental Protection. It is also the location of Aboriginal heritage sites, and National Estate listed heritage sites. During the 1990s, a large international corporation (then called BHP) wanted to build a plant there to produce methanol.

BHP conducted a hazard identification exercise. The first step was to develop designs for the project, and to collect baseline information on heritage values, local communities, and the natural and physical environment. The project was broken up into commissioning, operations and decommissioning phases. The baseline information provided background against which potential impacts of human activities could be assessed, using a hazard matrix to identify causal factors associated with potential effects.

Consequence tables were constructed separately for the marine environment, the terrestrial biological environment, the social environment, heritage values and traditional Aboriginal cultural values.

The design of the operations phase included a cooling tower. The tower would generate a plume of salt-laden spray, derived from the ocean water used in the cooling process. The engineering plan was to minimize impact through 'facility design' and to assess vegetation for potential impacts. That is, the engineers had identified an environmental hazard and wanted to know the magnitude of the consequences.

The ecologist who did the assessment was not able to specify the likely extent of changes in the vegetation that would result from the deposition of several tonnes of salt per hectare per year from the cooling tower. Certainly, salt spray is a natural feature of the environment, but not in the quantities envisaged from the cooling tower. The spray plume would extend several kilometres. The impact was outside her (or anyone else's) experience.

The ecologist communicated her judgements by saying that there would almost certainly be at least a minor change in vegetation cover and composition, because several species were not especially salt tolerant,

Table 5.1. *Part of a HAZOP table used to elicit judgements about hazards resulting from wildfire from national park managers in Victoria, Australia*

Guide word	Interpretation	Causes	Consequences	Action

'None'	No fire	No ignition
		Too wet
		Suppressed
			Build up of fuel
			Aging of
			  vegetation
				Fuel
				  reduction

'More of'	Greater extent
	  of burn
		Ignition
		High temperatures
		High winds
		High fuel
			Water quality
			  decline
			Water yield
			  increase
			Change in
			  habitat
			  quality for
			  animals
				Suppression

'More of'	Additional fire	Ignition
		High temperatures
		High winds
		Residual fuel
			Elimination of
			  reseeding
			  plant species
				Suppression

while others were. She also judged that it was at least possible that a major change would occur in the vegetation within the salt plume, with many species and most of the cover lost because of the additional salt load. The consequences of a major change would accrue to both the environment and corporate reputation (visible, substantial damage to the environment adjacent to company operations in a socially sensitive area).

The issue was taken with sufficient seriousness for engineers to redesign the cooling tower to use fresh water. The costs of the solution were well worth the benefits of avoiding a substantial hazard (Zaunbrecher 1999).

5.2.4 Hazard and operability analysis (HAZOP)

Hazard and operability analysis (HAZOP) is a kind of structured, expert brainstorming session. It uses conceptual models and influence diagrams together with guide words such as 'more of', 'less of' and 'reverse flow' to prompt the thinking of a small team of experts (e.g. Table 5.1). The experts, guided by a facilitator, apply 'what if' type questions to each component of a system in a systematic manner (Kumamoto and Henley 1996, Kletz 1999, Lihou 2002). The words are designed to encourage a group of experts to interrogate a system and apply their expertise beyond their own experience (Hayes 2002a).

HAZOP has been used to assess technical risks for several decades (CIA 1977). It was pioneered in the chemical industry and spread to civil, pharmaceutical, food processing industries and, to a lesser extent, to nuclear power and defence.

The process generates a repository of information containing actions, responses, dates and details of implementation, and references to external information. The facilitated meeting involving people with broad knowledge and experience aims to reduce the chance that something is overlooked (Kletz 1999). It depends on a conceptual model of the system. It is open-ended, more likely to identify all potential hazard scenarios than checklists or unstructured brainstorming, as are HHM and FMEA (below; Hayes 1997, 2002b).

Box 5.1 · *Structure of HAZOP procedures (after Kletz 1999, Hayes 2002a).*

The process operates as follows:

1. A group of experts is assembled.
2. A list of key words is compiled that describes the system, its components and operational characteristics.
3. If the list is large (usually they are), the words are split into manageable sections associated with different subsections of the system.
4. The list is distributed to the experts. They discuss potential problems in the system.
5. A facilitator (or a computer program) prompts the use of keywords and guide-words to stimulate thinking.
6. Potential problems are recorded as they are discussed.
7. The group aims to reach consensus on hazards associated with each part of the system and to specify what needs to be done. These deliberations are summarized in an action sheet that summarizes cause, consequence, safeguards and actions for each hazard.
8. Action sheets including deadlines for implementation are distributed to relevant operational personnel.
9. Personnel are required to submit response files that document implementation, feedback and any recommended additional actions. These are available for review and audit.

HAZOP elicitation is tedious. Kletz (1999) recommended developing influence diagrams and conceptual models beforehand. Meetings should be restricted to half-day sessions to avoid fatigue. Assessing complex projects may take many weeks.

5.2.5 Failure modes and effects analysis (FMEA)

Failure modes and effects analysis (FMEA) is a deductive process that aims to reduce risk (FMEAInfo Center 2002, Haviland 2002). Failure Modes are categories of failure, describing the way in which a product or process could fail to perform its desired function, defined in terms of the needs, wants and expectations of people (shareholders, customers or stakeholders). An FMEA is defined as 'a systematic process for identifying potential design and process failures before they occur, with the intent to eliminate them or minimize the risk associated with them' (FMEAInfo Center 2002).

FMEA shares this primary objective with HAZOP procedures but does not rely on structured brainstorming. Instead, it examines the behaviour and interaction of individual components ('elements') of a system to enable the consequences of undesired events to be assessed. It provides a detailed examination of causal relationships between elements in a system, in addition to generating a list of hazards.

The process involves calculating a 'risk priority number' (RPN) for each hazard (Figure 5.4). The number is the product of three quantities, *Severity*, *Occurrence* and *Detection*. A cause is the means by which a particular element of the process fails.

- *Severity*: an assessment of the seriousness of the effect of the failure.
- *Occurrence*: an assessment of the likelihood that a particular cause will lead to a failure mode during a specified time frame.
- *Detection*: an assessment of the likelihood that the current controls (design and process) will detect the cause of the failure mode or the failure mode itself, thus preventing it from occurring.

The RPN is used to set priorities for action on hazards and to identify elements that require additional planning. Critical thresholds may be set, above which action is mandatory. Actions are usually attempts to lower severity or occurrence. Adding validation or verification controls can increase the chances that a problem will be detected (thereby reducing detection scores). Design or process improvements may result in lower severity and occurrence ratings.

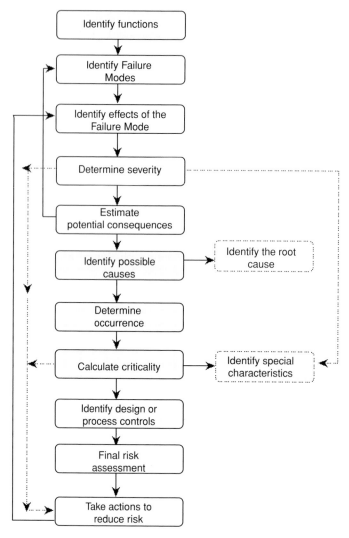

Figure 5.4. The elements of FMEA in a typical quality system (after Haviland 2002).

FMEA 'elements' are the individual components of a conceptual model. Typically, they are identified and analysed in the FMEA process. They appear as column headings in the forms resulting from the process. Commonly, the forms list Functions, Failure Modes, Causes, Effects, Controls and Actions.

Failure Modes are defined by the perspectives of the people involved in the risk assessment, or by those who bear the consequences of failures (see Haviland 2002). This recognizes the inherently subjective nature of hazard definition and would translate in a setting dealing with environmental risk to specifying the reasons why particular outcomes are considered to be hazards, from the perspectives of different stakeholders.

The method accommodates the fact that a single cause may have multiple effects, and that a combination of causes may lead to a single effect. Causes can themselves have causes (fish kills may be caused by elevated salinity, in turn caused by vegetation clearance; Figure 5.2). Similarly, effects can have downstream effects. Causes do not automatically result in a Failure Mode. The term 'potential' is often used to describe causes, to reflect this uncertainty.

Logic trees (described in Chapter 8) and influence diagrams (e.g. Figure 5.2) are complimentary techniques. They assist in describing and understanding causal relationships among system elements. FMEA input usually involves a multidisciplinary team with expertise in all aspects of the function and control of a system. The process by which information is elicited and gathered is not formally specified. Methods described in Chapter 4 may be adapted to this protocol.

FMEA has a great deal in common with the risk ranking methods described in detail in Chapter 6. It was developed originally by the United States Military in the 1940s and was later adopted and developed by automotive manufacturers (FMEAInfo Center 2002). In 1988, for example, the International Organization for Standardization issued the ISO 9000 series of business management standards. A task force including people from Chrysler Corporation, Ford Motor Company and General Motors Corporation developed QS 9000 to standardize supplier quality systems. QS-9000-compliant automotive suppliers must use FMEA in the planning process (FMEAInfo Center 2002).

FMEA (together with fault tree analysis (Chapter 8) and HAZOP procedures) is recommended by the international standard for risk management of medical device manufacture and operation (ISO 14971–1 1998). The US FDA (1998) evaluated 582 design control systems used by medical device manufacturers. A total of 285 firms used risk analysis techniques in their control systems. Of these, 71% of firms used some form of FMEA. A total of 15% included fault trees.

In a contrasting application, Hayes (2002b) used the approach to identify hazards when assessing the risks of invasive species. However, applications in environmental and natural resource management remain rare.

5.2.6   Hierarchical holographic modelling (HHM)

Hierarchical holographic models (HHM) recognize that more than one conceptual (or mathematical) model is possible for any system. They try to capture the intuition and perspectives embodied in different conceptual and mathematical models of the same system (see Haimes 1998, Haimes *et al.* 2000).

As noted above, sources of risk may be decomposed into different kinds, such as functional, operational, spatial, social and ecological risks. Each of these sources may be further decomposed hierarchically so that ecological risks may include marine, estuarine and terrestrial environments. Within the marine environment, it may be useful to explore benthic, pelagic and shore environments separately. Each submodel is a complete view of the system from a single perspective, a holographic submodel (Bier *et al.* 1999).

One modeller's view of a system carries with it biases, assumptions and simplifications that are peculiar to that modeller. Recognizing this means that it becomes impractical to represent any system by a single model.

HHMs aim to be holographic in the sense of embodying as many perspectives as possible. The broad perspective they bring should result in relatively comprehensive lists of hazards. HHM's are most effective when the analyst is able to identify and list all the important components and processes in a hierarchical fashion. HHM provides an ordered way of dealing with structural uncertainties in models.

HHM was designed to deal primarily with infrastructure (Haimes 1998) but has potential for much broader application (Hayes *et al.* 2004). It has been used to ensure more complete assessments of failure modes in a wide range of disciplines (Bier *et al.* 1999). The collective knowledge embodied in a number of models may provide an adequate summary of a range of opinions from different stakeholders and decision-makers (Haimes *et al.* 2000).

Hayes *et al.* (2004) developed an HHM to explore hazards associated with growing herbicide-tolerant Canola (Figure 5.5). The ecological system is large and complex. Many elements are poorly understood. Data are missing that would support a full, quantitative examination of risks. HHM deals with these uncertainties by identifying 'components' and

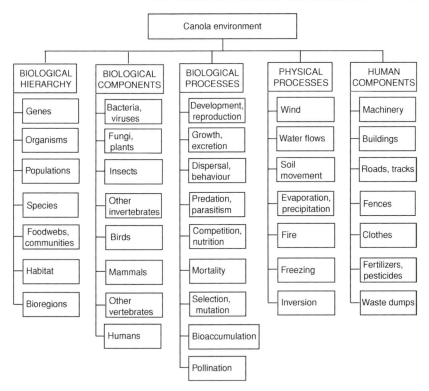

Figure 5.5. Hierarchical holographic model for hazards associated with herbicide resistant Canola (a subset of the elements explored by Hayes *et al.* 2004).

'processes' of subsystems. Tables are then used to suggest interactions and organize supporting information.

Each subsystem is a 'perspective', part of the hologram. They may be subjective conceptual models or detailed mathematical models. The analyst is responsible for finding appropriate perspectives, system boundaries and levels of detail/aggregation.

Hayes *et al.* (2004) avoided sociopolitical, economic and geographic (regional) perspectives, although they may have been legitimate. They restricted themselves to human, biological, chemical and physical components and processes of the environment (Figure 5.5).

A study team composed of technical experts considered 1356 potential interactions between these components and processes. They evaluated each one and ranked their potential to cause adverse environmental impacts. Team members scored a 'degree of concern' (high, medium, low) and their degree of confidence in the plausibility of the hazard (a value between 0 and 1). Scores were aggregated over individuals by summing concern scores and multiplying by average confidence (Hayes *et al.* 2004).

The ranks provided a basis for setting priorities for further analysis (Table 5.2). The process identified hazards that had been overlooked in checklists and unstructured brainstorming conducted previously.

Scenarios are a useful conceptual device for hazard identification and assessment akin to HHM, relying on verbal models rather than quantitative or qualitative models. They are hypothetical sequences of events that are put together to focus attention on causal relationships and decision points in a conceptual model (Cooke 1991, Van der Heijden 1996). They provide insight into precisely how an event may occur, and what management options exist to avoid the event or remediate its consequences. Kahn and Wiener (1967) developed this approach. It is outlined in greater detail in Chapter 12 because of its utility in risk management.

5.3 Discussion

Hayes (2002a) reviewed eight risk regulatory approaches. Only one – New Zealand's Environmental Risk Management Authority – identified a range of deductive and inductive hazard identification techniques, including brainstorming, checklists, logic trees and HAZOP analysis. The Organization for Economic Co-operation and Development (OECD) safety considerations recommended fault and events trees as a means to quantify probability but otherwise provided a simple checklist of potential environmental hazards.

Four of the remaining frameworks reviewed by Hayes (2002a) provided similar checklists. The UK Department of Environment, Transport and Regions and the Office for Gene Technology Research provided the most comprehensive lists. The European Community directive simply listed five generic hazards including toxicity, impacts on population dynamics, altered susceptibility to pathogens and effects on biogeochemistry. The Cartagena Protocol on Biosafety and United Nations Environment Program guidelines did not provide checklists or discuss any hazard identification techniques.

There is an opportunity to improve routine risk assessments for environmental management and conservation because there is a wide range of methods for hazard assessment that are rarely used. More complete lists of hazards should result from structured identification and assessment protocols. Confidence in the hazard assessment process would be enhanced

Table 5.2. *Ranked potential hazards (and benefits) associated with herbicide-resistant Canola. Hazards that were not identified in an equivalent checklist approach are noted by an X (hazards are from a list of several hundred, after Hayes* et al. *2004)*

Hazard category	Potential hazard	Rank	Checklist
Farming practice	Greater extent of Canola increases chance of weed tolerance development	14.8	
Dispersal	Off-site transport of pollen by insects	14.7	
Volunteers	Subsequent crop seed contamination	14.3	
Dispersal	Off-site transport by farm machinery	13.8	
Segregation	Consumer demand for segregation of harvest	12.9	X
Dispersal	Seed dispersal along transport routes	12.2	
Volunteers	Seed loss during harvest	12.0	
Segregation	Building and processing costs to maintain segregation	12.0	X
Volunteers	Need for alternative weed strategies to eliminate volunteers	11.3	
Unexpected expression	Gene expression in roots may modify exudation	11.2	
Volunteers	Additional monitoring requirements for volunteers and other herbicide-resistant weeds	11.2	
Farming practice	Inhibition of organic farming in region	10.7	X
Farming practice	Avoidance of region by beekeepers	10.2	X
Farming practice	Reduction of crop rotation options	10.1	X
Dispersal	Pollen spreads to conservation areas	10.1	

if lists of potential hazards and their interactions were continually updated to accommodate new knowledge and experience. Monitoring systems may influence risk assessments qualitatively by providing information to revise conceptual models and to identify novel causal relationships between components and processes in a system.

FMEA and HHM have properties that make them close relatives of the risk ranking methods outlined in Chapter 6. In particular, arithmetic is performed on subjective degrees of belief to generate ranks. These ranks may be used to set priorities for subsequent actions. The advantages and weaknesses of this approach are outlined in detail in Chapter 6.

Risk assessments should be tiered. For instance, Hill *et al.* (2000) described 'screening-level' and 'detailed' risk assessments. Different tools are useful in different contexts and at different levels of analysis. If hazards are unacceptable, the risk analysis may need to go no further than identifying hazards to be avoided or engineered out of existence. The benign nature of potential consequences may guide the analyst towards a simple qualitative assessment. More severe consequences may lead to monitoring and contingency plans, or to a more detailed exercise involving data collection and mathematical modelling. Everything is contingent on the premise that risk assessment has been guided at the outset by good conceptual models.

6 · Risk ranking

Risk ranking represents one of the most common forms of risk analysis. It is used extensively in engineering, mining, land development and industrial contexts in many countries. Yet, as currently practised, it is particularly susceptible to the vagaries of human perception and the inconsistencies of expert judgement outlined in previous chapters. Fortunately, there are some remedies to these problems.

This chapter describes how risk ranking is done. It explores some of the characteristics of the approach. It outlines ways of conducting risk ranking to evaluate more reliably the risk-weighted costs and benefits of environmental management options. These modifications will make risk ranking more useful in the context of the risk management cycle to guide managers to develop strategies that eliminate, reduce or mitigate risks.

6.1 Origins of risk ranking methods

Risk ranking is a risk assessment that relies on qualitative, usually subjective, estimates of likelihoods and consequences. It avoids the technical demands of more formal techniques and may use quantitative information where it is available.

In the late 1960s and 1970s, NASA followed examples set by the US military and adopted 'risk assessment tables' to assist analysts to quantify and set priorities for risks (Table 6.1). The objective was to create risk assessments that were more reliable than those conducted subjectively and without any guiding principles. Reliability was judged by the repeatability of the resulting relative risk assignments.

The tables had four components. The 'hazard probability rank' was a verbal description of the relative likelihood of the event (the hazard) (Table 6.1a). They ranged from frequent or continuous to improbable. Terms were based on Kent charts used to relate linguistic terms to frequencies (Chapter 4). It was not necessary to estimate frequencies of events, although frequency data could assist.

Table 6.1. *Elements of the NASA Risk Assessment Table (after Wiggins 1985, in Cooke 1991)*

a. Hazard probability rank

Level / scale	Description	For an individual item	For fleet or inventory
A	Frequent	Likely to occur frequently	Continuously experienced
B	Probable	Will occur several times in the life of an item	Will occur frequently
C	Occasional	Likely to occur at some time in the life of an item	Will occur several times
D	Remote	Unlikely but possible in the life of an item	Unlikely, but can be expected to occur
E	Improbable	So unlikely it can be assumed it may not be experienced	Very unlikely, but possible

b. Hazard severity categories

Level / scale	Description	Scenario / details
I	Catastrophic	Deaths or system loss
II	Critical	Severe injury, major system damage
III	Marginal	Minor injury, minor system damage
IV	Negligible	Less than minor injury or system damage

c. Risk management matrix

Hazard probability	Hazard severity			
	Catastrophic	Critical	Marginal	Negligible
Frequent	1	3	7	13
Probable	2	5	9	16
Occasional	4	6	11	18
Remote	8	10	14	19
Improbable	12	15	17	20

d. Risk acceptance index

Score	Category
1–5	Unacceptable
6–9	Undesirable
10–17	Acceptable with review
18–20	Acceptable without review

'Severity categories' were verbal descriptions of the importance (dreadfulness) of the consequences of the hazard (Table 6.1b). They were defined in terms of loss of human life, injuries, dollars or some other metric representing cost.

The method worked by assigning point scores to various combinations of hazard probability and severity (Table 6.1c). These scores represented the 'risk', although the values were not probabilities. Rather, they were weights reflecting judgements about the importance of the hazard for risk managers. The risks were compared to a table of thresholds describing acceptability (ranging from 'unacceptable' to 'acceptable without review'; Table 6.1d).

These tables carry assumptions. For instance, the time frames within which probabilities were judged were implicit. Despite being called 'quantitative' assessments, no attempt was made to quantify probabilities or consequences. Cooke (1991) thought this reflected distrust of subjective numerical representations of uncertainty, and concerns that numbers would be interpreted as though they were exact.

6.2 Current applications

Risk ranking procedures rely on experts to estimate qualitative categories. They are used to assess both technical and social risks in a range of institutions. They combine a subjective judgement of the relative likelihood that each hazard will be manifested with a subjective assessment of the magnitude of its consequences. The estimation procedure is subjective, implicit and undescribed. Risk is the product of the likelihood and magnitude of the consequences of a hazard. Figure 6.1 shows a general framework for risk ranking.

The framework in which risk ranking is done (Figure 6.1) has a great deal in common with the framework promoted by the US EPA (Chapter 3). There is little doubt that the compartmentalization of consultation and risk communication, for instance, is a result of the cross-fertilization of ideas.

This approach was recommended by the British Institution of Civil Engineers, the British Institute of Actuaries, and Standards Australia (ICE/FIA 1998, AS/NZS 4360 1999). It is an accepted part of the machinery by which corporations achieve compliance with International Standard ISO14001, a hallmark of environmentally sustainable practices. Evans (1999) called AS/NZS 4360 (1999) '...a welcome evolutionary development in the general application of risk management

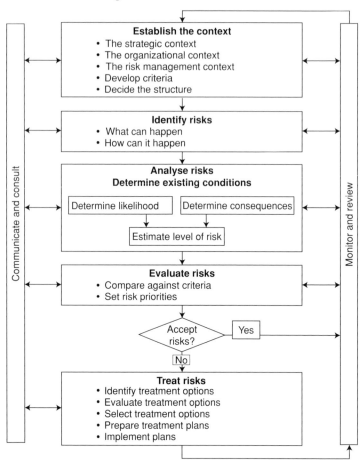

Figure 6.1. Risk ranking framework (after AS/NZS 4360 1999).

principles . . . '. It is an integral part of Failure Modes and Effects Analyses (FMEA) developed by the US military and the US automotive industry and recommended by the International Organization for Standardization (Chapter 5). It represents perhaps the most common form of risk assessment.

Despite the long list of approvals, the method has some serious weaknesses if applied carelessly. The following section outlines how risk ranking exercises are conducted, examines the extent and consequences of uncertainty in these kinds of analyses, provides an example of their application and recommends approaches that will maximize their reliability and utility for decision-making.

6.3 Conducting a risk ranking analysis

Risk ranking depends on a sound conceptual model and completion of a hazard identification and assessment phase. Usually, the list of hazards and a description of their consequences is the result of unstructured brainstorming. It could involve the techniques that are part of FMEA or Hazards and Operability Analysis (HAZOP) procedures, or even Hierarchical Holographic Modelling (HHM) if the problem is well defined and there are a number of conceptual or mathematical models available (see Chapter 5).

Once hazards have been uniquely defined and their consequences have been specified and described, estimates for both probabilities and severities (likelihoods and consequences) are made. However, usually quantitative estimates of these quantities are unavailable and there may be no quantitative expression for some consequences. For example, it may not make sense to try to quantify the outrage felt by people about exposure of infants to a carcinogen. Most often, likelihoods and consequences are classified subjectively into one of a few classes (e.g. Table 6.2a,b).

Once each hazard has been allocated to a likelihood and a consequence class, the scales associated with the subjective categories are multiplied together to give a nonprobabilistic 'risk' (Table 6.2c). This value is used to rank risks and to set priorities for evaluation, communication, treatment and monitoring.

Not much has changed in 40 years. Scores and 'probabilities' have been added to Table 6.1a. Otherwise, there is little difference between the recent set of tables and the earlier ones developed by the US aerospace industry. In other examples, the classes are defined with words tailored to suit the specific context of the risk assessment. There may be as few or as many classes as suit the purposes of the problem at hand. Other systems use as many as 10 categories for both consequence and likelihood although more than five to six categories may be unwieldy. When outcomes are unambiguous and data are available, the thresholds for likelihood are sometimes tied to specific probabilities.

Often, the consequence table is split into categories, each of which deals with a different dimension of effects (such as human health, social, ecological and economic risks). For instance, tables may be created for the influence of hazards on share prices or the public reputation of a corporation. Different tables may represent consequences for different segments of society, reflecting exposure, or attitude to an impact.

Table 6.2. *Elements of the British Risk Ranking Table (after ICE/FIA 1998. cf. AS/NZS 4360 1999; Burgman 1999)*

a. Qualitative measures of likelihood

Level / scale	Description	Scenario / details	Probability
16	Highly likely	Expected to occur in most circumstances	Over 85%
12	Likely	Will probably occur in most circumstances	50–85%
8	Fairly likely	Might occur at some time, quite often	21–49%
4	Unlikely	Could occur at some time	1–20%
2	Very unlikely	Not expected to happen	< 1%
1	Extremely unlikely	Just possible but very surprising	< 0.01%

b. Qualitative measures of consequence

Level / scale	Description	Scenario / details
1000	Disastrous / catastrophic	Deaths, toxic release off-site with substantial detrimental environmental effects, bankruptcy
100	Severe / major	Extensive injuries, release off-site with detrimental effects, serious threat to business
20	Substantial / moderate	Medical treatment required, toxic release on-site contained with outside assistance, or off-site release with no detrimental effects, significant reduction in profit
3	Marginal / minor	First-aid treatment, on-site release immediately contained, small effect on profit
1	Negligible / insignificant	No injuries, no important environmental effect, trivial effect on profit

c. Risk management matrix

Likelihood	Consequence				
	Negligible	Marginal	Substantial	Severe	Disastrous
Highly likely	16	48	320	1 600	16 000
Likely	12	36	240	1 200	12 000
Fairly likely	8	24	160	800	8 000
Unlikely	4	12	80	400	4 000
Very Unlikely	2	6	40	200	2 000
Extremely unlikely	1	3	20	100	1 000

d. Risk acceptance index

Score	Category
Over 1000	Intolerable
101–1000	Undesirable
21–100	Acceptable
Up to 20	Negligible

The information from such assessments usually is compiled and supplied to project planners, design engineers, quality control people, regulators and stakeholders. Reviews, updates and monitoring are sometimes implemented.

There is no need to be guided rigidly by the product of likelihoods and consequences. A likely hazard with a negligible consequence may be ranked lower than an extremely unlikely hazard with a severe consequence. These may be adjusted to reflect the attitude to risk of the assessor or the decision-maker. Some stakeholders may be more risk averse than others, necessitating different classifications for different elements of the community.

6.4 Pitfalls

6.4.1 Selection, elicitation and aggregation

Perhaps the most obvious difficulty is the use of (usually) expert subjective estimates. Most risk ranking exercises are composed almost entirely of expert judgement, 'supported' by various amounts of background information and technical detail.

Yet proposals for risk ranking do not provide any advice on how to select experts, elicit estimates or combine judgements. In most applications, rules for elicitation and aggregation are unspecified.

These deficiencies do not make subjective risk ranking exercises useless. At the least, risk ranking provides a means by which environmental hazards are treated by decision-makers in the same way as social and economic hazards. They are guaranteed a hearing. But the process could be more complete and honest.

The utility of different approaches to expert elicitation and aggregation will depend on the nature of the problem, the kinds of uncertainties involved, and the social and cultural context (Chapter 4). These issues should be resolved before a risk ranking exercise commences.

6.4.2 Discrete hazards

Another difficulty is that hazards are treated as discrete events with discrete consequences, whereas many events have a continuous range of potential outcomes.

For example, ADD (1995) assessed the possibility of a ship grounding and releasing oil in a new port facility (Chapter 1). We could draft a conceptual model for an oil spill but, as we noted in Chapters 1 and 2,

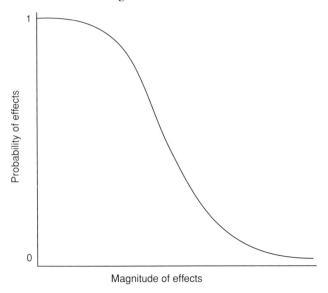

Figure 6.2. The likelihood and consequences of a hazard that would be best represented by a continuous range of outcomes (after Suter 1993).

there are many kinds of ships, different kinds of oil and a range of sizes of oil spill. Any single event could generate a continuous range of outcomes. Consequences of the greatest magnitude typically occur with low probability (Figure 6.2).

Some applications select the most likely consequence, or select a small range of relatively likely outcomes, and ignore the rest. One of the errors sometimes committed in risk ranking is to ignore events with catastrophic consequences, treating them as unrealistically unlikely and therefore trivial. This essentially truncates consideration of extreme events. It runs contrary to the spirit of risk assessment, which is to explore the extremes of possibility, the tails of event distributions.

Alternatively, the range of potential outcomes may be subdivided into a few discrete classes and each range treated as a separate hazard. Analysts define hazards in terms of their consequences, and then estimate the likelihoods of events with the specified consequences. This accommodates extreme outcomes and makes thinking about consequences easier, but it may create additional linguistic uncertainty.

6.4.3 Model complexity

The degree of disaggregation in a conceptual model (Chapter 5) often depends on the knowledge of the people involved. This problem is common to all methods for risk assessment but it is especially apparent in risk ranking exercises.

Engineers assessing the environmental risks of a new port facility will spend a great deal of time breaking down and evaluating the likelihoods of failures in things such as pumps, structures and storage facilities. A hazard such as a fire that starts in adjoining vegetation on-shore is likely to be treated as a single hazard.

If an ecologist joins the team, they are likely to lump engineering structures into a few crude baskets. They will, however, be preoccupied by whether the fire is a summer or a winter fire, whether it is a relatively hot fire, whether there was a fire the year before and so on, because these things determine its ecological consequences.

Arbitrary model complexity may be redressed by employing teams with broad technical expertise. Stakeholder involvement broadens the professional narrowness of experts. In addition, once conceptual models are complete, it is important to revise and simplify them in the light of the questions it is necessary to answer.

6.4.4 Susceptibility to risk perception

Risk ranking is particularly susceptible to the psychology of risk perception. Recall the things that affect judgement outlined in Chapters 1 and 4. They include the level of personal control, understanding of the issues, extent of personal experience, apparent dreadfulness of the outcome, equitability of distribution of the risk, cultural context, cognitive biases (judgement bias, framing effects, anchoring and insensitivity to sample size), motivational biases and advocacy.

The rich and extensive literature on the consequences of human psychology for judgements about risk has been ignored in the design of risk ranking protocols. There is an opportunity to design methods that anticipate biases and cognitive deficiencies, and thereby produce more reliable and transparent assessments. Some recommendations are outlined below.

6.4.5 Susceptibility to linguistic uncertainty

The idiosyncracies of human psychology that may affect subjective risk assessments are compounded by linguistic uncertainty (Regan *et al.* 2002a, Chapter 2). It is difficult to define terms and concepts in such a way as to ensure they are interpreted the same way by all those involved in the process. However, behavioural elicitation methods (Chapter 4) can eliminate elements of linguistic uncertainty.

Ambiguity may be reduced by careful attention to the definition of terms and is usually improved by allowing participants to discuss meanings and context. Vagueness may be dealt with most simply by defining arbitrary, discrete classes, although it is not always satisfactory to do so (Chapter 2).

6.4.6 Acceptability

Defining an acceptable level of risk is ultimately a social decision. It involves risk-weighted trade-offs between the prospect of unacceptable environmental damage and interference in relatively benign and productive activities (Fischhoff *et al.* 1981, Finkel 1995). Decisions about acceptability are tied to the thresholds in risk acceptance tables (e.g. Table 6.2d). The choices of these thresholds are rarely justified or explored. Often, acceptability depends on the way in which risks are expressed and communicated. This topic is addressed in greater detail in Chapter 12.

6.4.7 Sensitivity and validation

It is a striking feature of Figure 6.1 that it omits sensitivity analysis and validation. The general framework doesn't prescribe these things and they are almost never done. This creates a substantial gap in the risk management cycle (Chapter 3). Without them, the risk assessment is incomplete.

Risk ranking protocols could include a step in which an alternative set of hazards is identified and ranked, to determine the sensitivity of results to the way the conceptual model was interpreted. There is no reason why the reliability of ranks for a given set of hazards could not be assessed independently by different people or groups, and their results compared. The process could involve specifying conditions under which monitoring data could be used to evaluate the ranks for a set of hazards.

There is clearly room for improvement. Steps may be implemented to explore the sensitivity of management priorities and decisions to assumptions. Guidelines could be developed to validate estimates of likelihoods, consequences and risk ranks.

6.4.8 Unacknowledged uncertainty in results

Expert psychology, the complex nature of most ecological systems and the lack of attention paid to linguistic precision in risk ranking suggest that assessments of the ranks of environmental hazards will be unreliable. Despite the considerable uncertainty in all risk ranking exercises, ranks are presented as points, as though they were exact. This convention has the potential to mislead everyone involved.

In one example, UDMH, a breakdown product of the growth regulator Alar used on apple crops, was considered to be dangerous to humans. Ames and Gold (1989) explored the claim that Aflatoxin in a daily ration of peanut butter was 17.6 times more dangerous than UDMH in a daily ration of apple juice. The implicit rationale suggested that if you eat peanut butter, you should be prepared to eat treated apples.

Finkel's (1995) analysis revealed that the central tendencies (both median and mode) of the pdfs for Aflatoxin and Alar were nearly the same. However, one could be only 90% certain that the relative risk of Aflatoxin to UDMH lay somewhere between 300 : 1 in favour of Aflatoxin and 35 : 1 in favour of UDMH. The pronouncement of 17.6 might be both imprecise and of the wrong sign. Finkel concluded that such point estimates are only useful if they are accompanied by a statement of the degree of certainty one can have in the value. Interval arithmetic (Chapter 9) provides a way to combine uncertain likelihoods and consequences that could be applied routinely in risk ranking.

6.5 Performance

The claims of pitfalls above are based on inference. It remains to be seen if such problems result in important differences in ranks for hazards in real applications.

To explore the extent of uncertainties in risk ranks, groups of five or six assessors (senior undergraduate students, together with a few engineers and EPA scientists) were provided with information describing a range of projects. The projects included mining, engineering, transport and waste disposal proposals. The groups were asked to assess the environmental risks of the proposals, using a risk ranking table modelled after Table 6.2.

Each group selected a project (there were five projects on offer). Participants were provided with training in the psychology of perception, concepts of probability and uncertainty, and so on. Most had backgrounds in engineering or science.

The participants met and agreed on common definitions for terms such as likelihood and consequence, to eliminate some linguistic uncertainty. They independently created lists of hazards, and then compared and combined their independent lists into an agreed list of hazards. They then

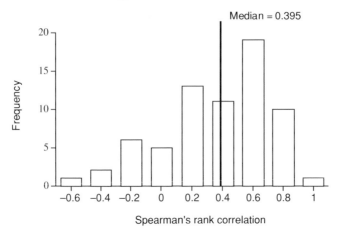

Figure 6.3. Disagreements between pairs of assessors ranking hazards. The *x*-axis labels are upper class limits. A total of 63 assessors were divided into groups of five, six or seven people, comprising as wide a range of disciplinary backgrounds as possible. Each group selected a project for which an Environment Effects Statement had been published. The information available to the assessors was a mixture of technical summaries, empirical data, expert judgement, and inferences based on other sites or similar projects. People compared their results to one another, within groups, using Spearman's rank correlation coefficient. These are plotted above (after Burgman 2001).

ranked the hazards independently. In Figure 6.3, a value of 1 represents perfect agreement between two assessors, and −1 represents complete disagreement.

The median value of 0.395 in Figure 6.3 reflects the extent of overall agreement over the ranks for each hazard. There were substantial areas of disagreement between assessors, represented by values near and less than zero. Strong agreement between assessors (represented by values of 0.8 or greater) was rare.

These results may underestimate the degree of uncertainty in most risk ranking exercises. The circumstances are artificially constrained and uniform. Differences in ranks are attributable to differences in risk perception and linguistic uncertainty because most people had identical information at their disposal. Furthermore, rank correlations ignore absolute values for likelihood and consequence.

The groups of assessors met a second time and discussed the basis for discrepancies. Sometimes this involved the introduction of new evidence, as people contributed knowledge about the process and new data. Most discussion centred on time horizons, resolution of context, and

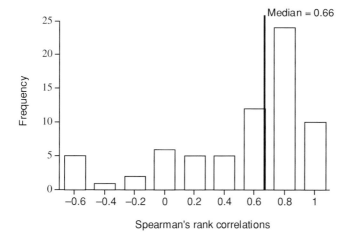

Figure 6.4. Disagreements between pairs of assessors conducting subjective risk assessments after repeated attempts to define terms and context.

redefinition of hazards, likelihoods and consequences. The groups then repeated the ranking exercise independently, and again compared the results (Figure 6.4).

The situation improved after the groups reconsidered definitions of terms and interpretation of the context of the project (Figure 6.4). The median value of the Spearman rank correlation coefficients increased from 0.395 to 0.66, reflecting substantially better agreement.

The same patterns of disagreement and change emerged when assessors were asked to judge human health or social and financial risks associated with the same projects. The changes were partly due to an improvement in shared understanding of the context and definitions of terms, reflecting the assessors coming to terms with some elements of linguistic uncertainty. In addition, some assessors were persuaded to change their estimates by argument, a result that is not necessarily due to linguistic uncertainty.

The majority of pairs of assessors improved their level of agreement after discussion, although a few disagreed more. When people's opinions diverged, it was because they found previous agreements were based on linguistic misunderstandings.

New information is added during discussion. Further resolution of differences may be possible given further discussion and redefinition. However, there remains a residual disagreement between assessors that is not resolvable by further negotiation. These are honest differences of opinion between assessors regarding the likelihoods and consequences of hazards. Often, they reflect different personal values.

These results are not due to the inexperience of the assessors. The experiment was repeated with other groups of senior undergraduate students, experienced professional ecotoxicologists, analytical chemists, engineers and ecologists, and each time gave the same qualitative results. Risk ranking exercises should anticipate this effect and deal with it using the approach documented above.

6.6   Examples

6.6.1   Risk ranks for the US National Ignition Facility

The US National Ignition Facility is working to focus laser beams on spherical targets containing deuterium and tritium. The intention is to make them implode, creating fusion energy in the laboratory for the first time.

Brereton *et al.* (1998) outlined the need to deal with a range of novel social and technical risks. They used a risk ranking procedure based on tables similar to those developed by NASA (Table 6.1a), although they employed only three likelihood categories.

Brereton *et al.* (1998) identified several hazards among the highest category of risk. 'High risk' hazards included successful legal challenges to stockpile stewardship, significant injury during construction, severe accident during start-up and subsequent operations, and performance shortfall. The analysis was used to implement risk management strategies and monitoring protocols across a range of social, construction and operation activities.

It is interesting that a facility devoted to quantitative understanding of physical systems did not report a process model for the construction and operation of the facility. Nor did it quantify uncertainties in judgements and present them in the results. There was no description of plans to evaluate the sensitivity of assumptions, or to validate the results. The analysis was incomplete and lacked transparency about uncertainty. However, it may have been a useful vehicle to tier detailed model development and monitoring.

6.6.2   The Paper risk rank calculator

The company Australian Paper uses a card with the relationships between likelihood and consequence printed on it (Figure 6.5; it is an example

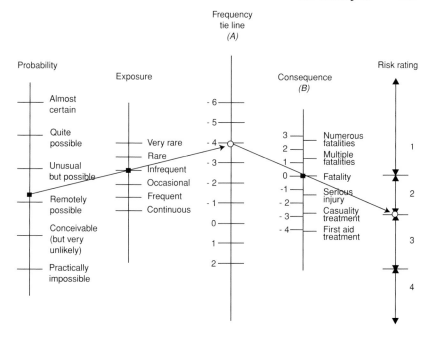

Risk score calculator

Figure 6.5. Australian Paper's risk score calculator showing an example application. The probability of a hazard is assessed as somewhere between remotely possible and unusual, with 'infrequent' exposures (black squares). This gives a score on the frequency tie line of −4. If it occurs, the event will cause a fatality (the black square on the consequence line), resulting in a classification on the border between a class 2 and a class 3 hazard.

of tools developed for occupational health and safety). Anyone in the company can estimate risk using the same standard.

The way it works is:

1. Estimate the probability that an event (an incident / hazard) will occur.
2. Estimate exposure to the hazard.
3. Mark points on the 'probability' and 'exposure' lines (represented by black squares in Figure 6.5); connect them with a straight line and extend it across to the 'frequency tie line' (the open circle).
4. Estimate the consequence of the hazard (the black square).
5. Draw a line from the intersection of the frequency tie line, through the point on the consequence line, extending to the 'risk rating' line.

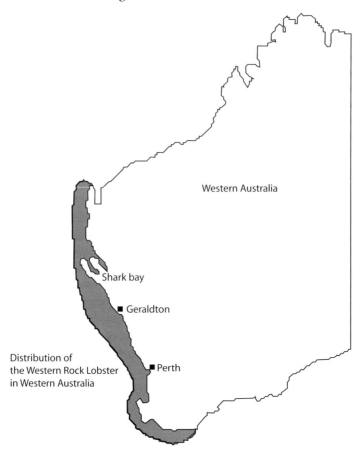

Figure 6.6. Distribution of the Western Rock Lobster (from IRC 2002).

6. The resulting score (1–4) is the risk priority rating (the example gives a value on the border between categories 2 and 3).

The method is simple, transparent and repeatable. Obviously, the subjective judgements make it susceptible to the linguistic and psychological uncertainties that affect other risk assessment, but no more so. It clarifies the order of terms describing probabilities, exposures and consequences. The visual interpretation of scales is a particularly useful feature.

6.6.3 Western Rock Lobster ecological risk assessment

The Western Rock Lobster (*Panulirus cygnus*) is distributed on the continental shelf of Western Australia (Figure 6.6, IRC 2002). The animals

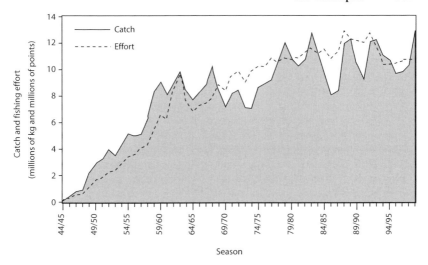

Figure 6.7. Catch and an index of effort in the Western Rock Lobster fishery since it began in the 1940s (from IRC 2002).

can live for more than 20 years, maturing at about 6 years and reaching weights of 5.5 kg. Larvae spend 9–11 months as plankton before moulting, and settle in shallow inshore reefs. Recruitment success depends on the strength of ocean currents and winds. About 4 years after settlement, the lobsters migrate to deeper reefs on the shelf.

The fishery began in the 1940s. The catch has stabilized at about 10 000 tonnes per year but varies depending on recruitment (Figure 6.7). The annual catches in the commercial fishery have varied between 7000 and 14 000 tonnes. The variation in the number of post-larval animals that successfully return to the shallow reefs each year is used to estimate the number of lobsters recruiting to the fishery, and consequently the catch, three to four years later (Caputi *et al.* 1995). The catch in 1999/2000 was worth about US$200 million, fished by about 600 boats (IRC 2002).

In March 2000, it was the world's first fishery to be certified as sustainable by the Marine Stewardship Council. Part of the certification process involved undertaking a 'comprehensive, and scientifically defensible ecological risk assessment' (IRC 2002).

A risk assessment of the Western Rock Lobster fishery in Australia (IRC 2002) employed the principles of risk ranking based on expert judgement. Problem formulation involved segmentation of activities into functional groups associated with different kinds of ecological consequences (e.g. Table 6.3).

Table 6.3. *Subset of the segmentation of activities associated with the Western Rock Lobster fishery (after IRC 2002)*

Ecological consequences	Activities
Removal / damage of organisms	Physical impact on benthic communities
	Bait collection
	Rock lobster fishing
	Ghost fishing
	Physical impact on coral
Addition / movement of biological material	Stock enhancement
	Discarding
	Displacement
	Bait
Other	Air, water, substrate quality
	Bird interactions
By-catch	Sealions
	Moray eels
	Turtles
	Whales
	Manta rays
	Dolphins

The assessment commenced with an expert group that constructed conceptual models using influence diagrams (e.g. Figure 6.8). These summarized functional relationships between human activities, elements of the fishery and the broader ecosystem. They were then sent to participants to support brainstorming and further refinement of ideas about causal relationships.

These diagrams were used together with other background information to support a workshop. They formalized the rationale for the design of monitoring systems and performance reviews.

The workshop had 15 participants selected for their involvement in the industry or in conservation groups, and for their scientific expertise. Thus, the sample of experts was stratified to represent a number of stakeholder positions, although the basis for the stratification was not explicit. The qualifications and experience of the participants were presented with the workshop findings and were used to justify the composition of the workshop participants, a useful feature of any process that uses experts.

Expert assessments were used to avoid the need for 'time-consuming collection and review of data' (IRC 2002). A total of 11 out of the 15

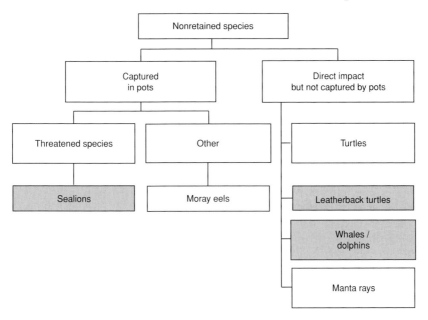

Figure 6.8. Influence diagram for the 'nonretained' species, termed a 'component tree'. Shaded elements were identified as warranting further investigation (from IRC 2002).

experts had training in science. Expert selection was not discussed beyond attribution of expertise.

The composition of the expert group may have influenced the resulting list of hazards. There were no social scientists, indigenous representatives, seasonal island residents, marine engineers or toxicologists, for instance. It is difficult to judge the importance of group composition, but the evidence from cognitive psychology suggests it is not a trivial consideration.

The classification of hazards was conducted by judging consequences and likelihoods for each hazard (both on scales of 1 to 5). If the product of the numbers was between 5 and 14, the hazard was classified as 'moderate'. Thus, the threshold of 15 was anchored to the notion of an acceptable risk by consensus following inspection of the table of consequences of the hazards. Moderate risks were judged to be acceptable as long as management was implemented to reduce them to be as 'low as reasonably practicable' (see Chapter 12 for more on acceptability of risks).

Unfortunately, no justification was provided for selecting this value. It is difficult for someone looking at the results of the assessment to know what values and standards the expert group used. The absolute level of risk classification is important because it affects public perception and how managers react.

The experts identified 33 hazards by brainstorming and iterative refinement of component trees, supported by generic checklists. A total of 4 hazards were ranked as 'moderate' and 29 were classified as 'low risk', based on a table of likelihoods and consequences similar to Table 6.2. Moderately risky hazards included:

- mortality of sealion pups in lobster pots,
- direct damage to coral by pots,
- entangling of leatherback turtles in fishing ropes, and
- dumping of domestic waste at islands within the fishery.

Facilitators sought group negotiation to a consensus (Chapter 4). This group approach is susceptible to psychological artefacts such as anchoring and hazard visibility. The process is also susceptible to the motivational biases resulting from vested interests of the participants. Deference to authority and the hierarchical relationships between individuals in the group may also be problematic. It is difficult to know how much influence any of these factors has in any instance, but elicitation procedures should anticipate them. One weakness, common to most risk ranking exercises, was that the summary of the elicitation process did not report the full breadth of opinion.

The Rock Lobster risk assessment is a good example of an application of risk ranking. It had several innovative features including the stratification of experts, detailing their credentials, using trees to structure the hazard elicitation process, and specifying tiered responses to monitoring and subsequent model development for the most serious hazards. For instance, the fishery is an excellent opportunity to develop population models for the sealion population, and to use it to identify sensitive life-history parameters. This would allow managers to estimate biologically important impacts on variables affected by the fishery. The risk management cycle could then be closed by creating monitoring strategies capable of detecting important changes, and by selecting management options that allow for unexpected outcomes.

The risk assessment could have been made more honest and complete by using a more broadly based group of participants, providing a more detailed report of uncertainty, and anchoring values for unacceptable risks to something tangible or external to the group of experts. It would have been useful to know if decisions were sensitive to uncertainty in the assessments.

6.7 Discussion

Risk ranking has the advantages of operational simplicity and transparency. Given a little training, most people can perform one and achieve useful results. Ecosystems and other systems are too complex to model explicitly. This approach allows ecologists (and others) to integrate complex qualitative and quantitative information and generate assessments without extensive data and full understanding.

Perhaps the greatest single advantage in operational circumstances is that risk ranking allows environmental managers to communicate with stakeholders, board members, financial staff, engineers and others in a common language. The same language communicates social, financial and human health risks, so that environmental risks are seen as equally worthy of consideration. As a result, environmental issues are treated in the same way as other corporate and public issues, and are included in discussions about the costs and benefits of actions.

Stakeholders should be involved in the process of building, testing, revising and interpreting risk assessments (Beer 2003). The most important skill of the analyst in these circumstances is to communicate the detail of alternative conceptual models and the consequences of their assumptions. Risk ranking creates a natural framework for including stakeholders. They may be part of the group that drafts a conceptual model and estimates likelihoods and consequences. This gives people ownership of the solutions to environmental risks and enhances the chances of acceptance of the outcome.

It is a relatively common strategy, once the list of hazards is known and ranked, to eliminate the most important hazards by re-engineering the system, or by selecting management options that avoid some hazards altogether. Some corporations direct all 'extreme' risks to the board or the managing director. Often, more detailed evaluation of technical risk is not warranted because boards decide that all extreme risks are, by definition, unacceptable, and projects will not go forward until all extreme risks have been eliminated from the design.

Risk ranking provides a natural basis for tiered risk assessments. Risks that are (confidently) ranked as acceptable may be given relatively little

Box 6.1 · *A method for risk ranking*

1. Set context and compile background information.
2. Identify a pool of potential experts and verify their substantive and normative knowledge.
3. Survey stakeholders, assessing their influence and importance.
4. Set stratification criteria for potential participants (experts, stakeholders).
5. Approach potential participants in a random sequence within strata until a sufficient number can participate (the number will depend on the breadth of scientific and social issues, time frame and budget).
6. Employ models for elicitation (Chapter 4) that suit the context.
7. In consultation with stakeholders and experts, set rules for elicitation and aggregation of opinions and estimates.
8. Anchor thresholds for the acceptability of risk on other broadly accepted probabilities and outcomes.
9. Distribute information to participants.
10. Meeting 1:
    a. create conceptual models,
    b. identify hazards (using structured brainstorming, FMEA, related methods),
    c. agree on numbers of categories and on definitions of likelihood and consequence,
    d. agree on acceptability thresholds,
    e. train participants to distinguish subjective and objective probability, and to estimate objective probabilities, and
    f. test abilities to estimate against known standards.
11. Individually, rank hazards using the risk ranking tables, including best estimates and bounds for estimates of likelihood and consequence, and for final risk ranks.
12. Meeting 2:
    a. discuss differences in ranks,
    b. introduce new information,
    c. refine conceptual models,
    d. disaggregate elements of the model,
    e. redefine hazards, and
    f. eliminate ambiguities, sharply define vague concepts, clarify underspecific terms.

13. Individually, re-rank hazards using the risk ranking tables and re-estimate bounds.
14. Combine information on ranks from all participants.
15. Present aggregated ranks and full detail of the remaining uncertainties and disagreements.
16. Attribute opinions to identified individuals.
17. Assess the sensitivity of ranks and resulting decisions to:
    a. choice of acceptability thresholds,
    b. exclusion of individual participants (one by one), and
    c. exclusion of participant strata (if there is more than one representative per stratum).
18. Specify data collection that will verify ranks and audit management decisions.
19. Identify hazards that will be subjected to more detailed analysis (tiered response).
20. Nominate a time, place and participants to meet and revise the conceptual model and risk ranks, based on data collected and analyses conducted in the interim.

attention. Others may warrant further data collection and analysis using more formal tools described in Chapters 7–12.

However, the framework fails to acknowledge the uncertainties surrounding assessments. There are no conventions for communicating the degree of certainty that may be placed in an assessment, or for measuring the reliability or repeatability of an assessment. There are no conventions for assessing the validity of ranks, or for exploring the sensitivities of decisions to arbitrary assumptions.

Because decisions about the use of the environment almost always involve conflicts with other priorities, they are laden with subjective values and interpretations. Risk ranking conventions fail to acknowledge linguistic uncertainty, personal values, the role of cognitive psychology and other sources of bias. They provide no advice or any mechanisms to deal with them, although they could do so.

The consequences of failing to recognize values, context and linguistic imprecision is that people who use risk ranking will make more mistakes than they would had these issues been taken into account. It is difficult to measure the frequency or the costs of these mistakes because they are submerged in the operational detail of project management and are confounded by stochastic processes. In the absence of a detailed monitoring programme, the feedback that would highlight the flaws in risk ranking is unavailable.

Because risk ranking suffers from the weaknesses outlined above, the results typically present a less than honest and complete picture to those who decide on whether to tolerate, avoid or mitigate a risk. From this perspective, it seems irresponsible not to employ strategies for stakeholder involvement, elicitation and aggregation, sensitivity analysis, validation and representation of uncertainty. If risk acceptability levels could be independently established, there would be robust grounds for developing a tiered analytical process. These changes are essential not because they guarantee a more accurate answer, but because, if properly employed, they result in an answer that is relatively internally consistent, transparent and free of linguistic uncertainty.

The insights offered by expert elicitation, the risk management cycle, taxonomies of uncertainty and observations from psychology create opportunities to make risk ranking more robust. The method in Box 6.1 is designed to take advantage of these advances.

7 · Ecotoxicology

Ecotoxicological risk assessments have their roots in the social activism of the 1960s (inspired by such things as Rachel Carson's *Silent Spring*, 1962). Governments in many countries created protocols that matured over the period from the mid 1960s to the mid 1990s and that continue to evolve. These methods have a particular focus: the assessment, approval and auditing of pollutants and toxicants in a regulatory system.

Circumstances demanded the rapid development of conventions, experimental techniques and standards for interpretation of evidence. These things coalesced into the ecotoxicological paradigm (Chapter 3). The system rests on the foundations summarized by Suter (1993): management and policy goals, assessment endpoints as well as indicators and measures of effect. This chapter examines the system's relationships with epidemiology and toxicology, and with broader concepts of environmental risk assessment.

To reiterate the ecological hierarchy outlined in Chapter 3, management goals encapsulate the spirit of a management or monitoring programme. As such, these goals have a social mandate and are ecologically relevant. Assessment endpoints are formal expressions of the environmental values to be protected. They provide a means by which management goals may be measured and audited.

Measures of effect (measurement and test endpoints, US EPA 1998) are quantitative biological responses, such as toxic effects on survival and fecundity. Indices are created from field measurements or laboratory tests. They represent sensitivities of ecosystem components to toxic substances and act as surrogates for other elements of the ecosystem.

Ecotoxicology provides a basis for assessing whether chemicals are likely to have adverse effects on ecosystems and to provide a basis for managing those effects. The broad objectives are to protect ecosystems, usually by protecting individual species and avoiding irreversible ecological changes (Calow and Forbes 2003).

Table 7.1. *Approximate acute LD$_{50}$ values for animals exposed to some chemicals (from Klaassen 1996)*

Chemical	Dose (LD$_{50}$) mg / kg body weight
Sodium chloride	4000
Ferrous sulfate	1500
Strychnine sulfate	2
Nicotine	1
Dioxin (TCDD)	0.001
Botulinum	0.00001

This chapter provides a sketch of the ecotoxicological approach to risk assessment. It begins with dose–response relationships. It outlines methods for estimating and extrapolating toxicological responses in populations, species and ecosystems, documenting places where assumptions make the methods susceptible to uncertainty. It then describes methods for establishing 'safe' exposures, and for developing models to estimate exposure from transport and fate models. Examples illustrate the role of uncertainty, the value of accounting for it explicitly and the need to carry uncertainty through chains of conventional calculations.

7.1 Dose–response relationships

The evolution of ideas in ecotoxicology was motivated by the need to regulate potentially damaging chemicals. This led to the definition of functions relating measured effects (proportion dying, mean weight, proportion germinating) to exposure (dose, concentration, duration). Some of the more common measures are:

- LC$_{50}$: median lethal concentration.
- LD$_{50}$: median lethal dose.
- EC$_{50}$: median effective concentration.
- LC$_{01}$: lethal threshold concentration.

L stands for 'lethal', E is 'effective', C is 'concentration' and D is 'dose'. An effective concentration is the concentration that elicits the specified response in that percentage of test organisms. It may be the concentration at which a percentage of organisms fails to produce eggs, germinate or achieve a specified height.

Typically, these measurement endpoints are acute: short-term and assumed to be severe or lethal. An LD$_{50}$, for instance, is the quantity of

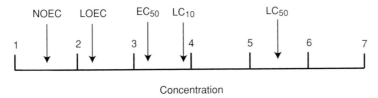

Figure 7.1. Toxic effects as a function of concentration. Usually, concentration is represented on a log scale.

the toxicant imbibed by organisms at which 50% die within a specified time, usually 96 h (e.g. Table 7.1). An EC_{10} is the concentration at which 10% of individuals exhibit the effects of exposure to a contaminant within a specified time. The tests are replicated under standard conditions, with increasing quantities or concentrations of the chemical, resulting in a concentration or dose that causes a specific outcome (Figure 7.1). LC_{10} values must be less than LC_{50}s for the same organism. EC_{50}s are less than LC_{50}s because they measure an effect other than death. LOEC (lowest observed effect concentration) and NOEC (no observed effect concentration) are lower again (see below).

In laboratory conditions, it is relatively easy to supply a constant concentration of a toxicant to the air, water or soil in which a test organism lives. When exposures require doses (inhalation, dermal, intravenous, intraperitoneal or intragastric), by convention animals usually are dosed once or several times per day. That is, the pattern of dose exposure is not continuous.

When these tests are carried out on fish, it is usual to use a 96-h LC_{50} test. It has been a standard procedure in the past to subject vertebrates, such as laboratory rodents, to acute, oral doses designed to estimate the median lethal dose.

It is possible to examine sublethal effects. For example, tests may measure the growth response of plants to different levels of a chemical (e.g. Figure 7.2), resulting in an EC_{50} value.

A full life-cycle (egg to egg) test may examine the time for a cycle to complete, or the sensitivity of different stages to changes in temperature, as a function of the concentration of the contaminant. Relatively comprehensive tests are rare because of the time and money involved, and the difficulty in replicating experimental conditions for more complex designs. Survival analysis (e.g. Newman and McCloskey 1996) more completely characterizes risk than LC_{50} and related statistics.

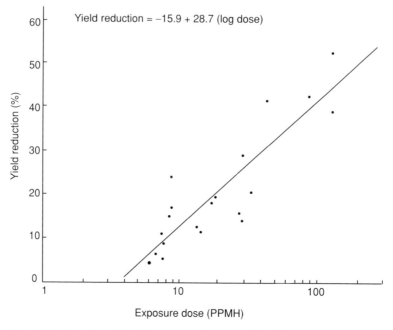

Figure 7.2. Chronic, sublethal test measuring growth reduction in the yield of beans (after McLaughlin and Taylor 1988, in Suter 1993).

7.1.1 NOELs and LOELs

A no observed effect level (NOEL, or concentration, NOEC) is the largest dose at which there is no statistically significant increase in the fraction of an exposed population who exhibit some effect (Crawford-Brown 1999, p. 130). A LOEL is the smallest dose at which a statistically significant effect has been demonstrated.

Crawford-Brown (1999) argued that it is important to reject increases that might be accounted for by random fluctuations, 'since such increases might not be due to exposure to the substance.' (p. 131). In contrast, he argued that confidence intervals used to interpret the significance of NOELs and LOELs need not necessarily be 95%, but could be, say, 90%, to give greater weight to the cost of false-negatives.

A no observed adverse effect level (NOAEL) is the level at which a statistical hypothesis test failed to reject the null hypothesis of no effect. NOAELs typically drive regulatory decisions. They usually result in higher levels of pollutants than NOELs because they encompass both statistical significance and biological importance. The 'true' threshold lies somewhere between a LOAEL and a NOAEL (Crawford-Brown 1999,

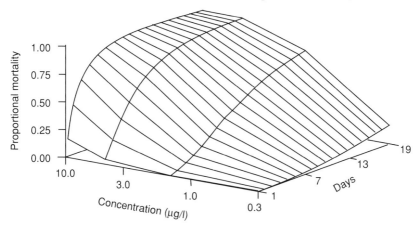

Figure 7.3. Three dimensions of the toxic effects: concentration (on a log scale), duration and proportions of test organism populations that die (after Nimmo *et al.* 1977 in Suter 1993).

p. 134). However, these arguments run the risk of being confounded by statistical power (or a lack of it).

NOELs, NOAELs and related statistics are a curious invention because their value depends on experimental design and sample size, rather than potency alone. Very large samples may detect significant differences between treatments and controls that are biologically unimportant. More usually, small samples and substantial experimental variation inflate NOAEL thresholds, decreasing the apparent toxicity of the chemicals (Laskowski 1995, Chapter 11). If a test is underpowered it may be nonsignificant despite important effects. Nevertheless, they are used throughout the world to set regulation guidelines and provide a basis to infer ecosystem effects from single-species data.

When the logarithms of chemical concentrations are plotted against the percentage of organisms exhibiting an effect, typically the curve is s-shaped. If the *y*-axis is transformed to a probability scale, the relationship between concentration and effect is approximately linear.

Clark (1933) noticed this pattern and pointed out that it is consistent with the relationship expected for the formation of a drug–receptor complex according to the law of mass action. The sigmoidal curves have the characteristics of a normal curve, suggesting an underlying lognormal distribution of sensitivity among individuals in the population (see Byrd and Cothern 2000 for more detail).

When tests are extended over time, they result in logarithmic relationships between effect and time (Figure 7.3). Often, the relationship

Table 7.2. *Hypothetical data on leukaemia occurrence resulting from occupational exposure to EMF radiation*

	Cancer	No cancer	Prevalence
Exposed	A	B	$A/(A+B)$
Unexposed	C	D	$C/(C+D)$
Ratio	A/C	B/D	
Exposed	2	9 998	0.0002
Unexposed	10	99 990	0.0001
Ratio	0.2	0.1	

between exposure or concentration and effect is estimated using logistic regression (Box 7.1).

'Equivalence' of the toxic effects of two chemicals is sometimes inferred by comparing replicates of LC_{50} or NOEC test results for a species (with a *t*-test of their means, for instance). If the test returns a result of 'do not reject the null hypothesis', people infer that the potency of the chemicals is (roughly) the same. Instead of using a single potency value, Piegorsch and Bailer (1997) recommended comparing the slopes and intercepts of regression models, thereby making better use of available data. The latter approach is better but, other assumptions aside, both suffer from the same weakness as the interpretation of NOELs. That is, the inference depends on statistical power. Substantial differences between the potency of two chemicals may be masked by small sample sizes and large experimental variation.

7.1.2 Odds ratios and relative risks

Odds ratios express the probability of one outcome relative to the probability of its opposite. They are independent of the relative numbers of different kinds of events and can have any value between 0 and infinity.

Take an example in which there was a suggestion that exposure to electromagnetic radiation may cause cancer in electricity workers. You collect some new data (Table 7.2). The 10 cases in the unexposed population represent the background rate, the risk of getting leukaemia from causes other than EMF radiation. Assuming all other things are equal, we could make the inference that the prevalence of leukaemia is twice as high in the exposed population as it is in the unexposed population (ignoring the tiny sample size).

Box 7.1 · *Logistic regression for toxicity estimation and odds ratios*

The binary nature of toxicity tests is an example of a wide class of relationships in environmental risk assessment in which a binary outcome (e.g. alive / dead, present / absent, true / false) is a function of one or more variables (e.g. chemical concentration, habitat quality).

In logistic regression, the response variable is binary. A condition or event (present, alive) is coded as 1, and the alternative condition (absent, dead) is coded as 0. The logistic model is:

$$\text{Logit}(p) = \ln\left(\frac{p}{1-p}\right) = a + bx$$

where p is the probability that the condition or event occurs. Taking the exponent of both sides and solving for p gives:

$$p = \frac{e^{(a+bx)}}{1 + e^{(a+bx)}}.$$

Because the response can take only two values (presence / absence, alive / dead), the errors are assumed to follow a binominal distribution. Maximum likelihood is used to estimate parameters. The model may be used to compare the number of predicted versus observed occurrences of an event, or to fit a dose–response model. The y-axis (symptom) is whether the organism is alive or dead, and the x-axis is concentration of the chemical. Even if a logistic regression model is correct, classification ability is limited by how clearly responses are separated by the causal variable.

Odds ratios are a natural part of logistic regression. Let $x = 1$ if an event occurs and $x = 0$ if it does not. We can write the logistic regression as:

$$\frac{p}{1-p} = e^{(a+bx)}.$$

Since $x = 1$ for an event, the odds become

$$\frac{p}{1-p} = e^{(a+b)}$$

and the odds for a nonevent are

$$\frac{p}{1-p} = e^{a}.$$

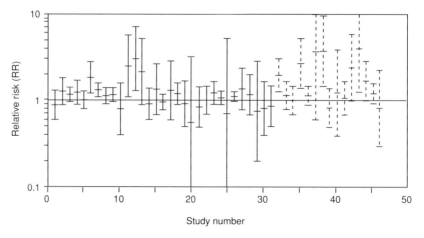

Figure 7.4. Risks of leukaemia through occupational exposure to EMF radiation. The data are from 31 studies in which combined relative risks for all leukaemia was reported (solid bars) and 15 studies in which the relative risks of acute myelogenous leukaemia were reported (dashed error bars) (after Oak Ridge Associated Universities 1992, in Shlyakhter 1994).

The odds ratio (OR) is the chance of exposure among diseased cases relative to the chance of exposure among healthy cases. The ratio is calculated from the table most easily as

$$\text{OR} = \frac{A/C}{B/D} = \frac{A \times D}{B \times C}$$

which gives an odds ratio in the case of Table 7.2 of 2 : 1 (after Byrd and Cothern 2000). The odds are often scaled so that the second number is 1. Relative risks (RR) are a closely related measure. They express the chance that a disease occurs in an exposed population relative to the chance in the unexposed population. They are given by

$$\text{RR} = \frac{A/(A+B)}{C/(C+D)}.$$

The relative risks of the event in Table 7.2 are 2 : 1. When events are rare, OR and RR are approximately equal. The terms are used more or less interchangeably in epidemiological literature (L. Flander, pers. comm.).

Relative risks are a common measure in public health and ecotoxicology. Figure 7.4 shows the relative risks measured in a number of studies

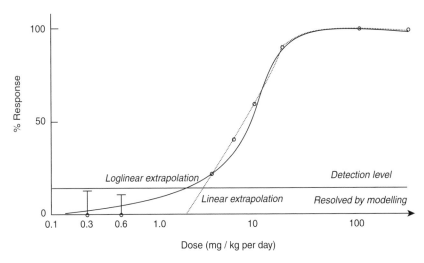

Figure 7.5. Alternative models for dose–response relationships below the level of detection (after Beer and Ziolkowski 1995, Adams 1995).

into the effects of exposure to large doses of electromagnetic radiation, compared to people not habitually exposed. The relative risks of acute myelogenous leukaemia are somewhat higher than background expected levels, even though the individual results from several studies are not significantly different from 1.

7.2 Extrapolation

Once a dose–response test is complete, the challenges remain to extrapolate the results to other taxa, and to decide on what a 'safe' level for a toxic substance might be.

7.2.1 Extrapolating to low concentrations

A concentration in a laboratory test that results in zero mortality (or no greater mortality than the controls) is not necessarily safe. If a safe level is, say, less than 1% mortality in a 96-h test, sample error may give results that are not significantly different from zero, or from the controls.

Extrapolations below the detection level are usually resolved by modelling the response of the species to the contaminant. There are numerous possible models. The choice could make a substantial difference to recommendations for 'safe' levels, depending on attitude towards and tolerance of the risk involved (Figure 7.5).

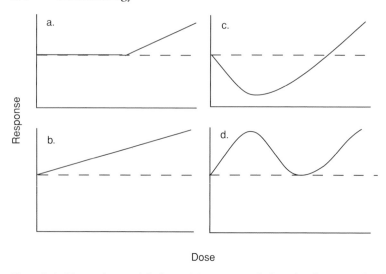

Figure 7.6. Alternative models for toxicity response below the detection threshold. a. Threshold response; b. linear response; c. hormesis; d. elevated low-dose response. Increasing response indicates greater damage.

The different conceptual models in Figure 7.6 emphasize the assumptions in extrapolating from high to low doses. Linear and loglinear extrapolations have been conventionally accepted since the 1930s (Figure 7.6b). It is easy to imagine responses in which a small increase in the concentration of a toxicant elicits little measurable response up to a threshold, beyond which another small increase causes a large response in an individual (Figure 7.6a).

Calabrese and Baldwin (2001) argued that benefical responses may be observed to some toxicants at very low exposures, termed hormesis (Figure 7.6c). They explained it as an adaptive response to low stress levels resulting in improved fitness. That is, hormesis occurs because response systems have beneficial effects until the system becomes overloaded at higher concentrations.

In striking contrast, Olson *et al.* (1987) and Cavieres *et al.* (2002) found higher mammalian responses at very low exposures to insecticide and herbicide mixtures (Figure 7.6d, Figure 7.7). They attributed the response to different physiological, neurological, endocrine and immune system responses across different ranges of chemical concentration.

The possibilities for responses at low doses place a burden on the analyst to make an explicit choice of a model and justify it with available data, where possible. In addition, irrespective of the choice, a complete risk

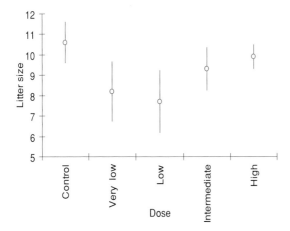

Figure 7.7. Effect of a chemical mixture on mice litter sizes (after Cavieres *et al.* 2002). The chemical was a mixture of 2,4D, mecoprop and dicamba, applied at levels well below regulatory guidelines, applied in the summer months (means and 95% confidence intervals).

assessment should put in place monitoring systems designed to detect biologically important effects below conventional thresholds.

7.2.2 Structure activity relationships (SARs)

The kinds of tests and the expected levels for safety of a new chemical may be established by inferring its mode of action and toxicity from similarly configured chemicals. SARs represent the assumption that similar chemical structures tend to interact in the environment through similar mechanisms. They are used to predict qualitative effects and quantitative exposure–response relationships.

Interactions between a toxicant and its receptor may be predicted accurately from knowledge of the structure of a chemical and its biological receptor. Metabolism and expression of toxic effects are not as easily predicted. Sometimes, regression equations are used to describe the relationship between the structure of a compound and its toxicity, resulting in quantitative structure activity relationships (QSARS). Uncertainty arises when the analyst makes a choice about how to describe the structure (how to interpret the chemical 'space'). The quality of data describing toxicity often is poor (Warren-Hicks and Moore 1995).

Some responses to toxins are continuous and some are discrete. For example, organophosphates interact with an enzyme site in nerve synapses.

The toxin blocks the normal function of the enzyme. The mode of action is common to all organophosphates. Toxicity depends on the rate of the reaction and the formation of the enzyme-inhibitor complex. This knowledge allows analysts to predict toxicity effects of new chemicals in this class. There has been some work on the reliability of such extrapolations (see Warren–Hicks and Moore 1995).

7.2.3 Extrapolating from single tests to species sensitivity

Usually, the toxicity threshold for a species is assessed using tests of a single attribute such as adult or juvenile mortality. Hanson and Solomon (2002) recommended determining effective concentrations (EC_X) for a number of endpoints for a single species. They assumed the EC_X values for a species would be lognormally distributed and replaced NOEC values with a low quantile of this distribution in hazard quotients and other calculations.

For example, they monitored plant length, total biomass, root number, root length, node number, chlorophyll a, chlorophyll b and carotenoid content in the water plant *Myriophyllum*. They estimated EC_{10}, EC_{25} and EC_{50} values for 4-, 7-, 14- and 28-day exposures to monochloroacetic acid. The thresholds from this approach are more conservative than single attribute tests and give a fuller picture of potential ecological responses.

7.2.4 Extrapolating species sensitivities

Extrapolating test results to a level that is deemed safe for all species requires additional assumptions and uncertainty. The results of toxicity tests on a small group of organisms are used to infer the effects that might be expected on a much larger group, assuming a few species provide a useful guide to the sensitivities of a much larger range of taxa.

For instance, fathead minnows are a standard test organism. LC_{50} values for fathead minnows may be a reasonably reliable guide to the responses of a range of fish, amphibians and invertebrates (Figure 7.8). LC_{50}s for a set of test species are assumed to represent a sample from the distribution of all possible LC_{50} results from all species. EC values for a range of attributes (Section 7.2.3) would be more complete. Typically, a model is selected to represent the variation in LC_{50} values over all species. Its parameters are estimated from the sample data. Lastly, a quantity is estimated (sometimes called the final acute value or FAV) such that most of the

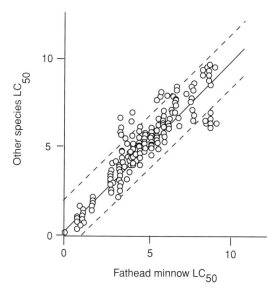

Figure 7.8. The LC$_{50}$ response values for fathead minnows plotted against the LC$_{50}$ responses of other organisms. The lines show a linear regression and 95% prediction intervals for the relationship (after Suter 1993).

(unsampled) LC$_{50}$s are expected to be greater than the estimated quantity (e.g. Figure 7.9).

The simplest extrapolation is to assume the logs of LC$_{50}$ tests from a set of organisms are normally distributed. A FAV may be selected that is, for instance, two standard deviation units less than the mean LC$_{50}$ (e.g. Figure 7.9). This implies that 97.5% of LC$_{50}$s from unsampled taxa will be greater than the threshold. Concentrations as high as the threshold will kill at least 50% of the individuals from 2.5% of the species exposed within 96 h.

One of the problems with these protocols is that LC$_{50}$s, FAVs and related values are treated as though they are exact. Of course, each one will include measurement error. In addition, natural variation will perturb values around any central tendency, affecting estimates of a 'safe' threshold (Figure 7.9d).

Hart *et al.* (2003) recommended an approach that accommodates variability in the likelihoods and the consequences of events. If information on variation in the contaminant is available, then a threshold may be selected that satisfies a level of safety with a given degree of probability,

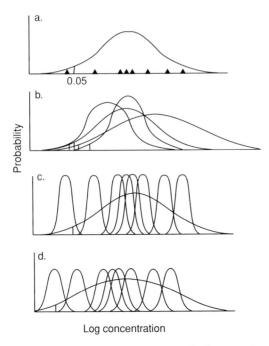

Probability

Log concentration

Figure 7.9. Uncertainties in setting a final acute value, a safe threshold for a toxicant based on extrapolations among species (after Suter 1993). The vertical bars show the lower 5th percentile of the distribution. a. Distribution fitted to LC_{50} values for eight species; b. distributions fitted to four samples of eight LC_{50}s; c. distribution fitted to eight LC_{50}s, each of which shows measurement error; d. distribution fitted to LC_{10}s for eight species extrapolated from LC_{50}s, each of which shows measurement error.

for a given proportion of the biota. The 'consequence' curve is the cumulative distribution of LC_{50}s for a large set of species. The 'likelihood' curve is the cumulative distribution of contaminant concentrations in the environment, measured over time. These curves may be plotted on the same concentration axis, giving, for example, the concentration at which the tolerance of about 5% of species is expected to be exceeded about 90% of the time (Figure 7.10).

The supervising scientist is an environmental regulator charged with monitoring the effects of uranium mining in Kakadu National Park in northern Australia. One of their concerns is the potential for uranium to contaminate the Magela Creek downstream from mining operations and tailings dams (Figure 7.11).

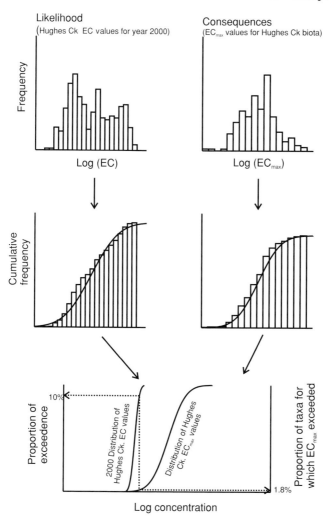

Figure 7.10. The 'likelihood' distribution represents the distribution of salinities experienced by species in Hughes Creek, in the Goulburn-Broken Catchment in Victoria, Australia in 2000. The 'consequences' distribution represents species sensitivities to salinity, based on field observations of species presence and salinity measurements in a range of systems (from Hart *et al.* 2003; see also Verdonck *et al.* 2003). The dotted line in the bottom panel indicates that 10% of the EC estimates for the site exceed the EC_max for about 1.8% of the species.

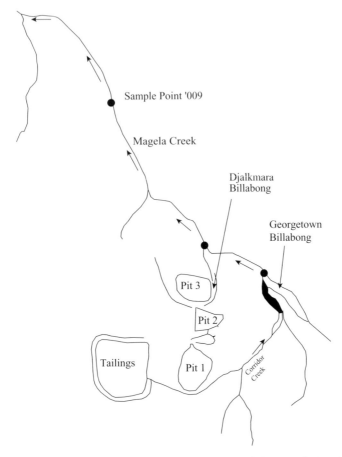

Figure 7.11. Tailings dam, mining operations and the Magela Creek at the Ranger Uranium site in northern Australia (after Bayliss, P. and van Dam, R., in prep., pers. comm.).

Data from sample point 009 suggest that 'Action Levels' are exceeded occasionally, a consequence of flow from the mine site during the wet season. The spikes in uranium concentration are attributed to human activities (Figure 7.12).

If the uranium concentration reaches 5.5 µg/l, then 99% of species will be 'protected' and 1% will be at risk, with 50% certainty (Figure 7.13). 'Protected' implies that the species' response will be less than the critical threshold in a toxicity test. In this case, sublethal effects (changes in reproductive output and population growth) were documented. The chance that uranium concentrations will exceed 5.5 µg/l is about 1 in

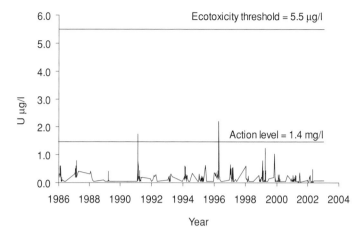

Figure 7.12. Uranium concentration at Magela Creek sample site 009, taken at weekly intervals over the wet season when the risk of contamination is high due to high rainfall runoff (after Bayliss, P. and van Dam, R., in prep., pers. comm.).

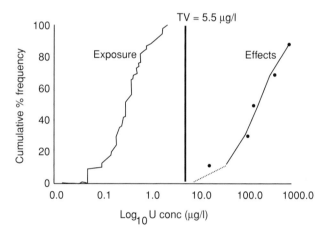

Figure 7.13. Exposure and effect distributions for aquatic organisms in the Magela Creek (after Bayliss, P. and van Dam, R., in prep., pers. comm.).

16 million, assuming the samples taken to date are a random sample of all concentrations and that the mean and variance of the process do not change in the future.

Exposure data were instantaneous whereas effects were measured over 72–148 hs, depending on the species. Direct comparisons (Figure 7.13) probably overestimate exposure. Extrapolations among species are difficult. It may be that the threshold is not sufficiently conservative or that it is too conservative. Extrapolated tests do not say anything about long-term (chronic) or other, more subtle, sublethal effects. They give little indication of population or ecosystem-level consequences. Laboratory bioassays do not predict responses under field conditions. Laboratory-based tests ignore the potential for toxins to interact with other chemicals or environmental variables. These issues are rarely discussed or resolved in ecotoxicological studies.

Lastly, the taxa selected are, in theory, a random sample of all exposed and susceptible taxa. In practice, they are a small set of vertebrates and invertebrates used conventionally in toxicity tests and do not sample the full spectrum of physiologies, life histories and behavioural traits.

Verdonck *et al.* (2003) outlined methods to calculate probabilistic risk from exposure to toxins. They described how several methods compare distributions improperly. They outlined methods for assessing risk that take into account exposure, environmental concentration and species sensitivity distributions. For example, they recommended interpreting the joint probability distribution formed from the environmental concentration distribution and the species sensitivity distribution. Routine toxicological protocols employ much cruder approaches to account for a range of uncertainties. They are outlined below.

7.2.5 Extrapolating from acute to chronic effects

Ecotoxicological studies may evaluate mortality, changes in life-cycle, reproduction or the survival of different stages. In the environment, species are exposed for extended periods, or continually, and the analyst has to infer long-term (chronic) responses from short-term (acute, usually 96-h) tests.

Regression relationships have been used to predict chronic NOEC values from acute LC_{50}s based on large numbers of tests for a few test organisms (e.g. Suter 1995). Because there are so few species for which these analyses are available, 'assessment factors' (commonly $= 10$) are used to extrapolate acute effects into chronic NOECs:

$$\text{Chronic NOEC} = LC_{50}/10.$$

In a few instances, the responses of test organisms to short- and long-term exposures have been compared. There is substantial variation between acute and chronic responses (Table 7.3), with variation among the responses of different species ranging over three orders of magnitude.

Table 7.3. *Ratios of chronic to acute toxicities for different animal species exposed to groups of chemicals with different modes of action (after Calow and Forbes 2003). The ratios are arranged from highest to lowest*

Metals	Nonpolar organics	Polar organics	Specific chemicals (e.g. pesticides)
125	5.6	5 012	126
15.8	3.2	112	20
6.3	1.6	10	7.1
1.8	1.3	2	4
0.9	1.1	1	0.8

Observations such as those in Table 7.3 make simple assessment factors unreliable. More reliable extrapolations will depend on better characterization of the responses of populations to short- and long-term exposures.

7.2.6 Extrapolating from toxicity tests to ecological effects

If a toxin affects only one life history stage, it may have almost no effect on the dynamics of a population, even given substantial mortality. Similarly, small changes in life history parameters (fecundity, survival, dispersal and so on) may have important effects on the chances that a population will persist. The ecological effects of exposures to toxins may be assessed by propagating the effect of the toxin through a population model.

Caswell (2001) summarized the effects of changes in individual elements in a life history table by measuring their effect on the long-term growth rate of the population. An alternative is to represent changes in a population through a model that accounts for uncertainty, resulting in changes in the probability of decline or loss of the population (see Burgman *et al.* 1993).

For example, we may represent the dynamics of a population with the expressions:

the number of adults next year = the number of adults that survive
+ the number of young that survive
and mature into adults,

and

the number of young next year = the number of offspring per adult
× the number of adults,

or more succinctly,

$$\text{Adults } (t + 1) = \text{Adults } (t) \times s + \text{Young } (t) \times s$$
$$\text{Young } (t + 1) = \text{Adults } (t) \times f$$

where s is the proportion of adults and young that survive from one year to the next in the absence of the toxin, and f is the average number of young produced by each adult.

Given LC_{50} data, the dose–response curve converts to additional mortality in a population through the slope of the mortality curve at the LC_{50} point

$$x = \frac{\ln(1 - m)}{T}$$

where m is the reduction in biomass (mortality), T is the test duration (in units of the time step of the model) and x is additional mortality (/ unit biomass per unit time) (see Spencer and Ferson 1998).

If the toxin kills only the young, to include mortality due to the toxin, we calculate

$$\text{Young } (t + 1) = \text{Adults } (t) \times f \times e^{-x}$$

Bartell (1990) estimated the chances of 25% and 50% reductions in the annual production of fish in a temperate lake from exposure to chloro-paraffins (Figure 7.14) using calculations such as these. The probabilities were calculated using Monte Carlo simulation (Chapter 10) of a bioenergetic population model. A food web model incorporating elevated mortality from a contaminant is developed in Chapter 10.

Statistical extrapolations are an alternative to model-based extrapolations. They depend on extensive data on laboratory and field testing of the responses of populations to exposure to a contaminant. The risk management cycle would suggest that the population models should guide the development of field measurements and experiments. The results should be used to revise model structures and parameter estimates.

7.3 Deciding a safe dose

The regulation of toxicants, pollutants, carcinogens and other substances involves the extrapolation of exposure–response tests to determine safe circumstances. In the USA, the FDA accomplishes regulation by publishing lists of 'allowed' substances and tolerances. Tolerances are maximum amounts of approved substances allowed in different circumstances (in foods, in the air and so on).

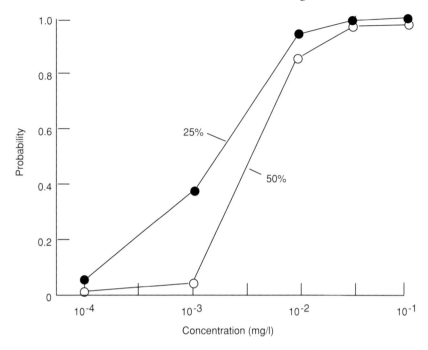

Figure 7.14. Chances of decline of a fish population resulting from exposure to chloroparaffins (after Bartell 1990).

Deciding on a safe level involves building conceptual models for exposure, and calculating the exposure experienced by an average person in a population when they consume normal amounts of a food (for instance). The allowable amount from all foods is called an acceptable daily intake (ADI, termed a reference dose by the EPA). It is the amount ingested every day without experiencing much likelihood of an adverse effect.

Politicians write legislation for safe food. No chemical, food or food-additive is risk free. The ADI concept rests on the concept of 'reasonable certainty of no harm', a term invented by the FDA which Byrd and Cothern (2000) call bureaucratic code for 'following FDA's traditional practices'. These are as follows (Byrd and Cothern 2000):

1. The proponent submits data to the regulator, following guidelines published by the regulator, and usually requiring animal toxicological studies and exposure scenarios that match the anticipated exposure (acute for acute, chronic for chronic).
2. Scientists working for the regulator review the data for quality and completeness.

3. The regulator derives a NOEL, (or a NOAEL in mg/kg·day), the highest dose that does not induce an adverse effect in any of the submitted studies. This threshold is determined by biological judgement, rather than statistical criteria.
4. The regulator converts the NOEL to an ADI by dividing the NOEL by a safety factor of 100. The implication is that regulators believe that an average person could consume this much of the substance every day of their lives without experiencing a toxic effect.
5. The regulator publishes tolerances for foods, label directions, conditions of use and so on. Usually, the proponent remains liable for adverse effects. The regulator compares the ADI to the exposures expected from the uses intended by the proponent and will set conditions for use if anticipated exposures exceed the ADI.

The use of safety factors in human health assessments embodies an undertaking to protect the normal range of human behaviour. Byrd and Cothern (2000) claimed that no substance approved by the FDA had adverse effects on humans, out of more than 1000 that were listed.

7.3.1 Reference doses, benchmark doses and uncertainty

Species sensitivity distributions (Figure 7.9) often are based on a small and unrepresentative set of species. Even so, the data are relatively expensive and time consuming to collect. Building models and collecting additional data to estimate population and ecosystem-level effects are more difficult still. It is likely that point estimates of potency, fixed extrapolation factors and acute tests will continue to play a role in setting standards for ecosystem protection (Calow and Forbes 2003).

A benchmark dose is the dose that corresponds to a predetermined level of response, such as the dose at which, say, 10% of the population exhibit an effect. If only a LOAEL / LOAEC has been recorded, then dose–response models may be used to extrapolate to a benchmark dose. The EPA depends on available scientific literature, rather than requiring specific studies and uses an extensive list of 'uncertainty' factors as denominators, to translate test results into no-effect levels.

If all doses above a zero dose cause an adverse effect, the LOAEL is used as the threshold (because it represents an upper bound on the threshold) together with a safety (uncertainty) factor (Crawford-Brown 1999).

Uncertainty factors account for interspecies variability (UF_A), intraspecies variability (UF_H), uncertainty in the duration of exposure and the duration of the study (UF_S), the use of a LOAEL, incomplete data bases, age dependence of thresholds (such as the greater susceptibility of children) and additional modifying factors (MF). The intent is to generate confidence that the true threshold for the most sensitive members of a population is greater than this value, reflecting application of the precautionary principle (Crawford-Brown 1999).

The reference dose (RfD) is given by:

$$RfD = \frac{NOAEL}{UF_A \times UF_H \times UF_S \times MF \times \ldots}$$

The factors are assumed by default to have values of 10, usually making the denominator in the above equation range between 1000 and 100 000. Reference concentrations (RfCs) are used for inhalation pathways. RfDs are used for ingestion (Crawford-Brown 1999).

Uncertainty factors are a crude way of providing assurances that actual doses and responses are no greater than those specified. They are designed to be protective. Unfortunately, the level of protection is unevenly applied between substances, reflecting the level of knowledge rather than the potential for harm. Furthermore, the absolute level of protection is unknown. This has led to accusations that regulatory thresholds for chemicals are hyperconservative to the point of disallowing many substances that might, on balance, be beneficial (Breyer 1993, Burmaster and Harris 1993, Finkel 1995). We explore this conundrum in detail in Chapters 9 and 12.

7.3.2 A bootstrap estimate of RfDs

It is a standard procedure to regress ranked, transformed LC_{50}, NOEC and EC_{50} data on concentration for a range of species to estimate a specified level of protection (see the example below). Suter (1993) and Newman et al. (2000) raised a number of concerns, including:

- Standard toxicity tests such as LC_{50}s do not give an adequate measure of effects on populations in the field. In particular, in situ exposures depend on life stage, behaviour, dispersal dynamics and so on.
- Any species loss may be unacceptable.
- The approach does not discriminate among species and it may be more important to protect ecologically important species (keystone species, dominants).
- The data ignore many sublethal effects.
- The assumption of a lognormal distribution usually is not verified.
- The sample sizes and the taxa used in tests are arbitrary.

Box 7.2 · *A bootstrap estimate of uncertainty factors (after Newman et al. 2000)*

The data are composed of a set of LC_{50}s or related information for a set of species (usually 30 or more). Their procedure works as follows:

1. The data (the species) are sampled randomly, with replacement, generating a set of 100 observations.
2. The taxa are ranked from lowest to highest (LC_{50}).
3. The value at the 5th percentile is taken to be the allowable threshold (the hazardous concentration).
4. The first three steps above are repeated, say, 10 000 times, giving 10 000 estimates of the hazardous concentration.
5. The 10 000 concentrations are ranked and the median value is taken to be the hazardous concentration.

The 2.5th and the 97.5th percentiles are the 95% bootstrap confidence intervals. The concentration corresponding to the 5th percentile gives the concentration that protects 95% of species with 95% certainty. The main cost of this approach is the computational overhead, a modest price for the benefits.

The choice of concentrations and the scale implied by the interval between concentrations can affect interpretations. For instance, when log scales are used, analysts are more likely to specify low–dose thresholds because the scale places greater emphasis on low concentrations compared to a linear scale (see Figure 4.6 and Nabholz *et al.* 1997).

The uncertainty factors outlined above are intended to compensate for these uncertainties. Newman *et al.* (2000) proposed an alternative analytical procedure based on a bootstrap (Efron and Tibshirani 1991, Box 7.2).

7.4 Transport, fate and exposure

Once a reference dose has been established, it must be translated into a concentration that is allowed in the environment, usually a concentration that will produce a dose below the threshold, expressed as the average daily intake (in units of mg/kg·day). The ways in which chemicals enter and move through the environment determine, to a large extent, what is allowed by regulators.

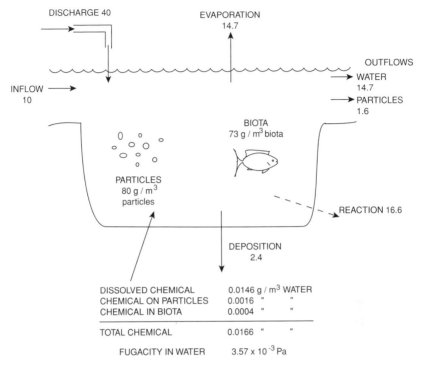

Figure 7.15. Conceptual model for the steady-state mass balance of a chemical in a lake (from Suter 1993).

Exposure modelling involves the collection of data on things such as discharge rates and concentrations of the chemical or its structurally similar alternatives within various environmental compartments. Defining important elements and pathways at an appropriate scale is an exercise in conceptual model building (Chapter 3). The model summarizes component species, the locations of people relative to emission sources and plumes, as well as hydrology, soil characteristics, or other elements and processes with which the chemical may interact.

The conceptual model for exposure may be formalized mathematically, or may be represented as a diagram (e.g. Figure 7.15), an influence diagram or a hazard matrix (Chapter 5). It is used to estimate amounts and concentration of the chemical in different parts of the environment, its transportation rates and estimates of the time for which it will remain.

The various elements in a model such as that in Figure 7.15 may be reduced to a set of reasonably well-understood physical and chemical processes (a few are summarized in Box 7.3).

Box 7.3 · *Chemical processes important in many exposure pathways*

- *Advection:* the transportation of a chemical from one medium to another by a carrier unrelated to the presence of the chemical. Such vehicles include dust, rainfall, food, or sediment particles suspended in a water column.
- *Diffusion:* occurs when a chemical migrates between media because it is in a state of disequilibrium.
- *Partitioning coefficient:* the name given to the ratio of concentrations of chemicals in two phases at equilibrium. This is also known as *fugacity* (a measure of the chemical's tendency to escape from the phase it's in).
- *Volatilization:* processes in which a chemical changes from a solid or liquid form to a vapour.
- *Bioconcentration factor:* the steady-state ratio of chemical concentration in organisms relative to the concentration of the chemical in the media in which the organisms live. It is the net result of the uptake, distribution and elimination of a substance, and depends on a variety of physicochemical and physiological factors.
- *Transformation:* describes the set of processes and reactions by which chemicals change composition, including physical processes such as oxidation and biologically mediated transformations.
- *Hydrolysis:* the term given to chemical reactions in which organic compounds react with water to produce other compounds. The hydrolysis rate depends on the pH of the medium, temperature and the presence of anions and cations. The rate is often expressed as a half-life.
- *Photolysis:* refers to reactions caused by light, such as sunlight photolysis of organic chemicals in surface waters, on soil and in the atmosphere. The photoreaction rate depends on sunlight intensity, UV absorption tendencies of the organic compound, and the efficiency with which absorbed light causes the reaction.

The chemical processes in Box 7.3 reflect the degree of maturity of the ecotoxicology paradigm for risk assessment. Conceptual modelling in this domain has been routine for several decades. In contrast, in some arenas, there is no consensus on how to write appropriate models, or even if models are useful.

7.4.1 Exposure assessment

Exposure assessment measures the exposure of components of a population or ecosystem to a contaminant. The assessment answers questions such as: 'Given output of fate models, which media are significantly contaminated?', 'To which contaminated media are the endpoint organisms exposed?', 'How are they exposed (i.e. routes and rates of exposure)?', 'Given an initial exposure, will the behavioural response of exposed organisms modify subsequent exposure (i.e. attraction and avoidance)?'.

Estimation of exposure relies on creating exposure scenarios, descriptions of the circumstances in which a chemical is likely to be used. The scenarios determine the equations that are used to estimate exposure and, subsequently, the level of the chemical that will be tolerated by regulators.

Exposure pathways reflect the way a chemical reaches a target, usually an assessment endpoint. For example, respiration provides opportunities for hazardous substances to reach organisms through inhalation of contaminated air, intake of contaminated water through gills, dermal sorption, or uptake of a gaseous contaminant through leaves. Solid media may provide opportunities for ingestion of contaminated sediments and sediment-associated food, vegetative uptake of contaminants through soil, or inhalation of contaminated dust particles.

For example, Suter (1993) noted that if you eat or drink methylmercury, it may damage your central nervous system, impair your hearing and cause tunnel vision, severe mental abnormalities, headaches, fatigue, death and deleterious effects on a developing foetus. Methylmercury may be found in sediment and suspended sediments in water. It is taken up by aquatic organisms via their gills and reaches humans through food and drinking water contaminated by suspended sediment particles.

The last step in the chain of logic used to set regulatory conditions is to assess the dose a person (or other endpoint) is likely to receive from different pathways. The typical approach is to use the 'dose' equation (see, for example, Crawford-Brown, 1999, p. 84):

$$\mathrm{DOSE} = \frac{C_{\text{pathway } x} \times I_{\text{pathway } x}}{\text{bw}}$$

DOSE = total daily dose (mg/kg$_{\text{bw}}$·day),
$C_{\text{pathway}x}$ = concentration of chemical in medium x (mg/kg),
$I_{\text{pathway}x}$ = intake rate, total daily intake of this medium (kg/day),
bw = average human body weight (70 kg).

Figure 7.16. A model for bioconcentration.

This equation summarizes the thinking in the exposure scenarios. In the case of methylmercury, the dose would include the sum of exposures from consumption of fish and shellfish, and from drinking contaminated water. It's then up to the regulator to set conditions for use of the chemical such that the concentrations do not result in doses that exceed the ADI.

7.4.2 Modelling transport, fate and exposure

Creation of exposure scenarios is essentially a process of constructing conceptual models. The dose equation translates the qualitative ideas into a quantitative framework. The models we build to understand and predict the consequences of human activities can be more complex than the dose equation, and should be if the problem warrants a more detailed treatment.

For example, a two-compartment submodel of some of the bioconcentration processes may be that represented in Figure 7.16. If we assume that the amount of material excreted by the fish has a tiny effect on the concentration in the water (because of the relative sizes of the fish and the lake) or because the excreted material is not dissolved, then the rate of change of concentration depends on the rate at which it is absorbed and the rate at which it is excreted.

In continuous time (a reasonable assumption for continuous diffusion processes), the simplification above may be further abstracted into an expression

$$\frac{dC_f}{dt} = k_u C_w - k_e C_f$$

where

C_f = concentration of the chemical in the fish (mol/kg),
C_w = concentration of the chemical in the lake (mol/kg),
k_u = uptake rate of water (l/(kg·day)),
k_e = elimination rate (1/kg·day).

The bioconcentration factor (BCF) $= C_f / C_w = k_u / k_e$.

This equation is an example of many that may be written to represent the way we think the system works. As in all models, the level of detail is a compromise between the questions we need to answer, the data and our understanding of the system. There may be other equations that are just as plausible, or that fit the data equally well. A complete analysis would consider structural alternatives. In addition, both variability and incertitude affect our knowledge of the parameters. In Chapter 11, we will see methods that allow us to include uncertainty in models such as these.

7.4.3 Hazard quotients and mixtures

The goal of the regulator is to prevent exposures that will produce doses above the benchmark. If the expected dose (the average daily rate of intake, ADRI) exceeds the reference dose (RfD), the regulator may be judged to have failed in its duty to protect the exposed population.

A hazard quotient (HQ) is the estimated exposure divided by a toxicity threshold. Typically,

$$HQ = \frac{ADRI}{RfD}$$

in which an HQ above 1 indicates the RfD has been exceeded in the exposed population. The job of the risk manager is to ensure that even the most heavily exposed members of the population have an HQ less than 1.

A hazard index is the sum of the HQs for all of the substances to which an individual is exposed, and that act by a similar mechanism. A value of the ratio above 1 does not mean that the effect will certainly occur. The chance of the event is unknown but it is assumed to be acceptably small for hazard indices less than 1.

This approach assumes the effects of different substances are additive. If substances with different modes of action are experienced, they are treated and regulated separately. Thus, there is no allowance for synergistic effects between substances. Just as importantly, even when using 'reasonable maximum exposures', point estimates for hazard quotients typically are conservative and the level of conservatism is variable (Cullen 1994).

Table 7.4. *Parameter estimates for the state variables and flows in the conceptual model represented by Figure 7.15 (Suter, 1993)*

Volume of water in lake, $V_w = 10^7$ m^3
Flow into and out of lake, W_I, $W_O = 1000$ m^3/h
Inflow of suspended sediment, $I_S = 0.05$ m^3/h
Deposition to bottom of lake, $D_L = 0.03$ m^3/h
Sediment flowing out of lake, $S_O = 0.02$ m^3/h
Hydrolysis rate, $K_h = 10^{-4}$ h^{-1} (half-life = 289 days)
Evaporation rate of water, $E_w = 10$ m^3/h
Air-water partition coefficient, $K_{aw} = 0.01$ (i.e. C_a/C_w)
Particle-water partition coefficient, $K_{pw} = 5450$
Biota-water partition coefficient, $K_{bw} = 5000$
Volume of particles, $V_p = 200$ m^3
Volume of biota (including fish), $V_b = 50$ m^3
Discharge rate of chemical, $D_c = 40$ g/h
Concentration of chemical in inflow, $C_I = 0.01$ g/m^3

7.4.4 Model-based assessments

We could continue to build the submodels into a cohesive whole. One of the benefits is to make it plain just how complex dynamics such as these can be, even for an extremely simple abstraction such as that in Figure 7.15. For example, Suter (1993) provided parameter estimates for a one-compartment aquatic model (Table 7.4).

This list is dauntingly long and complex. Usually, most of the parameter estimates will be missing, or based on expert judgement or on estimates made for other species and other lakes.

Note, in addition, that each parameter is represented by a best estimate alone. The analyst did not bother to report bounds, or any other measure of uncertainty in the parameters or the model structure. This is not a reason to disavow an explicit model. Rather, the analyst has the opportunity to make the lack of data clear by providing bounds for each parameter. At least, the point estimates mean that terms in the model are defined unambiguously. If we take care with units and the construction of equations, the logic of the problem may then be internally consistent. These properties are difficult to obtain in a model based on language and concepts.

This model represents just one kind of abstraction of the problem of evaluating the exposure and consequences of the release of a chemical. It is possible to write a set of equations for all of the pathways, and then to solve the equations to generate expected outcomes and their

Table 7.5. *Threshold values for classification of chemicals (after Beer and Ziolkowski 1995)*

	Very toxic (mg/l)	Toxic (mg/l)	Harmful (mg/l)
96-h LC_{50} for fish	≤ 1	≤ 10	≤ 100
48-h EC_{50} for *Daphnia*	≤ 1	≤ 10	≤ 100
72-h EC_{50} for algae	≤ 1	≤ 10	≤ 100

associated probabilities using Monte Carlo simulation or other methods (Chapter 10).

7.5 Examples

7.5.1 Atrazine

Atrazine is a chemical used to control weeds in crops. Tarplee (2000) provided an example of the thinking behind estimation of reference doses in a re-evaluation of Atrazine. Measured LOAELs and NOAELs are combined with uncertainty factors in a qualitative process to generate an acceptable concentration. Atrazine has been shown to disrupt the reproductive development of rats, raising concerns that it may affect the development of children, although there is no direct evidence for this effect. Tarplee (2000) settled on an acute RfD of 0.1 mg/kg, '...based on delayed or absence of ossification found in a rat development study (NOAEL 10 mg/kg; LOAEL 70 mg/kg) and an uncertainty factor of 1000...Three other studies were considered to support this RfD'. They included two other studies on rats and one on rabbits that showed similar effects using similar concentrations of the chemical.

7.5.2 OECD protection thresholds

OECD policy aims to protect the structure and component species of ecosystems and, by doing so, protect ecosystem functions. It uses risk limits of chemical substances in soil, air and water. Chemicals are classified as very toxic, toxic or harmful on the basis of standard laboratory toxicity tests (Table 7.5).

Maximum permissible risk concentrations are an extrapolation of toxicity tests. They are intended to protect 95% of species. For human health, thresholds are set at 10^{-6} additional deaths per year for noncarcinogens, and at 10^{-4} for genotoxic carcinogens (based on NOEL). Negligible risk concentrations are calculated to be 100 times below the maximum permissible risk concentration.

Thus, like the Atrazine example, the approach uses evidence of toxic effects together with uncertainty factors in a qualitative process to arrive at an acceptable concentration.

7.5.3 Cotton pyrethroid risk assessment

Cotton pyrethroid insecticides are chemicals used to control insects. They interfere with ion channels in insect nervous systems.

Companies seeking to register pyrethroid products have used farm pond monitoring, contained field experiments, runoff monitoring and research into bioavailability to establish toxicity levels. Registrants are required to recommend no-spray distances that will buffer nontarget species from the effects of the insecticides.

Cotton pyrethroids have a half-life of about four days and low toxicity to mammals and birds. The main environmental concern is their potential to affect aquatic organisms. These concerns led in the USA to studies of toxic effects on organisms in several trophic levels and with different ecosystem functions and physiological characteristics.

Solomon *et al.* (2001) obtained all available data for aquatic species and created a cumulative frequency distribution. They combined the results of LC_{50} studies with EC_{50} studies because EC studies of insects usually record immobility, which leads to death. There were few LC_{10} or EC_5 results available. They used acute assays (over 24–96 hs) because uptake is rapid and the half-life of the chemicals is short. For multiple studies of a single species, they used the geometric mean of the results, giving a conservative estimate of the species' value.

Transformed ranks of species were calculated as $(100 \times i)/(n + 1)$ where i is the raw numerical rank of a species and n is the number of points (species). These transformed ranks were plotted on a lognormal transformed axis against the log of the concentration (Figure 7.17).

Solomon *et al.* (2001) fitted a linear regression to the points and used it to estimate the 10th percentile of the distribution. This value was used in calculating reference doses to regulate the chemicals. When discussing extrapolation uncertainties, Solomon *et al.* (2001) noted acute to chronic ratios ranged from 2 to 415, with an average of about 44.

The analysis ignores interactions between chemicals. It does not investigate responses of species under a range of field conditions. In common with all such assessments, it is simply too costly and time consuming to pursue all potential lines of enquiry.

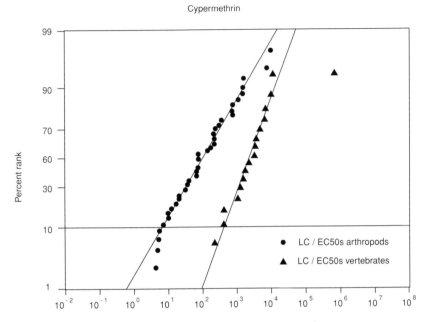

Figure 7.17. Distribution of acute toxicity values for cypermethrin, a cotton pyrethroid, in aquatic arthropods and vertebrates. The outlier represents a very insensitive species. This data point was omitted from subsequent analysis (after Solomon *et al.* 2001).

Unfortunately, the uncertainties involved in the analysis were not represented in the final value of the threshold. Instead, a conservative bound is offered that conveys the sense that real concentrations and levels of harm will be less than the threshold implies. Like the examples above, there is no way of knowing what the level of protection is, or how it compares to the levels applied to other chemicals.

7.5.4 Integrating pesticide risks in Italy

A variety of classification systems for pesticide risk have been developed by European countries. Some select a pesticide with the least impact among a candidate set. Others guide farming practices. All systems measure risk to water organisms (such as EC_{50}s) and most consider risks of groundwater contamination. Despite these similarities, Finizio and Villa (2002) noted that when the various methods were applied to a set of chemicals, the ranks generated were very different.

Finizio and Villa (2002) took elements from the various systems to create a rating system for pesticides on behalf of the Italian Environmental

Table 7.6. *Risk classification intervals and scores for nontarget organisms in surface water (after Finizio and Villa 2002). Scores were determined for each chemical, for each of algae (A), Daphnia (B) and fish (C), based on the results of tests. The scores were combined into a final score using* $P = 3A + 4B + 5.5C$. *The coefficients (weights) were determined subjectively. PEC is the predicted environmental concentration calculated from dilution, transport and fate models (see below). The toxicity exposure ratio (TER) is the* EC_{50} *(or* LC_{50}*) divided by the PEC*

TER	Score
>1000	0
1000–100	1
100–10	2
10–1	4
<1	8

Protection Agency. For example, they used measurement endpoints for algae, *Daphnia* and fish and assigned scores to different toxicity intervals. They then used a weighted linear combination of these scores to create a final score (Table 7.6). This score was used to rank pesticides, and to classify them as low-, medium- and high-risk hazards (Figure 7.18).

The purpose of the Italian system was to provide a basis on which to create incentives for more sustainable agricultural practices. People who used chemicals at the low end of the index were to be rewarded with financial incentives.

Like the risk assessments above, the analysis did not carry uncertainties through the chain of calculations. Several parameters were subjectively estimated, while other parameters such as LC_{50}s came with statistical bounds. Thresholds used in various jurisdictions were arbitrary. It would have been useful to conduct analyses that evaluated the sensitivity of ranks to uncertainties. The absolute values of the indices and the associated classifications may be less important because hazards are treated in a relative way.

7.5.5 Human health thresholds in Australia

Typically, health experts arrive at a consensus based on toxicity results, epidemiological data and any other available information. In theory, these standards integrate all exposure pathways and take into account individual

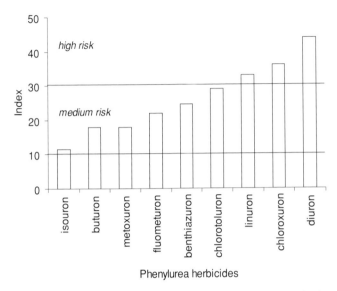

Figure 7.18. Integrated toxicity index scores for phenylurea herbicides in Italy (after Finizio and Villa 2002). The scores and classes are used to support financial incentives for farmers to reduce the polluting effects of agriculture.

sensitivities. However, the detailed reasoning used to arrive at such thresholds is rarely explicit and there are no standard equations.

For example, the Australian National Environment Protection Measures (NEPC 2000) provided ambient air quality standards and goals (Table 7.7). The standards specified details for the endpoints including the periods over which measurements should be taken, and the frequency with which the regulatory standards may be exceeded.

Like the reference dose thresholds set by the EPA, there is no way of knowing how protective each standard is, or how the level of protection varies between substances.

7.5.6 Methylmercury in the u'Mgeni River

A large mercury incinerator was commissioned near the u'Mgeni River in South Africa in 1986. In the early 1990s, effluent from a holding pond overflowed into a tributary that flows into the u'Mgeni River and the Inanda Dam, in a densely populated catchment in which the people rely on the dam for drinking water, watering livestock and fish (Oosthuizen 2001).

Table 7.7. *The Australian national criteria for ambient air pollution for carbon monoxide, nitrogen dioxide, photochemical oxidants, lead and particulate matter*

Pollutant	Averaging period	Maximum concentration	Goal within 10 years for maximum allowable exceedances
Carbon monoxide	8 h	9.0 ppm	1 day a year
Nitrogen dioxide	1 h/1 year	0.12 ppm/0.03 ppm	1 day a year/none
Photochemical oxidants (as ozone)	1 h/1 day/ 1 year	0.20 ppm/0.08 ppm/ 0.02 ppm	1 day a year/ 1 day a year/none
Lead	1 year	0.05 μg/m³	none
Particles as PM_{10}	1 day	50 μg/m³	5 days a year

Mercury poisoning causes neurological and developmental disorders. The World Health Organization (WHO 1994) estimated that a reference daily intake of methylmercury of 0.48 μg/kg body weight would cause no adverse effect in humans. Mercury concentrates in human hair follicles in a predictable way and the WHO estimated that the daily intake would, at equilibrium, equate to hair mercury concentration of 11 μg/g of hair.

The US EPA (1997c) specified a benchmark dose of methylmercury of 0.1 μg/kg body weight, based on the lower bound of the 95% confidence interval of the dose that produces a 10% prevalence of adverse effects, and an uncertainty factor of 10. Their calculations equated this dose with methylmercury concentrations of 44 μg/l in blood or 2.3 μg/g in hair.

Oosthuizen (2001) collected data on consumption and methylmercury concentrations in fish, and estimated the dose from the standard dose equation as:

$$\text{DOSE} = \frac{C_{\text{fish}} \times I_{\text{fish}}}{\text{bw}}.$$

He collected hair samples from a small population of 'high-risk' individuals, young boys who swam, played, drank and fished in the u'Mgeni River daily ($n = 9$), and from a control group of people with similar characteristics upstream from the plant ($n = 5$).

Hazard quotients are the calculated daily dose divided by the reference value representing a tolerable daily dose (see above). Hazard quotients relative to the WHO standard ranged from 0.8 to 2.7 with a median value of 2.3. Hazard quotients for the US EPA benchmark ranged from 4 to 13, with a median value of 11. These results suggested that the population of exposed children were at risk from elevated methylmercury. In contrast, the analysis of hair samples from the same children returned concentrations less than 0.5 μg/g, below the measurement detection limit for all cases in the control and the exposed groups, and well below concentrations that are typical of people who exhibit effects.

The samples sizes were small but the results raise some interesting issues. Oosthuizen (2001) questioned the effects of seasonal variation in diet, the reliability of dietary information, the possibility that chelating agents in the diet help eliminate mercury, the assumed relationship between blood and hair mercury concentrations, the hyperconservative assumptions in US EPA and WHO calculations, and the fact that previous studies on hair mercury levels were conducted on people predominantly of European and Asian origin. The physiology of mercury in African hair and blood may be different.

7.6 Discussion

The ecotoxicological paradigm has evolved ways of dealing with uncertainty that make it difficult to estimate just how protective the system is. Some say it is hyperconservative (Breyer 1993). Others say it is about right (Finkel 1995). The main problem is that the level of conservatism is uncontrolled and probably varies substantially between applications. Other methods share this problem (Chapter 9).

At least some levels of protection are higher than may be warranted by the effects of individual toxicants. Swartout et al. (1998) explored the extent to which uncertainty factors are biased in correcting a dose to a safe level. They used data for substances for which there were both human and animal studies of effects. They limited the studies to those in which data on variability in human responses were available, where studies of different durations had been conducted, and where both LOAELs and NOAELs had been estimated. These data allowed them to estimate directly, unbiased correction factors for each of these elements (between-species variability, within-species variability, duration of the study, and use of LOAELs). They combined lognormal distributions for each of these elements using Monte Carlo techniques (see Chapter 10).

They estimated that a correction factor of 234 for three uncertainty factors would produce a threshold that is 95% certain to provide sufficient protection for a randomly selected person from the exposed population. The default protocols would have created a denominator of 1000 for three factors, providing a confidence level far in excess of 99%.

But these estimates ignore interactions between chemicals, undiscovered pathways and so on. The degree of conservatism is intended in the USA to reflect a 'margin of safety'. Judge Bork made rulings (reported in 1994, recounted by Crawford-Brown 1999) over the effects of benzene and vinyl chloride (both carcinogens) that established several precedents. 'Reasonable protection' against risk does not require zero risk. Instead there is some social and legal threshold of acceptable additional risk. So long as the EPA established exposures that would keep the risk below an acceptable level, even for the most exposed individual, the regulated activity could continue.

The Judge ruled that the EPA must first establish a goal for risk (e.g. keeping added annual risk of dying below a value of 10^{-6}), and then recognize that the actual risk to individuals is subject to considerable uncertainty. If regulatory decisions ensure that the individual with the highest risk is below the target risk, even if the upper confidence limit is used to assess exposure, then the regulations may be said to provide an 'ample margin of safety'.

Ideally, to satisfy this stipulation, it would be necessary to specify the risk target, the acceptable fraction of people above the risk ceiling, and the confidence with which it can be stated that the actual fraction of the population is above the ceiling (Crawford-Brown 1999). The levels of conservatism in toxicological risk assessments and ways for dealing with them are explored in more detail in Chapters 9 and 10.

Measures of effect are difficult to relate to specific ecological outcomes when viewed in isolation. It is difficult to relate the objective of protecting sensitive elements of an ecosystem to an allowable dose that kills less than 50% of the adults of 5% of the species in a system. Ignoring population or higher-level effects, focusing only on individual-level endpoints can lead to biased judgements about risk (Pastorok *et al.* 2003). Ecological models of population processes may be used to translate the individual-level effects of toxicological tests into more concrete terms. The output of the ecological model will correspond to one or more of the assessment endpoints (see the examples documented in Pastorok *et al.* 2003). Models of this kind will be discussed in more detail in Chapter 11.

8 · *Logic trees and decisions*

Logic trees are diagrams that link all the processes and events that could lead to, or develop from, a hazard. There are two approaches: a fault tree works from the top down, linking chains of events to the outcome (fault tree analysis); an event tree takes a triggering event and follows all possible outcomes to their final consequences (event tree analysis).

These approaches are best developed in engineering where they are used to formalize conceptual models. They are sometimes called cause-consequence diagrams (Hayes 2002a). Logic trees use the same structured reasoning that appears in diagrams of arguments in informal logic (Walton and Batten 1984) although argument diagrams are not usually causal (Korb and Nicholson 2003). Logic trees are also intimately related to expert systems.

This chapter describes event trees and their extensions into decision trees, decision tables and related methods. It introduces classification and regression trees, probabilistic trees and Bayes' networks. Lastly, it describes the structure and function of fault trees. Logic trees are pervasive, even if sometimes the trees are not drawn. Examples from conservation biology, toxicology, human health and safety, engineering, freshwater ecology and invasive species risk assessment illustrate their utility and limitations.

8.1 Event trees

Event trees link possible outcomes following an initiating event. They are constructed as a series of dichotomies (yes/no). Each node is an event (or a decision). The tree represents a model of the causal pathways for the system. The focus is on a primary (initiating) event and consequences are traced forwards from it.

Most event trees have been constructed for engineering systems. Figure 8.1 shows a typical application outlining the possible outcomes resulting from a crack in a container containing a toxic substance. The branch

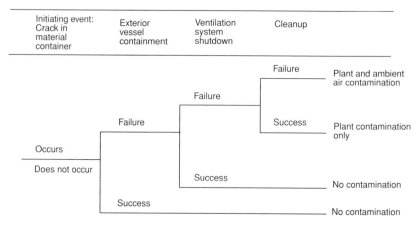

Figure 8.1. Event tree for toxic material container leak (after Stewart and Melchers 1997).

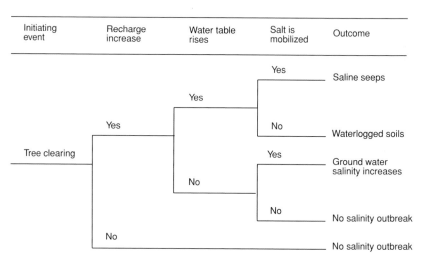

Figure 8.2. Event tree analysis for assessing risk of salinity effects after tree clearing (Bui 2000).

points are logical statements about the system, simplified into binary questions. Release into the atmosphere will occur only if there is a crack in the container AND the exterior containment vessel fails AND the ventilation system does not shut down AND the cleanup procedures are not executed properly.

The same kind of structure can be used to understand ecological consequences of land management decisions (Figure 8.2). This example makes some of the assumptions embedded in the logic tree clearer. Like hazard identification, the discrete nature of a logic tree is awkward when applied to continuous processes. It forces the analyst to apply sharp definitions to vague boundaries. For example, recharge is the volume of water that flows into an underground aquifer. If trees are cleared, the amount of water reaching the aquifer may increase because the transpiration rate of the vegetation is reduced. But what qualifies as an increase? The volume of water flowing into the aquifer is a continuous variable. The question is whether the recharge rate has increased substantially, compared to any increase that may have been expected in the presence of the trees. Water table changes and salt mobilization are likewise continuous variables. It can be useful to subdivide the range of a continuous response into categories, and treat each as a separate outcome.

In the container example (Figure 8.1), it is easy to accept the implicit assumption that the events either occur or do not occur. The tree clearing example illuminates the assumption. So, for the problem in Figure 8.1 we could ask, how much of a cleanup represents 'proper execution'. A cleanup may involve many operations and the outcome is likely to be the containment of a proportion of the released substance (a continuous variable).

Similarly, what represents ventilation system shutdown? Not all ventilation systems are airtight, and there may be a delay between the event and the shutdown that allows some emissions. There may be simple and sufficient answers to these questions, but they are worth asking.

8.1.1 Decision trees

Decision trees are event trees in which one or more of the branch points are decisions. They are not necessarily causal or binary. A famous example is of a person who wants to locate a gas valve in a dark room. If the person lights a match to locate the valve, there is a risk of an explosion if the valve has been left open. But finding it in the dark is difficult.

The decision-maker can decide to light a match, or not (the decision node). There is gas in the room or not (two possible states of the world). This gives four possible combinations of circumstances: no gas + no match, no gas + match, gas + no match, gas + match. Only the latter leads to an explosion (Figure 8.3).

This example is deceptively simple and the embedded assumptions are not easy to see. Of course, there may be gas in the room, but

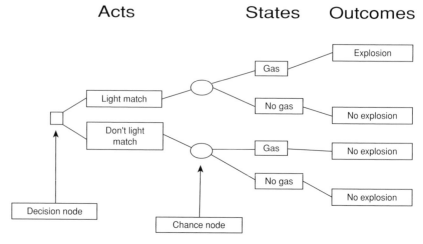

Figure 8.3. The gas valve problem represented as a decision tree.

not enough to cause an explosion. So the gas / no gas dichotomy is a sharp threshold through a continuous variable, and actually represents the condition 'enough gas to cause an explosion'. Similarly, there may be an explosion but it may be an acceptable one, small enough to cause no harm.

The structure of the decision is an example of many problems that people are asked to solve every day. It is a benefit-cost analysis, in which the cost of failing to light a match (crawling around in the dark for an interminable period) is weighed against the cost of lighting a match (an explosion). Imagine that you are a telephone operator and your job is to field calls to an emergency line. You are also required to dismiss 'obvious' cases of hoax calls. The structure of this problem is identical to the gas valve problem (Figure 8.4).

False alarms are costly. Ambulances and police are expensive. False alarms divert scarce resources from other urgent cases. But the cost of failing to report an urgent case, because it is thought to be a hoax, is also considerable and is highly visible, making the perception of this outcome an important consideration (Figure 8.5; Chapter 1).

It is possible that the person who incorrectly dismissed the call alerting them that someone was in trouble nevertheless made the right decision. The 'expense' in Figure 8.4 is more than just money. Hoax calls result in delays to other cases that may result in deaths. But they are less apparent and therefore the reaction to them is muted.

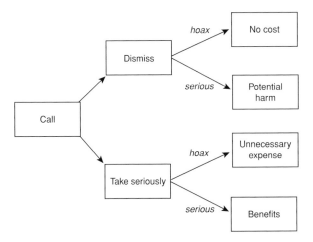

Figure 8.4. The hoax emergency call problem.

THE AGE
THURSDAY JUNE 7, 2001
BREAKING NEWS ✳ *theage.com.au*

OOO call woman left to die

By **DARREN GOODSIR**
and **LES KENNEDY**

A police phone operator dismissed as a hoax a 000 emergency call that an elderly woman had been left tied up alive after being bashed and raped in her home. Her body was discovered by chance 10 days later.

Police last week apologised to Joy Golbey Alchin's family, saying the 70-year-old probably would have been found alive if they had responded to the call.

Figure 8.5. The cost of a false-negative decision about an emergency call (D. Goodsir and L. Kennedy, '*The Age*', Melbourne, June 7, 2001).

If the person responsible for the error makes many such decisions and they are mostly right, then it might be worth an occasional error, in terms of human lives, to save the misdirected resources. Overall, more lives might be saved with this strategy, than by assuming all calls are serious. Unfortunately, such arguments are often lost when confronted with a stark press report of a single case.

Many decision-making circumstances are more complex than those involving two alternatives. A good example is provided by decisions about managing algal blooms in a river. The blooms contaminate drinking water, poison domestic animals, kill fish and halt recreational use of inland waters. They are a serious problem, but their effects can be substantially mitigated if there is early warning of an event. Domestic animals can be moved, people can be warned about drinking water, fish hatcheries can delay the release of hatchlings and towns can arrange independent water supplies.

A monitoring and decision framework was developed to react to measurement endpoints. Each week, a field officer travelled to several locations, took a surface water sample, measured water turbidity and recorded water flow rates over the preceding six days. The water sample was analysed for phosphorus (P) and nitrogen (N). The information was interpreted as follows (Figure 8.6):

- If nutrient levels were low (defined as concentrations in the sample of [P] less than 50 μg/l and [N] less than 500 μg/l) and sediment nutrient release was unlikely in the next six days, then the risk was classified as 'low'. There was no further action.
- If either nutrient condition was true (high levels or a likely release), light conditions were queried. If turbidity levels were high, it was considered that algal growth was unlikely and the risk was classified as 'low-medium'.
- If turbidity was low, meaning that light levels were high enough to support algal growth, flow conditions were queried.
- If flow was high for at least one of the last six days, thereby flushing developing algae downstream, the risk was classified as 'medium-low'.
- If flow was consistently low over the last six days, the risk was classified as 'high'.

The latter condition triggered a more detailed investigation of conditions, and may have led to advice to residents to move domestic animals away from the river, and to warnings to people about swimming and drinking conditions.

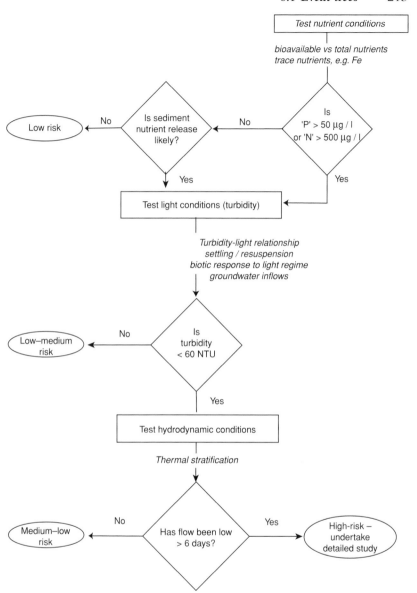

Figure 8.6. Decision tree for the risk of cyanobacterial blooms in lowland rivers caused by irrigation drainage (after Hart *et al.* 1999). NTU stands for nephalometric turbidity unit.

The advantage of Figure 8.6 over the dialogue above is that the train of logic is easy to see and understand. The decision tree is applied every week. Algal blooms are rare and, mostly, the tree leads to no action. The tree carries with it the assumption that cyanobacterial growth is controlled by nutrients, light and flow conditions (Hart *et al.* 1999).

It could lead to two kinds of mistakes: the analyst could advise low risk, and an algal bloom occurs; or, the analyst could advise high risk, and no algal bloom occurs.

Both mistakes have costs. Failing to warn about a bloom could result in the deaths of domestic animals and those who drink or swim may become ill. Rapid transport of water is relatively expensive and the impact on tourism may have far-reaching economic implications. Predicting a bloom that does not occur results in unnecessary costs associated with domestic animal movement, water transport and restrictions on recreation and tourism.

The thresholds for nutrient concentrations, turbidity levels and flow conditions are arbitrary, sharp boundaries applied to continuous variables. Different thresholds would lead to different frequencies for both kinds of mistakes. Increasing the frequency of one would decrease the frequency of the other. We will return to the issue of how to set thresholds to make optimal decisions in Chapter 11.

8.1.2 Probabilistic event trees

The critical elements in the gas valve problem and the hoax emergency call problem are that the decisions are weighted by a judgement about the probability of the true state of the world. In both cases, there is a fact: there is enough gas in the room for an explosion, or not; and there is someone in trouble, or not.

Decision-makers cannot affect the truth of the matter, or its likelihood of being true. Rather, they have a subjective belief in the truth of each condition. If their subjective belief is that the gas level is safe, or the call is a hoax, they will act accordingly. The consequences of a mistake are large, relative to the benefits, so that it only makes sense to light a match or ignore a call if the chance of a mistake is small.

An analyst or expert may estimate probabilities for each branch of an event tree subjectively. Sometimes, a model or data from similar cases support subjective estimates. For example, the engineers who designed the material container in Figure 8.1 may have data for failures in similar

Table 8.1. *The gas problem as a decision table*

		States	
		Explosive gas level	Little/no gas
Acts	Light match	Explosion	No explosion
	Don't light match	No explosion	No explosion

components, or the design specifications may allow the failure rate to be estimated from theory.

If it is possible to quantify the consequences in equivalent terms, then the decision-making problem reduces to one in which the expected benefit can be calculated by the product of the probability times the expected gain (or loss).

8.1.3   Decision tables and expert systems

One problem with logic trees of all kinds is that they become unreasonably large and unwieldy, even for modestly sized problems. There is a one-to-one correspondence between decision trees and decision tables. The latter are not visually appealing and do not communicate the problem as easily as decision trees, but they are much more compact (Table 8.1).

Like decision tables, expert systems may have a one-to-one correspondence with a decision tree. They were invented so that the thinking behind the tree would be more accessible to the person using it.

They differ in that the branch points of the logic tree are expressed in natural language. The rationale behind the branches is expressed as dialogue and may be accessed by the person using the tree. This provides the opportunity, for instance, to include information about the uncertainty associated with each branch, at least in narrative form.

Starfield and Bleloch (1992) described a protocol for building decision tables and translating them into expert systems:

1. Write relevant statements (states of the world) in the left-hand side of a table (e.g. Table 8.2).
2. Write potential acts (management options) across the top of the table.
3. Take the first column (the first decision): ask what combinations of true (T), false (F) and irrelevant (X) conditions will lead unambiguously to that decision.

Table 8.2. *A decision table for the question of how to manage fire in an African nature reserve (after Starfield and Bleloch 1992)*

States	Acts			
	Do not burn	Do not burn	Defer for a year	Burn after 1st rain
Need to remove old vegetation	F	X	T	T
Area recently burned	X	T	F	F
Vegetation is degraded (and a fire would help)	X	X	F	T
Grazers need grass (fire stimulates growth)	X	X	F	T

4. Ask what other combinations of T, F, and X will lead to the same decision.
5. Keep adding columns after the first decision until you have included all sensible routes.
6. Move to the next decision.

An expert system, in its simplest form, is a system for communicating the logic and evidence that supports a decision table. For instance, in support of a decision to burn, the expert system should communicate the fact that a fire is expected to restore degraded vegetation by stimulating the germination of the soil-stored seed bank and to encourage grass germination and growth by providing a release of nutrients from woody vegetation in ash. We would decide to burn after the first rain this year, for instance, IF we need to remove old vegetation AND the area has not been recently burned AND grazers need grass AND the vegetation is degraded.

There is no way of knowing, after the event, if a decision was 'right'. Sometimes, we can know it was wrong (as in the case of the phone call). Even then, people who make routine decisions must occasionally get them wrong, even if their decision strategy is optimal. The best we can do is to update decision criteria as experience and monitoring data accumulate, taking into account the costs of wrong decisions.

8.1.4   Classification and regression trees

Classification and regression trees (sometimes called CART analyses) provide a way of summarizing the knowledge of experts so that it may be used by less expert people. They create event trees in a repeatable and transparent way.

Explanatory variables are treated as 'classification' variables. The endpoint of a tree is a labelled partition of the set of possible observations and may be a factor or attribute (in a classification tree) or a predicted value (in a regression tree). The methods work by separating observations of the response variable into subsets, so that the response data in each subset are relatively uniform (Venables and Ripley 1997). Branch points are like chance nodes although they may be interpreted as decisions.

At each step, the data are partitioned in a way that is likely to produce good classifications or predictions in the future. The distributions of classes usually overlap to some extent so there is no unique way of partitioning that completely describes the classes. The easiest way to compare alternative partitions is to count the number of errors, and to choose the one that minimizes them.

One set of data called the 'training set' is used to build the model. An independent 'test' set is used to measure how well the tree predicts. The training set is assumed to be an independent random sample from the full set of observations.

The modeller must decide if a node is terminal, or whether it should be further subdivided. In most cases, the stopping rule is a practical choice about the desired size of the tree. Sometimes, the analyst is guided by the decrease in improvement in predictive accuracy. The most common strategy for building a tree is to build a large tree (one with more subdivisions and nodes than is desired) and then to prune the tree back to an acceptable size, a strategy that is similar to variable selection procedures in regression modeling (Elith 2000).

Most tree construction methods look ahead just one step when choosing the next split. Some methods allow only binary splits. Some split linear combinations of continuous variables (Venables and Ripley 1997).

Classification trees have been applied to habitat modelling (e.g. Franklin 2002). Kolar and Lodge (2002) used a classification tree to predict the potential for introduced fishes to invade freshwater lakes in North America. They used data on 13 life-history characteristics, 5 habitat requirements and 6 aspects of invasion history, and human use from 24 established (defined as 'successful') and 21 introduced but not established (defined as 'failed') alien fish species in the Great Lakes to build the model (Figure 8.7). They claimed that classification trees built in this way were relatively transparent and repeatable, compared to expert systems based on categorical attributes and expert judgements.

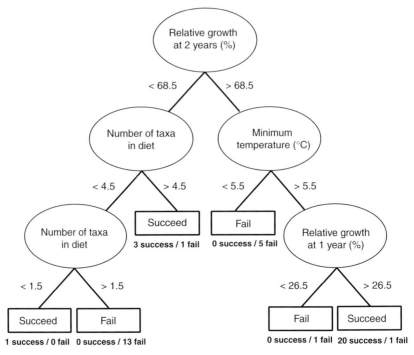

Figure 8.7. CART for successful and failed alien fish species in the Great Lakes. Ovals represent decision points and rectangles are the final classes. Each species is assigned to a class, depending on its characteristics. The tree based on minimum temperature threshold, diet breadth and relative growth classified 'failed' and 'successful' fishes with 94% accuracy (82% on cross-validation). The results of the predictions, applied to the species used to build the model, are shown beneath the nodes (after Kolar and Lodge 2002).

They pointed out that their model was likely to be unreliable when extrapolated to species with ecological characteristics outside the range of those used to build the model, and when applied to novel ecological circumstances such as tributaries and streams, rather than the lake environments. The cross-validation procedure (giving 82% accuracy) was also likely to be optimistic about performance, even for ecological conditions and species attributes equivalent to those used in model construction.

8.1.5 Bayesian networks

Bayesian networks (also called probability networks, influence networks and Bayes belief nets) are graphical models that represent relationships

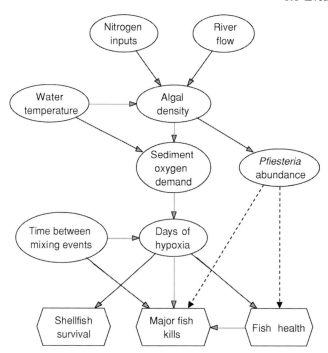

Figure 8.8. Probability network for the Neuse River Estuary (a subset of the relationships described by Borsuk *et al.* 2001a, 2003). Variables are shown as ovals and predictive endpoints as hexagons. Dashed lines represent relatively uncertain relationships.

among uncertain variables (Pearl 1988). Pearl (2000) recommended they be used to explain causal relationships and to integrate observations so as to eliminate models that are incompatible with data.

In the graph, nodes represent variables. A line between one variable and another represents a relationship (a dependency) between the variables. A node with no incoming edges can be described by an independent (marginal) probability distribution. A node with incoming edges depends on other variables and is described by a set of conditional probability distributions. The network describes the probabilistic relationships between all elements of a system. Probability distributions that depend on other distributions can be updated using Bayes' theorem.

Borsuk *et al.* (2003) used a probability network to model the ecological dynamics of an estuarine system in North Carolina (Figure 8.8). Building the model required each variable to be defined, reducing the chances that linguistic uncertainty would cloud issues (Borsuk *et al.* 2001a).

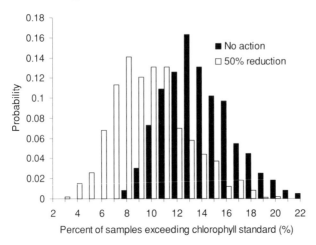

Figure 8.9. Anticipated effect of a 50% reduction in nitrogen inputs on the frequency of violations of the regulatory standard for chlorophyll a (40 µg/l) in the Neuse River Estuary (after Borsuk *et al.* 2003).

Each of the arrows represented a function that described how each element of the system depended on others (Borsuk *et al.* 2001b). Because functions were defined for all variables, the model involved some compromises in the level of detail. The endpoints in Figure 8.8 were adopted after public consultation (Borsuk *et al.* 2001a). They accord with the recommendations of Suter (1993) in that they satisfied ecological, operational and social demands.

Borsuk *et al.* (2001a) emphasized the need for subsequent monitoring to ensure the model is updated. Probability networks update new information through Bayes' theorem (see Section 4.9). Borsuk *et al.* (2003) used the model to explore 'what if' scenarios, an informal sensitivity analysis that displayed the consequences of predicted outcomes to potential management decisions (Figure 8.9).

Stow and Borsuk (2003) used networks to explore the relationship between fish kills and the presence of toxic algae *Pfiesteria*. Conventional wisdom was that *Pfiesteria* kill fish, implied by the dashed arrow in Figure 8.8. Toxic *Pfiesteria* are a subset of *Pfiesteria*-like organisms (PLOs). The conceptual model is that unknown environmental cues stimulate some PLOs to develop into toxic forms (Figure 8.10a).

The probabilities associated with each branch in a probability network are calculated using Bayes' theorem. Stow and Borsuk (2003) knew that the independent probability that toxic *Pfiesteria* would occur in a

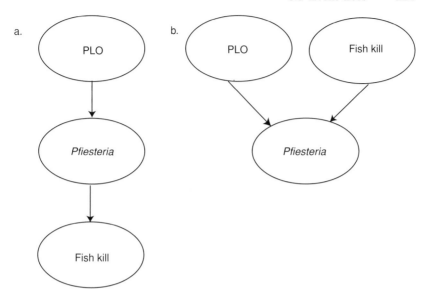

Figure 8.10. Alternative fish kill models. a. The model assumes a causal influence from *Pfiesteria*-like organisms (PLOs) to toxic *Pfiesteria* to fish kills. b. This model assumes fish kills stimulate the formation of toxic *Pfiesteria* from a population of PLOs that happen to be present at the site (after Stow and Borsuk 2003).

sample was 0.03. The probability that PLOs would be detected was 0.35. Because *Pfiesteria* are a form of PLO, if toxic *Pfiesteria* are present, by definition PLOs are present. Specifying the conventional model first:

$$P(\textit{Pfiesteria}|\text{ PLO}) = \frac{P(\text{PLO}|\textit{ Pfiesteria}) \cdot P(\textit{Pfiesteria})}{P(\text{PLO})}$$

$$= \frac{1.0 \times 0.03}{0.35} = 0.09.$$

The next step is to calculate the probability that a fish kill will occur, given that toxic *Pfiesteria* are present. Stow and Borsuk (2003) knew that if toxic *Pfiesteria* were detected, it was always in the presence of a fish kill:

$$P(\text{KILL} \mid \textit{Pfiesteria}) = 1.$$

The independent probability of a fish kill was 0.073 and the probability that toxic *Pfiesteria* are present when a kill occurs is 0.38 (*Pfiesteria* are not the only thing that can kill fish):

$$P(\text{KILL} \mid \text{no } \textit{Pfiesteria}) = \frac{P(\text{no } \textit{Pfiesteria} \mid \text{KILL}) \cdot P(\text{KILL}))}{P(\text{no } \textit{Pfiesteria})}$$

$$= \frac{(1 - P(\textit{Pfiesteria} \mid \text{KILL})) \cdot P(\text{KILL})}{(1 - P(\textit{Pfiesteria}))}$$

$$= \frac{(1 - 0.38) \times 0.073}{(1 - 0.03)} = 0.047.$$

Stow and Borsuk (2003) were uncomfortable with this conceptual model because they had noticed that toxic *Pfiesteria* were present only at fish kill sites, and never elsewhere. Furthermore, *Pfiesteria* were present at every fish kill site at which there were also other PLOs. What if, they asked, the fish kills cause the toxic *Pfiesteria*, rather than the other way around (Figure 8.10b)?

The Bayesian arithmetic changes. Since toxic *Pfiesteria* are present only if PLOs are present, then:

$$P(\textit{Pfiesteria} \mid \text{no PLO, KILL}) = P(\textit{Pfiesteria} \mid \text{no PLO, no KILL}) = 0.$$

Because *Pfiesteria* were never detected at sites where there were no fish kills,

$$P(\textit{Pfiesteria} \mid \text{PLO, no KILL}) = 0$$

$$P(\textit{Pfiesteria} \mid \text{PLO, KILL}) = 1,$$

and since PLO and KILL are unconditionally independent under the second conceptual model,

$$P(\text{KILL} \mid \text{PLO}) = P(\text{KILL} \mid \text{no PLO}) = 0.073.$$

This interpretation is consistent with all of the data. A particularly telling distinction is the expectation, under the model in which toxic *Pfiesteria* kill fish, that

$$P(\text{PLO} \mid \text{no } \textit{Pfiesteria, KILL}) = 0.33,$$

whereas under the model in which fish kills cause *Pfiesteria*,

$$P(\text{PLO} \mid \text{no } \textit{Pfiesteria, KILL}) = 0.$$

The data suggest that the probability is 0 of detecting PLOs at sites at which there was a fish kill and at which toxic *Pfiesteria* were absent (0 out of 35 such kills).

This may look like nothing more than the application of common sense. Hopefully, it is. But the presence of toxic *Pfiesteria* at 0.38 of the fish kills in the Neuse River Estuary had been enough to lead most researchers to believe in the first model. The advantage of the networks is that they make the logic of conceptual models plain, illuminating alternative explanations.

The assumption of causation illustrated by the fish kill example is an example of many embedded in most network models. When a cause, C, results in an effect, E, then C should not be positively or negatively correlated with (for instance) things that prevent E, other causes of E, or triggers for these (Cartwright 2003). There are so many possible networks, even for small numbers of elements (Table 4.1), that we settle on maps of causal interactions almost by intuition.

Bayesian networks may be built using continuous variables but it is easier to break variables into continuous classes (Korb and Nicholson 2003). Each value is a subrange of the original continuous variable. Most current software tools require it. Appropriate choices for class boundaries depend on the context of the problem and how sensitively the choice affects decisions or outcomes (Korb and Nicholson 2003).

8.2 Fault trees

Fault trees are built from hazard to consequence, and map what must occur for a hazard to be expressed. They express 'failure logic' and the contributing causes (and, optionally, associated probabilities for an unwanted event). Care is required to handle what is called 'failure modes', events outside the modelled system that affect different components simultaneously. Similarly, fault trees may not adequately model dependencies between system elements.

Fault trees are different to event trees because they are constructed around branch points that represent either a logical AND or a logical OR. They conclude with the primary (initiating) event with which event trees begin. Fault trees use some standard symbols (Figure 8.11). They have many of the same strengths and weaknesses as event trees. They are a versatile tool for mapping causal links between system components. Like event trees, they can become large and cumbersome even for modest

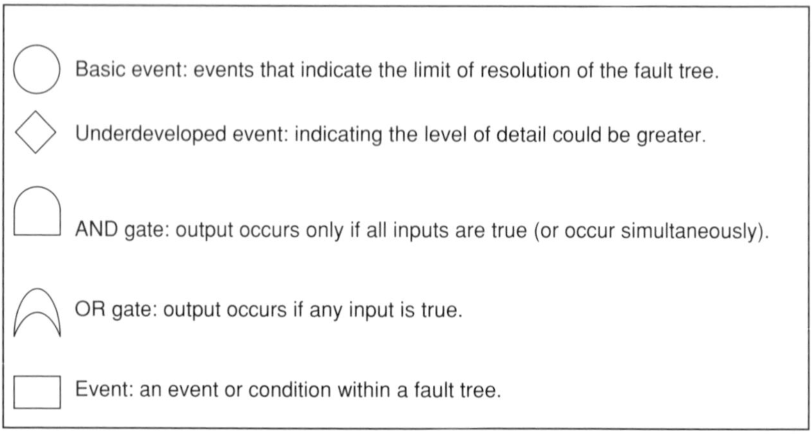

○	Basic event: events that indicate the limit of resolution of the fault tree.
◇	Underdeveloped event: indicating the level of detail could be greater.
⌂	AND gate: output occurs only if all inputs are true (or occur simultaneously).
⌒	OR gate: output occurs if any input is true.
□	Event: an event or condition within a fault tree.

Figure 8.11. Symbols used in fault trees (after Hayes 2002a).

problems. They are used most often to identify hazards and help design mitigation strategies (Hayes 2002b). Figure 8.12 is an example in which a fault tree was used to identify critical hazards in the management of the introduction of marine pests in ballast water.

The tree allows a risk manager to see readily if there are any system components that provide efficient monitoring and remediation strategies. Branches that depend on AND conditions, for example, are valuable because management needs to focus on only one component of the set linked by the AND statement. Different strategies are indicated on the tree, together with areas where managers need to conduct additional risk assessments.

Fault trees complement failure modes and effects analysis (Chapter 5). Trees use structured brainstorming to define each component of a system, and consider the causal links and consequences of failure of each component.

8.2.1 Probabilistic fault trees

The various AND and OR statements that make up fault trees are subject to ordinary probability calculus. Events may be mutually exclusive or independent. If they are mutually exclusive, then the probability that one or the other will occur is given by:

$$p(A \cup B) = p(A) + p(B)$$

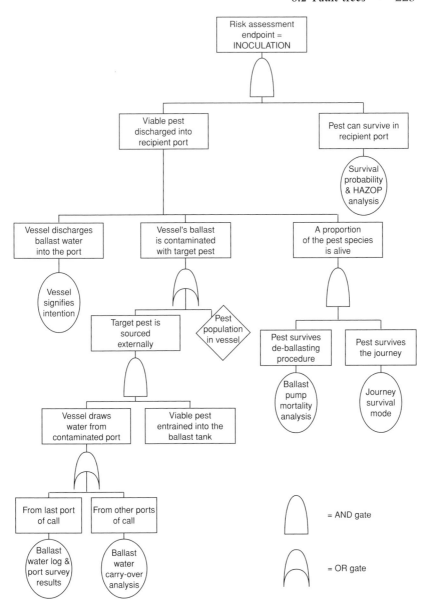

Figure 8.12. Part of a fault tree for marine pest introductions in ballast water (after Hayes 2002b). In this example, the assessor recommended a HAZOP assessment (Chapter 5) to further explore one of the branches.

and the chance that both will occur is, by definition,

$$p(A \cap B) = 0.$$

If two events are independent, then the chance that either one or the other, or both, will occur, is:

$$p(A \cup B) = p(A) + p(B) - p(A \cap B)$$

where,

$$p(A \cap B) = p(A) \times p(B).$$

For three events, A, B and C, the chance of $(A, B$ or $C)$ is given by

$$p(A \cup B \cup C) = p(A) + p(B) + p(C) - p(A \cap B) - p(A \cap C)$$
$$- p(B \cap C) + p(A \cap B \cap C).$$

Johnson *et al.* (2001) evaluated the probability that zebra mussels (*Dreissena polymorpha*), an invasive species of freshwater lakes in North America, will be transported overland by recreational boats moving from one isolated lake to another. They did not represent their analysis as a fault tree, but they could have. It would have looked like Figure 8.13.

Either adults or larvae may be transported, and Johnson *et al.* (2001) identified six pathways by which the event might occur. They interviewed boat owners, counted boats and sampled bilge water, cooling systems, hulls and anchors, producing the probabilities in Figure 8.13.

Employing the rules for combining independent events, the probability of transport of adults in macrophytes (entangled in trailers) or anchors is (ignoring hulls, because no adults were ever observed in their samples and the probability is assumed to be zero):

$$p(A \cup B) = p(A) + p(B) - p(A \cap B)$$
$$= 0.009 + 0.053 - (0.009 \times 0.053) = 0.0615.$$

Likewise, the probability of the transport of larvae in bait buckets or wells on board used for storing live bait (live wells) is

$$p(A \cup B) = p(A) + p(B) - p(A \cap B)$$
$$= 0.071 + 0.011 - (0.071 \times 0.011) = 0.0812.$$

The logic tree makes the rationale behind the conditional probability calculations plain. Fishing equipment will transport larvae if either wells or bait buckets are used. Adult mussels will be transported in macrophytes if the macrophytes are entangled and if they contain mussels. It is also plain

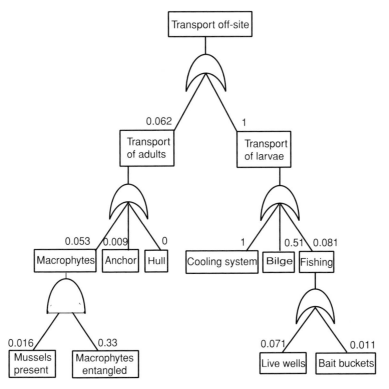

Figure 8.13. Fault tree for the transport of zebra mussels from infested lakes to uninfested lakes on recreational boats. The numbers next to the branches are the probabilities estimated from data by Johnson *et al.* (2001).

that the chance that adults are transported is 0.0615 while the chance of transporting larvae is 1. All boats transport larvae because all boats have engine cooling systems, all engine cooling systems use and store lake water and all lake water in infested lakes contains mussel larvae.

Of course, contamination of lakes also depends on the probability that a boat will move from an infested to an uninfested water body, that adults will live for the period during which the move takes place and that introduced animals are able to survive and reproduce in the new environment (see Johnson *et al.* 2001). Such assumptions should be made plain by the analyst when the results are communicated.

The operations above make the general assumption that events are un-correlated. This has been a substantial failing of logic trees, making them susceptible to a phenomenon known in engineering as 'common failure mode'. If events occur that influence several branches simultaneously,

Box 8.1 · *Calculating dependencies in logic trees*

Ferson (1996a) suggested using Frank's (1979) operators and labelled them AND_F and OR_F. If $p(A) = p$ and $p(B) = q$, then,
$A \; AND_F \; B$

$$\begin{cases} pq \; \text{if} \; s = 1 \\ 1 - \log_s \left[1 + \dfrac{(s^p - 1)(s^q - 1)}{(s - 1)} \right] \; \text{otherwise} \end{cases}$$

$A \; OR_F \; B$

$$\begin{cases} p + q - pq \; \text{if} \; s = 1 \\ 1 - \log_s \left[1 + \dfrac{(s^{1-p} - 1)(s^{1-q} - 1)}{(s - 1)} \right] \; \text{otherwise} \end{cases}$$

where

$$s = \tan \left(\frac{\pi(1 - r)}{4} \right)$$

and r is a correlation coefficient that describes the dependence.
Above, we calculated the chance equals 0.0812 that larvae would be transported off-site in fishing gear from the independent combination of the use of live wells or bait buckets. Assume that people tend to use both or neither. However, their combined use is not guaranteed and data give the correlation between their use as 0.7.
Intuitively, the chance that at least one event occurs should be less if the events are correlated than if the events are independent because they will tend to be coincident more often than by chance. With $p = 0.071$ and $q = 0.011$, Frank's (1979) OR operator gives,

$$s = 0.0041,$$

and

$$A \; OR_F \; B = 1 - \log_s(0.0064) = 1 - \frac{\ln 0.0064}{\ln s}$$

$$= 1 - \frac{-5.0515}{-5.4968} = 0.0810.$$

Thus the tendency for people to have both live wells and bait buckets, or neither, does not have an important effect on the probability of transporting larvae off-site, reducing the estimate from 0.0812 to 0.0810.

their responses will be correlated. The probabilities on the tree will be wrong. One solution is to redesign the tree so that the various independent causes and their consequences are shown appropriately. The other is to calculate probabilities that account for dependencies (Box 8.1).

8.3 Logic trees and decisions

A decision is a choice between two or more acts, each of which will produce one of several outcomes. Decision trees render decisions into a logical structure that reflects understanding of the system. Economists developed decision theory and defined decisions under risk and decisions under uncertainty (Morgan and Henrion 1990).

When making decisions under risk, states of the world are uncertain and probabilities are assigned to states. When making decisions under uncertainty, probabilities cannot be assigned to states. In the latter circumstance, a decision tree can map the structure of the problem but cannot say anything about the chances of different outcomes. Sometimes, risk assessors will allocate equal values to unknown probabilities. The principle of insufficient reason suggests that each state is equally likely because there are no data on which to discriminate among them.

Risk assessments acknowledge that the objective of decision theory is to clarify decisions. The specification of states and acts itself requires decisions. Problem specification is not necessarily unique. Elements of subjectivity, choice and linguistic uncertainty ensure that the values and perceptions of the risk assessor affect the construction of something as superficially objective as a logic tree.

8.3.1 Decisions under risk

Table 8.1 may be generalized, and the benefits (or costs) of each act may be specified quantitatively, together with the chance that the world is in each of the specified states (Table 8.3 and Figure 8.14).

States must be mutually exclusive and exhaustive, such that $\sum p_i = 1$. The framework assumes that acts are independent and that the decision-maker is an ideal reasoner and will act to maximize utility.

A utility is a measure of the total benefit or cost resulting from each of a set of alternative actions (decisions). It represents a scale of preferences among outcomes. A utility function may be a continuous representation of utilities (see Resnik 1987). Calculations depend on a probability associated with each state.

Table 8.3. *Decision table for two acts (A) and three states (S), with utilities (u and v) for each state, given an act, and probabilities for each state (p and q)*

	S_1		S_2		S_3	
A_1	u_1		u_2		u_3	
		p_1		p_2		p_3
A_2	v_1		v_2		v_3	
		q_1		q_2		q_3

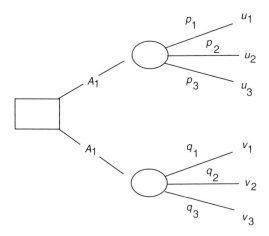

Figure 8.14. Decision tree for the decision problem represented by Table 8.3.

In this framework, the expected utility (E) of each act is the sum of the utilities for each state, A_i

$$E(A_i) = u_{i,1} \times p_{i,1} + u_{i,2} \times p_{i,2} + \cdots,$$

where p_i is the probability of the state, and u_i is the utility of the state. When $p_1 = p_2 = \cdots$ (decisions under uncertainty), expected utility is the mean of the utilities for each state.

When utilities are used to assist in evaluating problems, two assumptions are common and important.

- *Transitivity*: if X is preferred to Y and Y to Z, it implies X is preferred to Z.
- *Connectedness*: there are preferences for all outcomes.

A strategy is a series of acts. This formalism provides for some general strategies, such as the MaxiMin strategy (Morgan and Henrion 1990). Given an ordinal scale for outcomes, in which larger numbers represent greater utility, the procedure is to:

- Identify the minimal outcome associated with each act.
- Select the act with the largest minimal value.

  When the variable of interest is a loss not a gain, employ MiniMax:

- Identify the maximal outcome associated with each act.
- Select the act with the smallest maximal value.

The table helps to collect and order thoughts about the costs and benefits of management alternatives. It is most likely to be used to support thinking in a collective decision-making environment. It could be less than useful if used as a decision-making tool because it fails to account for the uncertainty inherent in the probabilities for states of the world, or the likelihoods and utilities of outcomes of management alternatives.

8.3.2 Bayesian decision analysis

Bayesian decision networks represent causal structure and also include prediction and the consequences of intervention. Decision nodes and utilities (values) are added to the network. The model may include a single decision, a sequence of decisions, or combinations of decisions (strategies). The latter are not available in many software packages (Korb and Nicholson 2003).

Robb and Peterman (1998) used Bayesian decision analysis to assist the management of the salmon (*Oncorhynchus nerka*) fishery on the Nass River in British Columbia, Canada. They used Bayesian statistics to calculate probabilities in a decision tree that accounted for uncertainty in annual recruitment, the timing of the salmon run and the catchability of fish. The analysis resulted in an optimal rule for timing the opening of the fishery. The result was sensitive to assumptions about the shape of the stock–recruitment model (Figure 8.15).

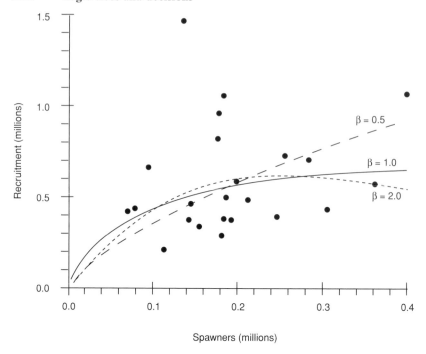

Figure 8.15. Alternative stock-recruitment models for a salmon fishery in Canada (from Robb and Peterman 1998; see also Varis and Kuikka 1999). β is a parameter of a density dependence model.

8.3.3 Gains from management of orange-bellied parrots

Orange-bellied parrots migrate annually from breeding grounds in Tasmania to spend winter in the coastal areas of mainland Australia. At the turn of the twentieth century tens of thousands of birds migrated to winter feeding habitat between Adelaide and Sydney. By 1990, there were about 200 birds and fewer than 30 breeding pairs remaining. The recovery team is not sure why the decline occurred or what the best recovery strategy is (Drechsler *et al.* 1998). There are several alternatives, and some of them are summarized in a decision table (Table 8.4).

In this case, utilities are defined as the percentage increase in the minimum expected population size within 50 years that results from implementation of the management option, as measured by a detailed population model for the species (of the kind developed in Chapter 10).

The expected result, E, from management option A is given by:

$$E(A) = \sum p_i u_{i1} = 0.2 \times 20 + 0.3 \times 5 + 0.4 \times 5 + 0.1 \times 0 = 7.5.$$

Table 8.4. *Decision table (utilities and probabilities) for three management options and four states. Utilities are the expected percentage increase in the minimum expected population size resulting from management aimed at alleviating the effects of the potential cause of decline*

System state (cause of decline)	Likelihood of each state (subjective probability that this is the primary factor inhibiting recovery)	Option A (predator control)	Option B (habitat rehabilitation)	Option C (reducing exposure to toxins)
	p	u_1	u_2	u_3
Feral predators	0.2	20	0	0
Grazing impacts	0.3	5	15	0
Loss of habitat area	0.4	5	10	0
Ecotoxicological effects	0.1	0	5	30
Expected utilities		7.5	9	3

Management options A and C promise a bigger response to individual causes, resulting in greater benefit if the diagnosis of the primary cause of decline is correct. But the best option is management option B, because the action provides a measure of recovery from two different potential causes, which together are the most likely. Management option A has collateral benefits for habitat but does not do enough to outweigh the advantages of option B.

8.3.4 Interpreting decision trees

Decision trees are presented as though they were exact. Interpretation relies on subjective judgements about likelihoods and utilities. There will always be considerable uncertainty associated with these values, not least because they are the product of expert opinion and represent subjective beliefs about the state of the world. In most cases, utilities will be value-based, especially when dealing with environmental issues. They are likely to change from one context to another.

Probabilities are required to obey the rules of probability calculus. Transitivity is a cornerstone of rational decision-making. It is easy to create examples in which preferences are not transitive. Keynes (1936, p. 161), one of the founders of modern economics, objected to expected utilities, saying that most decisions are instinctive, and are not, 'the outcome of a weighted average of quantitative benefits multiplied by quantitative probabilities'. More importantly, it is rare that we have reliable, measured estimates for probabilities of states of the world. These values are much closer to degrees of belief.

Decision tables and the associated arithmetic are a curious abstraction of real thought processes and should be treated cautiously. In Chapter 11 we will explore info-gap methods that accommodate uncertainty in decision problems.

In addition, interpretation cannot be limited to a single currency. The alternative management options in the parrot example have different costs. There are social costs associated with poisoning, shooting or trapping introduced predators. Excluding birds from sources of toxicants is an expensive business, far more expensive than rehabilitating habitat. The analysis performed above had an embedded assumption that the associated costs were equal.

All decisions involve trade-offs and usually the currencies of the elements of the trade do not match. There are no general solutions to the problem of weighing the value of, say, a 10% reduction in risk of extinction against the value of $100 000 or the value of public perception to an organization involved in shooting foxes at the boundary of a residential area. Issues such as the voluntary nature of a risk, visibility and potential for outrage (Chapter 1) become important.

8.3.5 Conservation status

The IUCN (1994, 2001) developed a protocol for assessing the risk of extinction of species. The intent is to classify each species as belonging to one of several categories. It is worthwhile examining in this context because it includes a decision tree, each branch of which ends at a fault tree. It results in a risk ranking for species.

The system is usually represented as a table (e.g. Table 8.5). The overall structure of decisions about the status of a species is given by Figure 8.16. This table is set in a broader context in which the adequacy of data are evaluated (Figure 8.16). Species may be assessed as *extinct, critically endangered, endangered, vulnerable, near threatened, conservation dependent* or *low risk*. Thus, the table forms part of a decision tree that ends in a classification of species.

Table 8.5. *Summary of the IUCN categories of 'critically endangered', 'endangered' and 'vulnerable' and criteria (after IUCN 2001)*

	Critically endangered	Endangered	Vulnerable
A. Declining population			
Population declining at a rate of: Using either			
1. Population reduction observed, estimated, inferred or suspected in the past, or	> 80% in 10 years or 3 generations	> 50% in 10 years or 3 generations	> 20% in 10 years or 3 generations
2. Population decline suspected or projected in the future, based on direct observation, an abundance index, decline of habitat, changes in exploitation, competitors, pathogens, etc.			
B. Small distribution and decline or fluctuation			
Either extent of occurrence (EOO): or area of occupancy (AOO):	< 100 km^2 < 10 km^2	< 5000 km^2 < 500 km^2	< 20 000 km^2 < 2000 km^2
and two of the following three:			
1. Either severely fragmented or known to exist at a number of locations	1 location	≤ 5 locations	≤ 10 locations
2. Continuing decline in habitat, locations, subpopulations or mature individuals			
3. Fluctuations of > 1 order of magnitude in extent, area, locations or mature individuals.			

(cont.)

Table 8.5 (*cont.*)

	Critically endangered	Endangered	Vulnerable
C. Small population size and decline			
Number of mature individuals:	< 250	< 2500	< 10 000
and one of the following two:			
1. Rapid decline of:	> 25% in 3 years or 1 generation all subpops < 50	> 20% in 5 years or 2 generations all subpops < 250	> 10% in 10 years or 3 generations all subpops < 1000
2. Continuing decline of any rate and either populations fragmented with:			
or			
all individuals in a single population			
D. Very small or restricted			
Number of mature individuals:	< 50	< 250	< 1000 or AOO < 1000 km^2 or locations < 5
E. Quantitative analysis			
Risk of extinction in the wild:	> 50% in 10 years or 3 generations	> 20% in 20 years or 5 generations	> 10% in 100 years

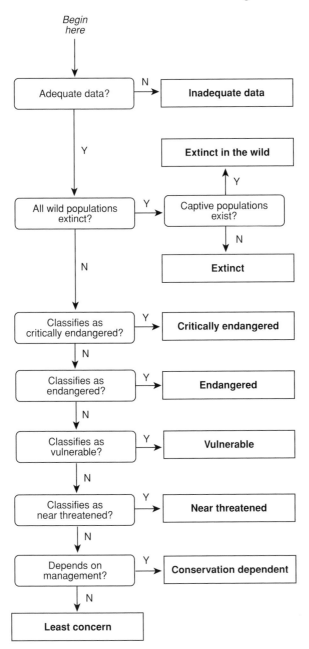

Figure 8.16. Structure of IUCN classification decisions (after Akçakaya and Colyvan, in Burgman and Lindenmayer 1998).

Each branch in the decision tree above asks a question. The answer to the question (yes / no) is determined by the conditions outlined in Table 8.5. The logic in this table may be reduced to a few rules. For example, a species is regarded as being critically endangered:

IF   Decline of ≥ 80% in 10 years or 3 generations
OR   Range < 100 km² or occupied habitat < 10 km²

AND
at least two of the following three conditions are met:

1. Severely fragmented or in one subpopulation
2. Continuing to decline.
3. Fluctuations > one order of magnitude.

OR   Number of mature individuals < 250

AND
at least one of the following two conditions is met:

1. ≥ 25% decline in three years or one generation.
2. Continuing decline and ≤ 50 per subpopulation, or a single sub-population.

OR   *< 50 individuals*
OR   ≥ 50% risk of extinction in 10 years or 3 generations.

This logic may in turn be represented as a fault tree (Figure 8.17). The 'fault' is the condition of being 'critically endangered' and it occurs if the various conditions are met. If the fault is true, the species is classified at that level.

Much of the IUCN (1994, 2001) classification scheme for threatened species is calibrated to apply most effectively to terrestrial vertebrate species. Alternative classification criteria have been proposed for plant and butterfly species so they can be considered in the context of relevant threats and demographics of plants and butterflies, using sharp boundaries for vague terms that make better sense for these taxa (Keith 1998, Swaay and Warren 1999).

The logic tree for classifying conservation status has the same weaknesses as other logic trees. It imposes sharp boundaries such as 80%, 10 years and three generations on vague criteria. The importance of the choices of the kind and level of thresholds and logical structures depends

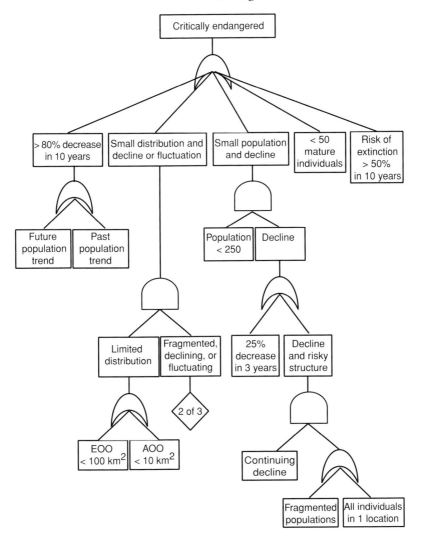

Figure 8.17. Logical structure of the IUCN (2001) fault rule set (resulting in the classification of a species as either critically endangered or not critically endangered).

on how they affect decisions. As with other trees, the best we can do is to accumulate outcomes by monitoring the consequences of decisions, and updating the thresholds and structures iteratively. Revision must account for the relative frequencies and costs of false-positive and false-negative judgements.

A complete risk assessment would be distinguished by monitoring outcomes, evaluating the costs and benefits of decisions and revising the logic tree tools based on the outcome. An honest risk assessment would be distinguished by operations that carry uncertainties through calculations and display them in the results. Chapter 9 outlines methods that can be used to include uncertainties in arithmetic calculations for logic trees, including the IUCN methods.

8.4 Discussion

When different people are confronted by the same problem, they will often draft different logic trees to represent it. This kind of structural uncertainty reflects different conceptual models, but it is rarely, if ever, considered in the building and reporting of logic trees.

Cooke (1991) reported a test in which teams of experts from 10 countries built fault trees for a feedwater system for a nuclear power plant. The teams made independent estimates of the probability that the system would fail. The test was conducted in four stages to examine the influence of different data and different fault trees on the estimates of failure.

Each team undertook a qualitative analysis of the data and made point estimates for the failure probability independently. They met and discussed their differences. Each group produced a fault tree and made a second estimate. In the third stage they agreed on a consensus fault tree but used the data they had collected independently to assess the problem. In the final stage, the groups used the same data and the same fault tree. Clearly, different ideas about the logical structure of a problem can contribute to substantial differences in risk estimates (Table 8.6).

Similarly, differences in data collection and interpretation may influence the construction of a logic tree. It is important to subject both the structure and the parameters in logic trees to sensitivity analyses by creating circumstances in which the data acquisition and tree construction are independently replicated.

To represent structural uncertainty, the analyst has the option to resolve linguistic differences and other misunderstandings. Remaining differences in the results may be represented in the form of two (or more) logic trees, or as a consensus tree.

Differences reflect different beliefs about dependencies and opinions about causes and their effects. Data can often resolve at least some of these differences. But, as in the treatment of risk assessments using

Table 8.6. *Differences in estimates of the probability that a feedwater system will fail (after Cooke 1991)*

	1. Independent qualitative estimate	2. Estimate based on independent fault trees and independent data	3. Common fault tree with independent data	4. Common fault tree and common data
Range of estimates	8×10^{-4} to 2×10^{-2}	7×10^{-4} to 2.5×10^{-2}	1.4×10^{-3} to 1.3×10^{-2}	$\sim 1.4 \times 10^{-4}$ to 1.4×10^{-4}
Ratio: max/min	25	36	9	1

subjective methods (Chapter 5), it would be a mistake to submerge residual, honest differences of opinion in a consensus or an average tree. Residual differences should be reported and their effects on decisions evaluated.

Sensitivity analyses might be undertaken for logic trees. Clearly, the tree structure can lead to different estimates of the probability of an event, or the calculation of utilities associated with decisions. In addition, decision thresholds often are arbitrary. The sensitivity of a classification or of a decision to the specification of thresholds should be a routine part of a risk assessment.

9 · *Interval arithmetic*

All steps in environmental risk assessments are uncertain. Risk assessments involve arithmetic, even if it is as simple as multiplying a likelihood by a consequence to generate a rank for a hazard (Chapters 5 and 6). Risk analysts choose whether to make the uncertainties apparent, or to submerge them in the assumptions of the analysis. Risk analysts should make all relevant uncertainties, and the sensitivity of decisions to these uncertainties, as plain and as accessible as possible.

Interval arithmetic provides an exceptionally simple tool that can be used on a routine basis to carry uncertainties through chains of calculations. It may be applied appropriately in all of the methods outlined so far, including risk ranking, ecotoxicological methods and logic trees. This chapter starts by describing some methods that share a few of the characteristics of interval arithmetic. It explores different kinds of intervals and how they may be estimated and elicited. Lastly, it describes the basic operations of interval arithmetic and applies them to some examples.

9.1 Worst case analysis

The people whose job it is to protect human health and the environment from the harmful effects of toxins have always faced substantial uncertainty. In many instances, they chose to be cautious and err in favour of health. This caution is applied through an approach known as worst case analysis (see Burmaster and Harris 1993).

Recall the exposure equation (Chapter 7),

$$\text{Dose} = \frac{C \times IR \times EF}{\text{bw}},$$

where

$C =$ concentration of chemical in medium (mg/l)
$IR =$ intake/contact rate (l/day)
$EF =$ exposure frequency
bw $=$ body weight (mg).

Each term in the standard dose equation is uncertain. Concentrations may vary from day to day depending on industrial activity, the direction of the wind and so on. Intake rate varies between individuals and between days, depending on activity levels. Exposure depends on the location, behaviour and body weights of individuals.

Despite these uncertainties, the regulator is required to set an acceptable target. In 1989, the US EPA (1989, section 6.1.2) stated that actions to clean contaminated sites '...should be based on an estimate of the reasonable maximum exposure (RME) expected to occur under both current and future land-use conditions.... Estimates of the RME necessarily involve the use of professional judgement... The intent of the RME is to estimate a conservative exposure case (i.e., well above the average case) that is still within the range of possible outcomes.' Setting aside the involvement of experts and their judgements, this definition is vague and ambiguous. The solution was to make a set of conservative assumptions that were likely to provide a level of protection at least equal to that required.

The process of worst case analysis may be summarized as follows:

- estimate an upper bound for potency,
- estimate an upper bound for exposure,
- estimate the risk to the most susceptible individual who receives the highest dose.

To calculate dose, for example, the EPA recommended that 95th percentile values may be used for each of the parameters in the numerator of the equation, and the 5th percentile values for body weight. Uncertainty is incorporated by replacing the point (best) estimates by more 'conservative' locations in the distributions. It assumes that daily concentrations, intake rates and exposure frequencies are all relatively high, and that they affect a relatively small or young (a susceptible) person. For example, while many calculations involve the number of hours per day a person is exposed to a pollutant, the common (but not universal) assumption is that exposure is 24 h/day.

Burmaster and Harris (1993) calculated post-clean-up concentration targets for carcinogens. They estimated that an exposed person had a risk of developing cancer in the range from 10^{-6} to 10^{-4} per year. The calculations were made such that the cancer risk goal was exceeded by no more than 5–10% of the potentially exposed population.

Sometimes, the median is used in place of the lower 5th quantile of the body weight distribution. The effect is to increase the tolerable

---

Box 9.1 · *Worst case calculations and their effect on tails of distributions*

If we have random variables $X_1, \ldots, X_n$ with corresponding probability density functions $f_1, \ldots, f_n$, with

$$P = X_1 X_2 \cdots X_n,$$

in general

$$C_{0.95}(P) \neq C_{0.95}(X_1) \times (X_2) \times \cdots \times C_{0.95}(X_n),$$

where $C_{0.95}$ is the 95th quantile of a distribution.

With $S = X_1 + \cdots + X_n$

$$C_{0.95}(S) \neq C_{0.95}(X_1) + C_{0.95}(X_2) + \cdots + C_{0.95}(X_n).$$

These inequalities apply whether the variables are dependent or independent. For example, if X and Y are independent, the sum of upper bounds $[C_{0.95}(X) + C_{0.95}(Y)]$ lies between the 90.25% and 99.75% quantiles of the distribution for $S = X + Y$. If the variables are dependent, a greater range of error is possible (see Cogliano 1997).

---

concentration of the contaminant. Combining upper-bound and mid-range exposure factors is intended to estimate an exposure scenario that is both protective and reasonable (and not the 'worst case'). The choices involve implicit trade-offs between the costs of exposure and the benefits of use of a chemical. Methods for trading the costs and benefits of assumptions explicitly are outlined in later sections of this chapter and in Chapter 11.

9.1.1 The trouble with worst case

The level of protection resulting from worst case analysis often is arbitrary and unknown. The reason is that the product of the 95th quantiles of a set of distributions is not equal to the 95th quantile of the product of the distributions. The degree to which these two quantities diverge, a measure of the conservatism associated with a regulatory target, depends on the number of arithmetic operations (Box 9.1) and the distributions involved.

The results of worst case analyses may be hyperconservative, much more protective of the environment than the data warrant. Hyperconservatism results in litigation, direct and indirect market costs, blocking potentially beneficial chemicals and the misdirecting of scarce resources for protecting the environment (Breyer 1993, Chapter 7; see Section 9.1.2).

Because the level of conservatism is uncontrolled, comparisons across studies are not meaningful, making it difficult to compare and rank management options. Ranking on conservative or worst case estimates of risk is unlikely to optimize protection of public health or the environment.

These problems are compounded by the fact that not all parameters are replaced by extreme values. The mix of upper and lower tails with means or medians makes it difficult to have even an intuitive feel for the degree of conservatism in the regulatory target.

Lastly, the result of worst case is a single number. Whatever information was available about the contributing distributions is lost. There is no information on how reliable the estimate is, or how sensitively it depends on the assumptions.

9.1.2 Arbitrary thresholds and acceptable levels of risk

Different perspectives on the consequences of worst case analysis created a fascinating debate in the 1990s. A US judge, Justice Breyer (1993), claimed that the practice of dividing a result from worst case analysis or the LOAEL by an 'uncertainty factor' (Chapter 7) to arrive at a safe human dose errs too far on the safe side. He suggested that the results were damaging, leading to a 'vicious circle' of biased technical methods, skewed public perceptions and haphazard political actions, leading to excessive regulation of some areas, and a complete lack of necessary regulation in others (paranoia and neglect).

More generally, Breyer's concern stemmed from his observations of risk assessment procedures including:

- the use of conservative values for uncertain parameters,
- the choice of conservative models to deal with 'scientific ambiguity',
- the practice of estimating risks for relatively exposed individuals,
- the choice of sharp thresholds between acceptable and unacceptable risks,
- an emphasis on probabilities of death, disease, and other losses perpetuating an overdeveloped sense of fear in the public,
- the fixation on achieving reductions in already trivial risks, leading to unnecessary constraints on economic activity, and

- risk protection measures that may result in circumstances more dangerous than the risks we are trying to prevent.

For example, concern for air safety results in more regular flight checks,
and shorter life-times for aeroplane components. The frequency of plane
crashes falls marginally. Air fares rise to accommodate the changes, and
a small proportion of people elect to drive instead of fly. Driving is
inherently more dangerous, the death rate in the population as a whole
increases, and is apportioned among those least able to pay for protection.
In a similar vein, Breyer (1993) argued that fear of cancer is 'the engine
that drives much of health risk regulation' but only 3–10% of all annual
cancer deaths in the US (i.e. 10 000–50 000 annual deaths) are caused by
pollution and industrial products. He claimed that only a small proportion
of these is likely to be reduced by legislation.

Finkel (1995) argued that Breyer's analysis was confused and jaun-
diced, that his diagnosis of risk assessment was incomplete and unduly
pessimistic. Defending conservative assumptions, he argued that they account implicitly for a broader set of unacknowledged uncertainties. He
outlined a more complete set of uncertainties, including:

- extrapolations of toxicity data into new environments,
- extrapolations to new taxa (data on most species are unavailable),
- the quality of toxicity data,
- interaction between toxicants,
- unanticipated exposure pathways,
- unanticipated speciation of toxicants,
- structural uncertainties in fate and transport models, and
- unpredictable bioavailability of toxicants.

His position was that current levels of protection are only conservative if,
for instance, each hazard is taken in isolation, it is assumed that all hazards
are evaluated, and that the extrapolations between laboratory and field
conditions, laboratory and field exposure levels, and between taxa are
reliable. Finkel argued that 'plausible conservatism' reflects public preferences to reduce health risks over unnecessary economic expenditure.

Leaving aside the social question of what is a 'small' number or a
trivial risk, a benchmark for an acceptable risk in the US from pesticide
residues on food is fewer than 10^{-7} additional deaths per year, a threshold
derived from studies of seven known carcinogens. Finkel (1995) argued
that there were at least 200 potential carcinogens, most of which had not

been tested. Typically, it is assumed that untested carcinogens have effects similar to those tested. Thus, the probability of cancer from pesticide residues on food is $200 \times 0.000\,000\,1 = 0.000\,02$. Essentially, we assume that the effects of these carcinogens are independent and additive, as well as approximately equal in effect. There are about 300 000 000 people in the US, making about 6000 additional deaths per year from this source alone.

If interactions between components give rise to cancer, then elimination of one of them may have a much larger effect than if the consequences are due to the additive influences of the carcinogens.

Uncertainty creates circumstances in which we choose a number to represent a distribution – the median will be too high or too low with equal probability. The 95th quantile has a 95% chance of being too high. Breyer (1993) and Finkel (1995) agree that the choice of thresholds should be conditioned by the balance between the cost of underestimating versus that of overestimating the risk. Levels of stringency reflected in intervals should be equivalent when risks are compared. Otherwise, inefficient risk reduction, and greater overall risk will result when limited budgets are directed to the wrong risks (Nichols and Zeckhauser 1988, Bier 2004). We return to this topic in Chapters 11 and 12.

9.2 Defining and eliciting intervals

Some of the problems of worst case analysis can be avoided by using interval arithmetic. Interval arithmetic is useful in circumstances far beyond the narrow confines of setting contamination thresholds for toxic substances. Before outlining the mechanics, we need to explore different kinds of intervals and how they may be obtained.

Intervals may be defined by their source. They may be calculated from data, estimated based on expert knowledge, result from optimistic and pessimistic assumptions about models, or elicited based on subjective individual judgement.

Intervals may be defined by their end use. They may be employed to make decisions, to reduce uncertainty about the distribution of data or parameters, to predict future observations, or as a heuristic device to explore the attitudes of stakeholders towards an issue.

Different kinds of intervals have different utilities in these circumstances. It may be best to use probability intervals indicating where 50%, 75% or 95% of the data are located. It may be necessary to indicate a range encompassing 100% of the data, so-called sure bounds. Sure bounds may be based on theory or judgement suggesting thresholds beyond which observations are impossible. Alternatively, circumstances may require us to estimate confidence intervals with a specified probability of enclosing an unknown numerical parameter. It may be best to view an interval as bounds on a random variable with its own distribution.

Intervals acknowledge uncertainty and are therefore more informative than point estimates. If a point estimate has been provided, it is also possible to provide an interval of one kind or another. The following sections outline some more formal approaches to defining intervals.

9.2.1 Confidence intervals based on measurements

The upper and lower confidence limits for the mean are

$$c = \bar{x} \pm t \frac{s}{\sqrt{n}}$$

where t is a value from Student's t-distribution for appropriate α and sample size, n, and s is the measured standard deviation, assuming normality. The interval, c, defined above has a $100(1 - \alpha)\%$ chance of hitting the true population mean (where t depends on α). In the long run, someone who computes 95% confidence intervals will find the true values of parameters lie within the computed intervals 95% of the time.

Prediction intervals provide an estimate of the confidence interval within which you expect the next single observation to lie. They are calculated knowing that the parameters are estimated from the data. Typically, confidence intervals are not used in interval arithmetic.

9.2.2 Probability intervals

A median divides an ordered set of data into two equal parts. A *quantile* is a point in the ordered data set below which a specified percentage of the data lie. The first *quartile* is the point below which 25% of the data lie. The third quartile is the point below which 75% of the data lie. The interquartile range is the interval between the first and third quartiles and it encloses 50% of the data. This thinking can be generalized to any quantiles, so that they enclose a pre-specified percentage of the data.

In a set of m ordered observations, $x_1 < x_2 < \cdots x_m$, the easiest way to estimate a quantile, $\hat{Q}\frac{i}{m+1}$, is to interpret the x_ith observed value as the $(\frac{i}{m+1})$th quantile (Morgan and Henrion 1990).

Table 9.1. *Values of d2 for estimating the standard deviation of a population from the range of a sample (Montgomery 2001)*

d2				n				
2	3	4	5	6	7	8	9	10
1.128	1.693	2.059	2.326	2.534	2.704	2.847	2.970	3.078

For example, if we have nine independent estimates of the population size for a threatened species (say, 78, 81, 89, 89, 93, 95, 101, 102, 103), the 50th quantile is 93, which is the 5th observation $(5/(9 + 1) = 0.5)$. The 90th quantile is 103, which is the 9th observation $(9/(9 + 1) = 0.9)$.

For a normal distribution, interquantile ranges can then be calculated as

$$\bar{x} \pm s.z_{(1-p)}.$$

They enclose a specified proportion (p) of a probability distribution.

These empirical approaches estimate the proportion of a distribution between two limits. They work best if the data have been transformed to conform with the normal distribution as closely as possible. They depend on the availability of at least some direct data.

Experts often can recall extreme values from a larger set of observations. The range of observations $(\max(x_i) - \min(x_i))$ depends on the sample size (the number of times the expert has observed the outcome) and the variation in the process. If the outcomes are independently and normally distributed, then:

$$\hat{\sigma} = \frac{\max(x_i) - \min(x_i)}{d2_n}$$

where $d2$ is a constant that depends on sample size, n (Table 9.1). The result can be used to estimate an interquantile range, using the formula above.

For instance, given an expert recalls only the smallest and largest estimates of population size from those in Section 9.2 above, the standard deviation is estimated to be $(103 - 78)/2.97 = 8.41$. Computed from the full set of nine measurements, it turns out to be 8.99. The range is an efficient estimator of σ when sample sizes are small ($n <$ about 7; Montgomery 2001).

9.2.3 Bayesian credible intervals

Bayesian intervals for a parameter take the view that the parameter varies randomly. A Bayesian credible interval can be defined in two ways. It may be the shortest interval that contains a specified amount of a (posterior) probability distribution. Alternatively, it may be the amount of a probability distribution contained within specified bounds (Jaynes 1976). An interval must be accompanied by 'some indication of the reliability with which one can assert that the true value lies within it' (Jaynes 1976, p. 179).

9.2.4 Imprecise probabilities

Betting rates can be used to set the bounds for a probability, an idea that dates at least to Laplace in the 1800s (Jaynes 1976). The procedure works as follows.

- Give a person a ticket and say that it is worth a reward (say, $100) if the event, E, occurs.
- Then, offer the person a reward (in cash, X), payable immediately, for the ticket.
- Begin with a small value for X and increase it slowly until a point is reached at which the person is willing to part with the ticket (this is called the point of indifference). The point of indifference is termed the selling price, X_S.

If the person is rational, $X_S < \$100$. Even if they view the event as certain, $100 now is more valuable than $100 at some time in the future. The person's estimate of the probability of the event is:

$$p(E) = X_S/100$$

Savage (1972) recommended betting analogies to elicit probabilities. However, people sometimes exhibit irrational behaviour when confronted by bets. A propensity to gamble can cloud judgement and people are willing to lose money to support what they believe to be true (Morgan and Henrion 1990).

de Finetti (1974) developed the reference lottery strategy that, theoretically, avoids problems with gambling behaviour. The objective is to elicit a judgement for a probability, p, that event E will occur. The reference lottery is in the form of a tree and the idea is to ask the expert to adjust the value of p to the point that they are indifferent to which lottery they could be involved in (Figure 9.1).

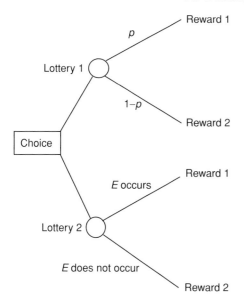

Figure 9.1. Reference lottery for elicitation of the probability that event E will occur. The respondent is asked to choose between Lottery 1 and Lottery 2. Usually, Reward 1 is less than Reward 2 when E is a low-probability event. The value of p is adjusted until the person is indifferent between the choice of the two lotteries (after de Finetti 1974, Morgan and Henrion 1990).

The value for p that results from application of the reference lottery is taken to be the probability that the event will occur. The same strategy has been used to elicit cumulative density functions by letting the event be the probability that the value of a parameter is less than a specified value.

Analogous systems can be used to elicit preferences (utilities) and ranks for nonprobabilistic quantities (see Chapter 12). It can be cumbersome to use in routine elicitation procedures and requires training (Morgan and Henrion 1990). People want to jump to an estimate. It is the analyst's responsibility to engineer circumstances that ensure estimates are reliable. The details of an elicitation protocol always involve trade-offs between the repeatability and reliability of expert judgements (Chapter 4), the number of parameters and time.

You may turn the questions for betting around, as follows:

- Say to the person that you have a ticket and that it is worth $100 if the event, E, occurs.
- Ask if they would be willing to give you X for the ticket.
- Start with a small value of X and gradually increase it.

A person should be willing to buy it at small values of X and will reach a point of indifference. This is the buying price, $\$X_B$. Interestingly, in most situations, $\$X_B < \$X_S < \$100$. The buying and selling prices provide bounds for the uncertainty the person feels about the event, E. Their answers can be used to construct an interval around the probability of the event,

$$p(E) = [p_{\text{lower}}, p_{\text{upper}}] = \left[\frac{\$X_B}{100}, \frac{\$X_S}{100} \right].$$

Walley (1991, 2000a,b) defined imprecise probabilities as a *gamble*. Subjective upper and lower probabilities based on buying and selling prices for gambles have comparable meanings (Walley and DeCooman 2001).

The probabilities $[p_{\text{lower}}, p_{\text{upper}}]$ embody a commitment to act in certain ways (a behavioural model; Walley and DeCooman 2001). For example, E might be the amount of rain that will fall next week, regarded as a reward. It is rational to pay any price smaller than $\$X_B$ for the uncertain reward E. It is rational to sell the reward for any price higher than $\$X_S$.

If the price is between $\$X_B$ and $\$X_S$, it may be reasonable to do nothing, to buy or to sell E. The buying and selling prices for risky investments offered by traders in financial markets are examples of such bounds (Walley 2000a). This approach differs from other forms of decision analysis (Chapters 8 and 12) in that it may produce only a partial ordering of the possible actions; we may be unable to determine which of two actions is preferable without regarding them as equally good (Walley 2000a, b).

9.2.5 Which intervals?

The choice of an interval may be determined by use. Sometimes, intervals are used to measure and convey attitudes (Chapter 12). When they are used to further investigate an issue, the analyst is responsible for ensuring that people share understanding about the meaning of the intervals. Eliciting judgements using gambling analogies is consistent. Resulting intervals should have comparable meanings.

If, instead, the parameters are subjectively estimated quantities, Bayesian credible intervals may be appropriate. If the purpose of a study is to detect outliers or implausible outcomes, sure bounds may be useful. If the purpose of the assessment is to provide probabilistic thresholds and data are available, the analysis will require quantiles of appropriate probability distributions.

In most environment risk assessments, most parameter estimates are composed of either subjectively estimated quantiles or plausible limits derived from experts applying optimistic and pessimistic assumptions. For example, Morgan and Henrion (1990) reported that in the 1970s, under the guidance of the statistician John Tukey, the Impacts Committee of the US National Academy of Sciences concluded that if the release of chlorofluorocarbons continued at 1977 levels, there was a 50% chance that atmospheric ozone depletion would eventually reach 10–23%. The committee was 'quite confident' (translated as equivalent to 19 chances in 20) that depletion would be in the range 5–28%.

Making such judgements is not easy. A few years later, a similar committee constituted without Tukey was unable to make the same kind of interval estimate (Morgan and Henrion 1990). In all such circumstances, analysts are obliged to consider the issue of elicitation (Chapter 4) and the use of imprecise probabilities.

Generally, upper and lower bounds are intended to provide an envelope that brackets the true value and the majority of possible outcomes. The locations of intervals are sensitive to the tails of a distribution. Even with measurement data, underestimation of the full extent of uncertainty will result in intervals that are too narrow. If the absence of data obliges the use of expert judgement, it is useful to know beforehand that experts often are optimistic and provide bounds that are too narrow. It is difficult to know what to do with this knowledge, other than to correct judgements arbitrarily. A better strategy is to measure the bias and correct accordingly (Chapter 4; Cooke 1991).

Many ecologists and environmental scientists refrain from making judgements about parameters because there are insufficient data. We are poorly trained to handle uncertainty. Without knowing exactly how much more data would be needed to make a reliable decision, we wait until a threshold of information is reached and then provide a point estimate. Instead, we should begin with bounds that capture the breadth of uncertainty, and successively reduce the intervals as knowledge and understanding improve. With training and coherent elicitation protocols, such as betting analogies, expert performance may improve.

It is important to discriminate between parameter estimation, prediction and decision making. Risk assessments involve all three. The methods outlined in this chapter are appropriate for parameter estimation and prediction. Other methods for dealing with uncertainty in predictions and decisions are outlined in later chapters. Irrespective of how they are constructed, intervals are a substantial improvement on point estimates that allow us to imagine there is no uncertainty at all.

9.3 Interval arithmetic

The objective of interval analysis is to carry quantitative uncertainties through chains of calculations in a way that is guaranteed to enclose an estimate with at least the surety required. The method is transparent and simple to calculate. It depends on extreme assumptions about dependencies. The value lies in the fact that if the results of the arithmetic do not straddle a decision threshold, dependencies and other uncertainties may be ignored because they do not affect the decision. If, however, the results do straddle a threshold, we need to find out more about sources of uncertainty and dependencies between variables, and make decisions that take into account the possibility of being wrong (Chapter 12).

For each parameter, we specify a range within which we are confident of the true value (to a certain degree). Typically, the same degree of confidence is assessed for each interval. Each operation ensures that the result encloses the true value with at least the level of confidence specified for the individual parameters. The operations in Box 9.2 are sufficient for most environmental risk assessment applications.

For example, an expert believes the probability that a ship will run aground creating a damaging oil spill is in the interval [0.6, 0.8]. The expert estimates that a proportion of the coastline in the interval [0.2, 0.7] will be substantially affected. If we take the conceptual approach outlined in Chapter 6, the 'risk' is the likelihood times the consequence (see Section 9.3.3 for another such example). In this case:

$$\text{Risk} = [a_1, a_2] \times [b_1, b_2] = [0.6, 0.8] \times [0.2, 0.7]$$
$$= [0.6 \times 0.2, 0.8 \times 0.7] = [0.12, 0.56].$$

To rank the hazards, we could compare this interval to others. The rules we adopt to rank intervals depend on our attitude towards uncertainty. If we are risk averse, for instance, we may decide to rank on the upper bound of the risk interval (the worst case for each).

More detailed background including the rules for back-calculation, dealing with zeros and other useful tricks can be found in Moore (1966), Neumaier (1990) and Ferson (2002).

Box 9.2 · *Interval arithmetic operations (for $a_1 < a_2, b_1 < b_2, 0 < a_1 < b_1$)*

Addition

$$[a_1, a_2] + [b_1, b_2] = [a_1 + b_1, a_2 + b_2]$$

Subtraction

$$[a_1, a_2] - [b_1, b_2] = [a_1 - b_2, a_2 - b_1]$$

Multiplication

$$[a_1, a_2] \times [b_1, b_2] = [a_1 \times b_1, a_2 \times b_2]$$

Division

$$[a_1, a_2]/[b_1, b_2] = [a_1/b_2, a_2/b_1]$$

Operations with a constant ($h \geq 1$)

$$h \times [a_1, a_2] = [ha_1, ha_2]$$
$$h + [a_1, a_2] = [a_1, a_2] + [h, h] = [a_1 + h, a_2 + h]$$

Powers ($b_1, b_2 \geq 1$)

$$[a_1, a_2]^{[b_1, b_2]} = \left[a_1^{b_1}, a_2^{b_2}\right]$$

9.3.1 Dependencies

Interval arithmetic gives assurances about the reliability of the results by making conservative assumptions about dependencies. It assumes that quantities that are added or multiplied are perfectly, positively correlated. Quantities that are subtracted or divided are assumed to be perfectly negatively correlated.

It may seem that these assumptions make the intervals as wide as possible. In fact, they are as narrow as possible, while remaining faithful to what you are willing to specify about dependencies. If you know something about two variables, the arithmetic should reflect it. Monte Carlo simulation is capable of handling dependencies if they are specified exactly.

For example, you may know from data that body size is positively, linearly correlated with intake rate. But if subjective knowledge tells you they are positively correlated but no more, then Monte Carlo will struggle to provide the full set of potential outcomes, and interval arithmetic will be too broad, ignoring the positive association. Other methods such as p-bounds may provide the appropriate solutions (Chapter 10).

9.3.2 Intervals for the dose equation

Assume the following data describe exposure of a crustacean to a contaminant (the example is from Ferson et al. 1999). The question is, what dose should be expected?

$C = [0.007, 3.30] \times 10^{-3}$ mg/l (5th and 95th percentiles from data)
$IR = [4, 6]$ 1/day (subjective range estimate by experts)
$EF = [45/365, 65/365] = [0.12, 0.18]$ (subjective range estimate by experts)
bw = $[8.43, 4.514]$ g (5th and 95th percentiles from data for a different population, assumed to have the same life history characteristics as the population under study)

$$\text{Dose} = \frac{C \times IR \times EF}{\text{bw}}$$

$$= \frac{[0.000\,007, 0.0033] \times [4, 6] \times [0.12, 0.18]}{[8430, 45\,140]}$$

$$= \frac{[3.36, 3564] \times 10^{-6}}{[8430, 45\,140]}$$

$$= [3.36/45\,140, 3564/8430] \times 10^{-6}$$

$$= [7.44 \times 10^{-11}, 4.23 \times 10^{-7}]\ (\text{mg/day})/\text{mg}.$$

The result spans four orders of magnitude. Providing a point estimate for the dose would have been misleading. The regulator is free to use the upper bound to set regulatory limits. This would be equivalent to using a form of worst case analysis, but the benefit is that at least some of the uncertainty is apparent.

9.3.3 Intervals for site contamination

Consider another example provided by Lobascio (1993, in Ferson et al. 1999). Groundwater contamination is an especially important issue in urban areas. Hydrogeologists use physical equations to estimate the time it will take a contaminant to move from the point of contamination through the soil to some target or sensitive site. For example, hydrocarbon contamination travelling time may be calculated from:

$$T = \frac{L(n + Df_{oc}k_c)}{Ki}.$$

A point estimate may be reported as 500 years. On that basis, you might be prepared to plan the development of a school in the path of the plume, thinking that the scheduled life of the school is a maximum of 100 years.

Consider the same problem recast in interval arithmetic. You are given uncertainty in each of the parameters:

L = distance to well = $[80,120]$ m
i = hydraulic gradient = $[0.0003, 0.0008]$ m/m
K = Hydraulic conductivity = $[300,3000]$ m/year
n = effective soil porosity = $[0.2,0.35]$
D = soil bulk density = $[1500,1750]$ kg/m^3
f = soil carbon fraction = $[0.0001,0.005]$
k_c = partition coefficient = $[5,20]$ m^3/kg.

Each interval represents a measured or subjectively estimated interval such that you are 90% certain it contains the true value. You have no knowledge of dependencies among the parameters. Then:

$$T = \frac{80 \times (0.2 + 1500 \times 0.0001 \times 5)}{3000 \times 0.0008},$$

$$\frac{120 \times (0.35 + 1750 \times 0.005 \times 20)}{300 \times 0.0003}$$

$$= [32, 234\,000] \text{ years.}$$

That is, the time it will take for the plume of contamination to move from the point of contamination to the school grounds could reasonably be anywhere between 32 and 234 000 years. An interval of this magnitude seems very unsatisfactory but it serves to illuminate the implicit deceit entailed in using point estimates.

Would you still recommend the school building to go ahead, knowing there is a chance the plume could arrive within 32 years, and a higher chance it could arrive within 100 years? Of course, the plume may never arrive, so the answer depends on the risk-weighted health costs of deciding it is safe, versus the social and economic costs of relocating the school when it is perfectly safe for it to be where it is planned.

This example illustrates the importance, and the difficulty, of estimating bounds. Taking a small part of this analysis further, Renner (2002) summarized reviews of the octanol–water partition coefficient. One review of 700 publications between 1944 and 2001 found estimates for the coefficient that varied over four orders of magnitude. There was no evidence that estimates over time were converging. The value of the coefficient in a given circumstance depends on many chemical and physical variables. Substantial understanding of the conditions at hand would be necessary before a range for k_c as small as [5, 20] used above could be justified.

9.3.4   Intervals for risk ranking

Risk ranking combines likelihoods (estimated on a scale of, say, 1 to 5) with consequences (on a scale of, say, 1 to 5) to produce a value (a 'risk'). The 'risk' is compared with threshold values (producing a classification of high, medium and low risks) and with the values for other hazards (producing a rank). Most ranking exercises use a range of judgements derived from expert opinion and quantitative data (Chapters 5 and 6).

When point estimates are used for likelihoods and consequences, information about the degree of certainty in judgements is lost. Interval arithmetic can be used in subjective risk assessments to preserve this information so that it may be considered when assessments and decisions are made.

For example, take a case of expert assessment of the ecological effects on local vegetation of the construction and operation of an industrial plant (Section 5.2). Construction and operation create the possibilities of fire, road construction, the introduction of invasive plant species and modification of vegetation caused by emissions from the industrial complex. A typical assessment would look like Table 9.2.

Often, the values are generated by a single assessor, making them susceptible to the values, perceptions and biases of the individual (Chapters 1 and 4). Better methods achieve a consensus from a range of experts, or use indirect or direct empirical data (Chapter 4). If the range of data or opinions are retained (Table 9.3) then the range of risk estimates may also be retained with interval arithmetic.

The analysis makes it plain that the experts are happier to agree about the risks of road construction than they are about the risks of changed fire frequencies. The basis for disagreement about fire lay in the estimation

Table 9.2. *Subjective risk assessment of four hazards associated with the construction and operation of a methanol plant*

Hazard	Likelihood	Consequence	Risk	Rank
Increased fire frequency	2	2	4	5
Road construction	5	1	5	4
Pest plant introduction	2	3	6	3
Emissions leading to minor vegetation change	4	2	8	2
Emissions leading to vegetation loss	2	5	10	1

Table 9.3. *Range of expert opinion about two hazards from Table 9.2, together with interval calculations for the resulting risk*

Hazard	Likelihood	Consequence	Risk
Increased fire frequency			
Expert 1	1	2	2
Expert 2	2	2	4
Expert 3	2	3	6
Expert 4	5	3	15
Interval	**[1,5]**	**[2,3]**	**[2,15]**
Road construction			
Expert 1	5	1	5
Expert 2	4	2	8
Expert 3	5	1	5
Expert 4	5	1	5
Interval	**[4,5]**	**[1,2]**	**[5,8]**

of its consequences, rather than the estimation of its likelihood. There is even a chance that the risk resulting from fire effects is greater than the risk resulting from emissions.

This wealth of information is lost from most risk ranking exercises. There is no reason to hide the uncertainties, unless it is to create an undeserved veil of certainty. Interval arithmetic provides a means for retaining them. The results provide an opportunity to rank the hazards on the basis of a best guess or consensus (Table 9.2), and on the basis of, say, the upper or lower bound from each assessment. The latter analysis would provide a measure of the sensitivity of risk ranks to underlying uncertainties in the assessments.

Box 9.3 · *Interval operations for logical operators when a and b are prob-*
ability intervals (limited to [0,1]; after Akçakaya et al. 2000)

Let $p(A) \in [a_1, a_2]$ and $p(B) \in [b_1, b_2]$ be likelihood intervals.
If A and B are independent,

$$p(A \cap B) = p(A)p(B)$$
$$= [a_1, a_2] \times [b_1, b_2]$$
$$= [a_1 b_1, a_2 b_2].$$

The probabilistic sum, $\oplus$, is:

$$p(A \cup B) = p(A) + p(B) - p(A) \cap p(B)$$
$$[a_1, a_2] \oplus [b_1, b_2] = [a_1 + b_1 - a_1 \times b_1, a_2 + b_2 a_2 \times b_2].$$

It is possible to make different assumptions about dependencies. The
following operators contribute to other interval versions of logical
conjunctions:

The minimum, 'min', is defined as:

$$\min([a_1, a_2], [b_1, b_2]) = [\min(a_1, b_1), \min(a_2, b_2)].$$

The maximum, 'max', is defined as:

$$\max([a_1, a_2], [b_1, b_2]) = [\max(a_1, b_1), \max(a_2, b_2)].$$

The envelope, 'env', is defined as:

$$\text{env}([a_1, a_2], [b_1, b_2]) = [\min(a_1, b_1), \max(a_2, b_2)].$$

Akçakaya *et al.* (2000) developed a method for dealing with uncertainty in the IUCN decision tree. They used fuzzy numbers instead of
intervals but the operators they selected work for interval operations.
The logic tree uses comparisons (greater than, less than) and logical
operations (*a* OR *b*, *a* AND *b*). They evaluated comparisons using
rules such as:

$$a \le b = \begin{cases} 1 & \text{if } a_2 \le b_1 \\ 0 & \text{if } b_2 \le a_1 \\ [0, 1] & \text{otherwise.} \end{cases}$$

Akçakaya *et al.* (2000) used operators that assumed dependence was
positive but otherwise unknown. For example:

$$a \text{ AND } b = \text{env} [a \times b, \min(a, b)],$$
$$a \text{ OR } b = \text{env}(\max(a, b), 1 - (1 - a) \times (1 - b)).$$

With two intervals, $a = [0.2, 0.5]$ and $b = [0.4, 0.9]$:

$$a \text{ AND } b = \text{env } [a \times b, \min(a, b)]$$
$$= \text{env } ([0.08, 0.45], [0.2, 0.5])$$
$$= [0.08, 0.5].$$

9.3.5 Intervals for logic trees

Logic trees outline the inter-relationships between basic events or conditions, to assist with decisions and risk evaluation. They include logical operations such as IF, AND and OR (Chapter 8). If information is available to estimate intervals for the probabilities of the basic component events in the tree, then interval arithmetic can be used to produce likelihood intervals for other tree nodes corresponding to the various logical combinations of basic events (Box 9.3). Intervals are defined for most logical operations (Neumaier 1990, Ferson 1996a), preserving uncertainties.

The bounds that assume independence are considerably tighter than those resulting from operators that do not assume independence. These two assumptions represent the extremes of normal assumptions about dependencies. Other operators provide the intervals for other assumptions about dependence (Ferson 1996a, 2002; Box 9.3).

Akçakaya *et al.* (2000) used the logic tree for the IUCN protocol outlined in Chapter 8 to assess the status of the Towhee (*Pipilo crissalis eremophilus*) in California. Some data for the species had been published (Gustafson *et al.* 1994, Laabs *et al.* 1995). Destruction of habitat resulted in a decline in the population (Akçakaya *et al.* 2000; see also Akçakaya *et al.* 1999). Some things were known with certainty, such as that there was only one remaining population in California. Some things were unknown, such as the rates of past and future decline. A few important variables were known imprecisely and were represented as intervals, including:

Population size $= [194, 300]$ mature individuals
Extent of occurrence $= [431, 100\,000]$ km^2
Area of occupancy $= [246, 430]$ km^2.

Interval operations applied to the logic tree generated a rank for the subspecies: [critically endangered, endangered]. Either rank is plausible. There is insufficient information to distinguish between them reliably. It is reasonably certain, however, that the species is threatened to some extent.

9.4 Discussion

Intervals may be able to cope with all types of numerical uncertainty (i.e. the uncertainty about a parameter value) simultaneously. For instance, uncertainty in the number of endangered species in a region that arises due to vagueness, measurement and systematic error, natural variation and subjective judgement can be subsumed within upper and lower bounds. Walley and DeCooman (2001) proposed an approach for translating linguistic uncertainty into imprecise probabilities with betting analogies.

One of the costs of using intervals is that they do not use all available information about a number. Using an interval loses information about the central tendency, standard deviation, sample size, distribution shape and so forth. In addition, intervals compound uncertainty and may introduce underspecificity where none existed before. Lastly, intervals are only appropriate for numerical uncertainty. Many instances of linguistic uncertainty are not numerical and should be treated in the most appropriate manner for their subcategory (Regan *et al.* 2002a).

Probabilistic and other methods may be combined. In such applications, the probabilities of vague events are quantified (see Gabbay and Smets 1998) and the sources of uncertainty are treated separately – probabilistic methods are used for epistemic uncertainty and fuzzy set methods for vagueness. This is different from an interval-based method where all the uncertainty is combined (see Ferson and Ginzburg 1996, Ferson *et al.* 1999).

Treatments of uncertainty in environmental science have focused largely on epistemic uncertainty. Few studies have acknowledged and dealt with both epistemic and linguistic uncertainty simultaneously. One notable example is the IUCN categories and criteria for the classification of threatened species. The IUCN categories deal with non–numerical vagueness in the terms *vulnerable, endangered* and *critically endangered* and accommodate natural variation in elements such as population size. Furthermore, a formal method exists based on intervals to deal with the various types of uncertainty in each of the parameters in the criteria (Akçakaya *et al.* 1999, 2000), as well as a method to deal with numerical vagueness in the categories (Regan *et al.* 2000).

Some authors have objected to intervals because they have sharp boundaries. It seems unreasonable to say nothing about a distribution within the limits, and yet to specify the boundaries as though they are exact. There are some solutions that are not outlined here. Fuzzy numbers are essentially stacks of intervals, each level of which represents a different degree of surety about the boundary (Kaufmann and Gupta 1985). Info-gap theory generalizes the problem of specifying what is not known about a decision, leading to a model of uncertainty that does not require assumptions about sharp boundaries (Ben-Haim 2001, Chapter 12).

10 · *Monte Carlo*

The name 'Monte Carlo' dates from about 1944 and the Los Alamos project in the US that produced the atomic bomb. The work involved simulating the random diffusion of neutrons. Von Neumann, Ulam and others worked on the bomb project and later disseminated the idea of Monte Carlo methods to solve both deterministic and stochastic problems.

Ulam (1976) related how the idea for Monte Carlo arose while he was in hospital after a bout of meningitis. While playing solitaire, it occurred to him that one could get an idea of the probability of an event such as a successful outcome in a game of cards, simply by recording the proportion of successful attempts. At the time, people were trying to estimate probabilities by following all chains of possibilities, a difficult task for all but the simplest cases.

This idea formed the foundation of what came to be known as Monte Carlo analysis (alluding to uncertainty in gambling). The methods originally focused on analytical solutions, but computers gave rapid solutions to complex problems by simulation.

Some of the earliest applications of Monte Carlo were to environmental problems. Hammersley and Handscomb (1964) noted a report in *The Times* of London from 1957 in which Monte Carlo methods were used to design controls of floodwater and the construction of dams on the Nile. The problem was inherently probabilistic because rainfall is unpredictable. The data consisted of weather, rainfall and water levels over 48 years. The problem was to see what would happen if certain dams were built and policies for water control exercised. There were a large number of combinations of dams and policies, each to be evaluated over a range of meteorological conditions, including typical and extreme scenarios. The model included details of dams, river hydrology, losses to evaporation and so on. The behaviour of the system had to be assessed in terms of engineering costs and agricultural, hydrogeological and economic outcomes. The problem was explored using Monte Carlo simulation.

Hammersley and Handscomb (1964) reported isolated and undeveloped instances from much earlier. In 1901, Lord Kelvin used Monte Carlo techniques to find solutions to mathematical equations. In 1908, W. S. Gosset (who worked for the Guinness company, did his mathematics in his spare time, and published as 'Student') used repeated random samples to help him to discover the statistical distribution of the correlation coefficient and to confirm his derivation of the t-distribution.

Monte Carlo methods provide a way to solve numerical problems. This is important for environmental risk assessment because explicit models provide a framework for the development of internally consistent ideas and a platform for exploring the consequences of intuition. They allow us to study systems in compressed time. They are often used where it is too expensive, risky or slow to test a proposed change in a real system. Such models encourage us to justify decisions, clarify problems, and identify important parameters.

10.1 The modelling process

In the risk management cycle (Chapter 3) at the point at which model building begins, the context has been determined, assumptions have been specified and hazards and their consequences have been identified. The purpose of the model determines the level of detail and the specifics of the modelling approach, within the constraints of time and money. Monte Carlo simulation models are built around the same conceptual models that support risk ranking (Chapter 6), exposure assessment (Chapter 7) and logic trees (Chapter 8).

The full value of model-based risk assessment is gained through an iterative process of model construction, calibration, sensitivity analysis and validation (Box 10.1). In this context, sensitivity analysis and verification of model predictions are among the most valuable parts of the exercise. The formalism of modelling ensures that terms are fully specified and relatively free from ambiguity. These attributes do not guarantee that linguistic uncertainties and biases won't arise during risk communication and management, but they go some way towards acknowledging and alleviating the problems.

10.1.1 Random variables

Random variation lies at the heart of Monte Carlo analysis. If a parameter in a model is uncertain, it is necessary to build a model for that uncertainty.

Box 10.1 · *A simple approach to stochastic model building*

The following steps summarize a useful strategy for building a model:

1. Develop a deterministic (mechanistic) model.
2. Add stochastic elements to represent uncertainties.
3. Add assumptions about dependencies.
4. Use the stochastic model to estimate the statistical distribution of the result.
5. Compare the result with reality and update the model.

In most cases, simulation is used to solve difficult or intractable mathematical calculations.

Some important elements of modelling include:

- *Calibration*: adjusting model parameters, structures and assumptions so that they fit available data and intuition, i.e. refinement of ideas.
- *Sensitivity analysis*: calculating the magnitude and rank order of responses of consequences as a function of model parameters, assumptions and model structure, i.e. exploring assumptions and judgements.
- *Validation*: comparing independent field observations with predictions, i.e. testing ideas.

It may require choosing a statistical distribution and specifying its parameters. Then, to solve the problem, a source of random variation is used to play the game over and over, providing estimates of the likelihoods of different outcomes.

Dice are a source of random variation. We let X (or Y) denote a random variable (formal definition comes later), and x (or y) the values the random variable can take. We may decide to use the dice because we are interested in estimating the number of offspring that a bird may lay in a clutch of eggs (between 1 and 6). The random variable (X) in this instance is clutch size, which may take any one of six values (x).

If empirical evidence suggests that each clutch size is equally likely, the dice are an excellent model for the variation in the system. The variable would be represented by a discrete form of the uniform distribution, in which each value has a 1/6 chance of occurring. The distribution of relative frequencies of different kinds of events is known as a probability density function. Formally, $f(x)$ denotes the probability density function.

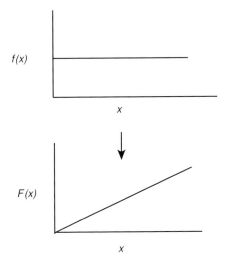

Figure 10.1. A continuous uniform probability distribution and the associated cumulative probability distribution.

It is often useful to record both the frequency of each event and the cumulative frequencies of events up to a given magnitude. A cumulative probability distribution gives the probability, p, that the random variable X will be less than or equal to some value x. It sums the value of the probability distribution from left to right.

The cumulative probability distribution for the dice is a step function that increases from 0.167 (the chance of throwing a 1), to 1 (the chance of throwing a number less than or equal to 6). This can be written as,

$$F(x) = p(X < x),$$

where $F(x)$ is the cumulative probability function and ranges from 0 to 1. We know the value of $p(x)$ because we can measure it through repeated trials. For example, if we didn't know the chance of throwing a 1 was 1/6, we could measure it by throwing the dice 1000 times. The number 1 would turn up in about 167 instances.

If the distribution was continuous instead of discrete, the interpretation would be the same (Figure 10.1). Instead of summing over values of x, the cumulative probability distribution would be:

$$F(x) = \int_{-\infty}^{x} f(x)\mathrm{d}x.$$

10.2 Kinds of distributions

Monte Carlo uses statistical distributions to represent different kinds of uncertainty, combining them to generate estimates of a risk. It is important to be familiar with the properties of a few of the most widely used distributions. Complete details of methods for estimating the parameters of these and other distributions are provided by Johnson *et al.* (1992, 1994, 1995; see also Morgan and Henrion 1990, Stephens *et al.* 1993, Vose 1996, Hilborn and Mangel 1997, Knuth 1981). The summary below presents a few simple ones.

It is useful to recall that the mean and standard deviation of any sample are given by the familiar formulae,

$$\mu = \frac{\sum x}{n}$$

and

$$s = \sqrt{\frac{\sum (x - \overline{x})^2}{n - 1}}.$$

10.2.1 Uniform

The discrete and continuous forms of the uniform distribution are described above. Specifying the parameters of a uniform distribution is equivalent to specifying an interval in interval arithmetic. Formally, a uniform distribution is defined as:

$$X \sim U(l, u),$$

where l is the lower bound, and u is the upper bound. The expression says that the parameter X is drawn from ($\sim$) the uniform (U) distribution with bounds l (lower) and u (upper). The probability that X lies between a and b is given by:

$$p(a < X < b) = p(X < b) - p(X < a) = F(b) - F(a)$$

where

$$p(X < b) = b/(u - l) \quad \text{and} \quad p(X < a) = a/(u - l).$$

The mean is $\frac{l+u}{2}$. The standard deviation is $\sqrt{\frac{(u-l)^2}{12}}$.

When uncertainties are unknown, it is common practice to assign equal probabilities to different events. The logic is that when there is no evidence to the contrary, the best assumption is that all events are equipossible, and therefore equiprobable. This assumption is attributed to Laplace around the end of the eighteenth century although it was in use at least a century earlier; Leibniz made this assumption in 1678 (Hacking 1975).

It is possible to argue that there is never 'no' knowledge about the probability density within an interval, that there will almost always be some central tendency. The tails of at least one end of the distribution are likely to occur less frequently than intermediate values (Seiler and Alvarez 1996). Even when representing expert opinion, it is rare that experts will have opinions about sharp boundaries for a parameter, but have no opinions about its central tendency.

The primary role of the uniform distribution is that it is a model for independent random variation from which other distributions may be constructed. This role only becomes important once the Monte Carlo engine starts to perform its calculations, outlined below.

10.2.2 Triangular

The triangular distribution accommodates a lower bound, a central tendency (a 'best guess'), and an upper bound for a parameter. It has no theoretical basis but is used particularly to represent expert judgement or belief. It tends to weigh the tails of a distribution more heavily than most other distributions (except the uniform). The notation

$$X \sim \text{Triang}(a, b, c)$$

says that the parameter is drawn from a distribution within the bounds a, c and the best estimate of the parameter, represented by the mode, is at b. The mean is $\frac{a+b+c}{3}$. The standard deviation is $\sqrt{\frac{a^2+b^2+c^2-ab-ac-bc}{18}}$. Its simplicity, intuitive definition and flexible shape have made it a popular tool, but it can generate biases for skewed data. The distribution of mass is such that when, for instance, the maximum value is large (the distribution is right skewed), the value of the mean will be sensitive to it, resulting in inordinately large estimates for the mean. The weight it gives to the tails is unrealistically large in most circumstances.

10.2.3 Normal

Parameters that result from the sum of a large number of independent random processes tend to produce normal distributions, no matter what the shape of the processes that contributed to them. The normal is used extensively, particularly for measurement errors and other random processes. When sampling a normal distribution, the notation

$$X \sim N(\mu, \sigma)$$

indicates that X is drawn from a normal distribution (N) with mean μ and standard deviation σ.

10.2.4 Lognormal

Parameters that result from the product of a large number of independent random processes tend to produce lognormal distributions. The notation

$$X \sim LN(\mu, \sigma)$$

says that X is drawn from a lognormal distribution (LN) with mean μ and standard deviation σ. The lognormal is a popular choice for quantities that are positive and right skewed, such as river flows, rainfall and chemical concentrations. It is also used commonly in situations in which expert judgement uses multiplicative assessments such as 'known to be within a factor of three', or 'to within an order of magnitude'.

10.2.5 Beta

The most direct use of the beta distribution is to represent the probability of a random event in a series of trials. It can take on a wide variety of shapes, including both symmetric and asymmetric forms (either left or right skewed), and horseshoe shapes. The notation

$$X \sim \text{beta}(\alpha_1, \alpha_2)$$

indicates that X is drawn from a beta distribution with parameters α_1, α_2. The uniform is a special case of the beta distribution with $\alpha_1 = \alpha_2 = 1$. One special use of the discrete form of the beta is to model the number of successes, r, from a given a number of trials, n. Then $\alpha_1 = r + 1$ and $\alpha_2 = n - r + 1$.

In general, the mean is $\frac{\alpha_1}{\alpha_1 + \alpha_2}$. If $\alpha_1, \alpha_2 > 1$ then the mode is $\frac{\alpha_1 - 1}{\alpha_1 + \alpha_2 - 2}$, and is 0 or 1 otherwise. The standard deviation is $\sqrt{\frac{\alpha_1 \alpha_2}{(\alpha_1 + \alpha_2)^2 (\alpha_1 + \alpha_2 + 1)}}$. It can be a useful model for uncertainty when a process is bounded and the mean and/or the mode are known. Because of its flexibility, it is a popular tool for representing expert judgement about statistical processes. To model a variable that is limited to an interval $[a, b]$, the beta distribution (X) can be rescaled (X^+) (Vose 1996) by:

$$X^+ = [a + \text{beta}(\alpha_1, \alpha_2) \times (b - a)].$$

10.2.6 Binomial

The binomial distribution is closely related to the beta. The binomial gives the number of successes, r, in n trials where the probability, p, of any single trial succeeding is known and constant. The notation

$$X \sim \text{binomial}(n, p)$$

says that X is drawn from a binomial distribution with a probability of the event, p.

For large n (roughly > 30), the binomial is close to the normal, $N\{np, [np(1 - p)]^{1/2}\}$. This fact is useful because it is often more efficient to sample a normal distribution than to sample a binomial distribution. The latter usually involves repeated trials and can be computationally expensive.

The mean number of successes in n trials is np. The standard deviation of the mean number of successes in n trials is $\sqrt{np(1 - p)}$.

The binomial is used routinely to model things such as the survival of individuals in a population, the failure of components in power plants and the presence of physical features at different locations in the landscape.

10.2.7 Exponential

When an event occurs at random and the probability of this event is constant, the time between successive events is described by an exponential distribution, sometimes called the negative exponential distribution. The occurrence of independent, random events is called a Poisson process. The notation

$$X \sim \exp(t)$$

says that X is drawn from an exponential distribution with a mean time until the next event, t.

The mean of the process is t. The standard deviation of the inter-event interval is also t.

The exponential can also be parameterized by the rate of the Poisson process, which generates it as a waiting time. In this case, $X \sim \exp(\lambda)$ with mean $1/\lambda$ and standard deviation $1/\lambda$.

It is important in risk assessment because it provides an excellent model for many processes, including the expected time until the next earthquake, the next failure of a tailings dam wall and the next forest fire. It forms the basis for the gamma (not described here) and the Poisson distributions. The exponential distribution is a special case of a more general distribution called the Weibull distribution (see Morgan and Henrion 1990).

10.2.8   Poisson

The Poisson distribution models the number of occurrences of an event that are likely to occur within a time, t, when the probability of an event occurring per interval of time is constant, and independent of any other events that may have occurred. The notation

$$X \sim \text{Poisson}(\lambda t)$$

says that X is drawn from a Poisson distribution with a mean of λt (i.e. the number of events in an interval of length t in a Poisson process with rate λ events per unit time).

Like the negative exponential from which it is derived, it has only a single parameter. The mean is λ and the standard deviation is $\sqrt{\lambda}$.

The distribution is used to estimate the number of failures in a repetitive process such as manufacturing, and the distribution of births per female in natural populations of plants and animals.

10.3   Choosing the right distributions

The shape of the gamma distribution is like the lognormal, although less of the gamma lies in the extreme tails. Like the lognormal, it is used routinely in some circumstances to represent rainfall and contaminant concentrations. The Weibull distribution is often used to model time to failure, material strength and related properties. It can take on shapes similar to the gamma distribution and tends to be less tail heavy than the lognormal.

In contrast, the popular triangular and uniform distributions make the tails of distributions much heavier than most other distributions and

Table 10.1. *An example of guidelines for use of probability density functions in stochastic models of nuclear reactor safety (after Stephens* et al. *1993)*

Type	Used for parameters	Examples
Constant	With a well-known, fixed value	Radionucleotide decay constants
Uniform	For which few data are available, but firm bounds are known	Rate constant
Normal	For a variable made up of the sum of a set of independent, random variables	Permeability of backfill in a toxic waste repository
Lognormal	For a variable made up of the product of a set of independent, random variables	Coefficients for element mobility in soil
Triangular	For which the upper and lower bounds and most likely value have been estimated, but little else is known	pH of groundwater in a repository
Beta	For which the upper and lower bounds and most likely value have been estimated, and other qualitative information is available on shape	Coefficient for calculations with redox potential of water in contact with used nuclear fuel

are hard to justify on empirical grounds (Seiler and Alvarez 1996). Vose (1996) used a modified beta distribution (the 'Betapert') because of the relative simplicity of interpretation of its moments, making it easier to use in elicitation processes.

Despite the importance in risk analysis of correctly estimating the tails of distributions, convention usually determines the choice. For example, Stephens *et al.* (1993) provided examples of rules of thumb for eliciting a choice of a distribution from experts (Table 10.1). Often, the normal distribution is used in cases where it is clearly unsuitable (for instance, where the data are strongly skewed or the tails are poorly fitted).

Theory may provide a basis for the choice. A process may consist of a series of effectively random, time-invariant, independent events. The beta, the negative exponential, the binomial or the Poisson may be appropriate, depending on what it is we need to estimate. If processes affecting the uncertainty in a parameter are known to be independent and additive, such as when estimating the magnitude of measurement error, then the normal distribution might be justified.

Often theory doesn't tell us what we need to know. Usually, we know so little about empirical phenomena that we cannot construct the expected distributions from their fundamental properties. We are then obliged to guess at the correct distribution by simply observing the relative frequencies of events or outcomes.

If direct measurements are available, the response variable is continuous and the values are right skewed, it is conventional to adopt the lognormal distribution. If the process is thought to be skewed and experts guess its shape and parameters, it is conventional to estimate parameters for a triangular or a beta distribution. In Section 10.10, a method called 'p-bounds' is outlined that does not require specific guesses about distributional shape. Alternatively, empirical data may guide choice through goodness-of-fit.

10.3.1 Goodness of fit

Several tests provide a p value, the probability that a fitted distribution would produce a test statistic as large, or larger, than the one observed, by chance alone. The Kolmogorov–Smirnoff (K–S) statistic compares the maximum distance between an empirical cumulative distribution and a theoretical, fitted distribution, with a table of critical values. The chi-square test compares the numbers of sample observations in discrete classes with those expected under the proposed distribution, perhaps best suited for discrete random variables.

The Anderson–Darling (A–D) statistic sums the squared vertical distances between two cumulative distributions. The A–D statistic is useful in risk analysis because it pays greater attention than the K–S statistic to the tails of the distribution. However, it is computationally more difficult and can't be done easily by hand or in a spreadsheet.

It is a sensible idea simply to plot the expected and observed cumulative distributions and examine them for patterns of deviation. Outliers in the main body of the distribution and in the tails may be informative. Plots can be made of cumulative probabilities, of probability versus probability, or of quantile versus quantile.

The main problem with goodness-of-fit tests is that calculating the probability that the fitted distribution could have produced the data is not the same thing as calculating the probability that the data came from the fitted distribution. Other distributions could have produced a similar outcome. Goodness-of-fit is a necessary but not a sufficient condition for selecting a distribution.

Usually, the problem comes down to one of choosing between a limited set of plausible alternatives. There are a few methods for choosing, including maximum likelihood.

10.3.2 Estimation and maximum likelihood

There are many ways of estimating the parameters of statistical distributions (see Johnson *et al.* 1992, 1994, 1995, Hilborn and Mangel 1997). Maximum likelihood is one of these. It may also be used to choose a distribution and to evaluate the goodness-of-fit of data to a model.

In many cases, the parameters of a distribution may be estimated using a simple equation. For example, if there are x observations in n trials, the probability of the event may be estimated as x/n. Some of the formulae in Section 10.2 give maximum likelihood estimates of the parameters of the distributions. For complete details of the maximum likelihood equations for these and other distributions, see Johnson *et al.* (1992, 1994, 1995). In situations where several unknown parameters must be estimated from data simultaneously, a brief introduction to maximum likelihood estimation may be helpful (Box 10.2).

10.3.3 Other selection criteria

Selection of appropriate distributions should draw on all available information. There are many relevant questions. Is the variable discrete or continuous? What are the bounds of the variable? Is the distribution skewed or symmetric? If the distribution is skewed, is it left or right skewed? What other properties of the shape of the distribution are known? Is there a mechanistic basis for choosing a distributional family? Is the shape of the distribution likely to be dictated by physical or biological properties or other mechanisms? Are there data from other, similar systems that provide support for one distribution or another?

One of the most serious difficulties in choosing a distribution is that normal and lognormal distributions (for instance) may look good over most of the range of the data, but deviate substantially in the tails, where risks are expressed most seriously (Hattis 1990, Hattis and Burmaster 1994). For example, Shlyakhter (1994) estimated the parameters of a normal distribution for human global population size estimates and for several physical constants. He then measured the deviations of subsequent

Box 10.2 · *An example showing how maximum likelihood estimation works*

Table 10.2 gives counts for diatoms in water samples taken downstream from a source of a chemical used on a farm to control insects. The diatom is an indicator of the presence of the contaminant.

The average natural 'background' frequency of the diatom in samples from these streams is about 0.5 cells per sample. In other studies, diatom frequencies of about two cells per sample reflect concentrations that led to increased risks of fish kills and changes in stream insect fauna. The first objective is to measure the average number of cells in the stream water, to see if the count is 'above background'. The second objective is to evaluate if there is any gradient in the effect of the contaminant downstream from the source, and about how far it extends.

The samples are made up of independent counts so we model variation with the Poisson distribution. We can calculate the likelihood of observing four cells, for example, from the probability density function for the Poisson,

$$L_1 = P(4|\lambda) = \frac{e^{-\lambda}\lambda^x}{x!} = \frac{e^{-\lambda}\lambda^4}{4!}.$$

Thus, if we guess that $\lambda = 1.2$, the likelihood of observing a cell count of four is 0.026 and the natural log of that likelihood (the 'log likelihood') is -3.65. For a count of three (again assuming $\lambda = 1.2$), the log likelihood is -2.44 (see the first two elements of Table 10.2, under the column for $\lambda = 1.2$).

If we calculate the likelihood of each observation for a given value of λ, the log-likelihood of the model (for example, a Poisson distribution with a parameter $\lambda = 1.2$) is given by the sum of the log-likelihood values for each observation,

$$\ln L = \ln(l_1) + \ln(l_2) + \cdots \ln(l_n) = -24.91$$

(see the values at the foot of Table 10.2).

The value of $\lambda = 1.2$ may not be the best estimate. We can then try a range of values of λ. The maximum likelihood estimate of λ is the value of λ that maximizes the value of the sum of $\ln L$.

Table 10.2. *The number of diatom cells per sample in 30 samples from a range of locations downstream from a source of an agricultural chemical (hypothetical data)*

Distance	Count	$\lambda = 1.2 \ln(l_i)$	$\lambda = 1.47 \ln(l_i)$	$\lambda = 2 \ln(l_i)$
1	3	−2.44	−2.11	−1.71
1	4	−3.65	−3.11	−2.41
1	2	−1.53	−1.39	−1.31
2	4	−3.65	−3.11	−2.41
2	2	−1.53	−1.39	−1.315
4	0	−1.20	−1.47	−2.00
4	0	−1.20	−1.47	−2.00
6	1	−1.02	−1.08	−1.31
6	1	−1.02	−1.08	−1.31
8	2	−1.53	−1.39	−1.31
8	1	−1.02	−1.08	−1.31
10	0	−1.20	−1.47	−2.00
10	0	−1.20	−1.47	−2.00
12	2	−1.53	−1.39	−1.31
12	0	−1.20	−1.47	−2.00
	$\ln L = \sum \ln(l_i)$	−24.91	−24.50	−25.67

The sum, L, of the log (likelihoods) for each observation, i, in Table 10.2 is maximized at a value of $\lambda = 1.47$, assuming the Poisson distribution (Figure 10.2). The average number of cells per sample, 1.47, is somewhat above background (0.5 cells) but not as high as the frequency associated with serious environmental change (two cells). So far, we have ignored the distance information. We expect the concentration of the chemical and the frequency of the diatom to decline as we move further from the source. It seems reasonable to expect the decline to approach zero gradually. An exponential decline will probably be a good approximation. We can use maximum likelihood to estimate the parameters of this slightly more complicated model.

We make the expected frequency of each observation equal to $\exp(a + b \times \text{distance})$, instead of λ, where a and b are the slope and intercept of the exponential decline of chemical concentration with distance. Using maximum likelihood estimation to fit the parameters, we get $a = 1.30$ and $b = -0.23$ (these details are not shown here, but the calculations were done using Microsoft Excel's 'solver'). At zero

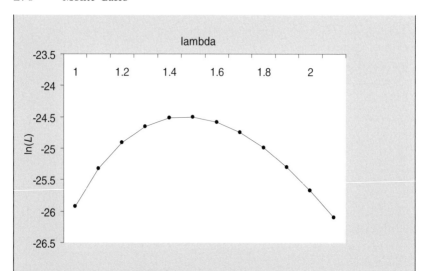

Figure 10.2. Log likelihood versus a range of values of λ for the Poisson model of diatom frequencies.

distance, we expect about 3.5 cells per sample, declining to an average of about 1 cell per sample at a distance of about 5 km downstream. We could choose another discrete statistical distribution and repeat the exercise, comparing the log likelihoods of the two models, adjusted for the number of parameters in each model. Akaike's Information Criterion provides one way of making such comparisons. Hilborn and Mangel (1997) and Burnham and Anderson (2002) give a number of other examples.

observations from the original mean, and compared the extreme values to the tails of the normal distribution (Figure 10.3).

The plots of the tails of four sets of observations, three for physical processes and one for human population estimates, demonstrated unexpectedly large deviations in successive measurements of the same quantities. Shlyakhter (1994, p. 480) described this as a 'pattern of overconfidence' in which the normal distribution seemed adequate but underestimated the frequencies of extreme events in subsequent measurements.

Sometimes a process is best described by a mixed distribution in which the parameter(s) of a distribution is itself a random variable. For example, Shlyakhter (1994) solved the problem of poorly fitted tails by assuming

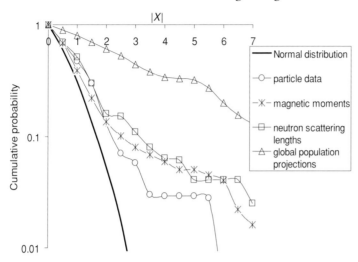

Figure 10.3. Probability of extreme (unexpected) results in physical measurements and population estimates. Cumulative probability plots that new measurements (*a*) are at least |*X*| standard deviations (*s*) away from reference values drawn from previous results (*A*), where $x = (a - A)/s$ (after Shlyakhter 1994).

that the normal distribution included a parameter that was a random variable.

The question will often remain, in the absence of underlying theory and high-quality, relevant data, what is the best assumption? The analyst has the option of averaging the expectations of different plausible models, weighted by their likelihood (see Section 10.6.3). If the context demands that the analysis be risk averse, it may be best to avoid strong assumptions about distribution type, and to choose the widest distribution consistent with the state of knowledge. But conservative assumptions about risk may have other costs, and the degree of conservatism should be made explicit and should be communicated when decisions are made.

10.3.4 Knowledge and inherent uncertainty

Parameters vary naturally and lack of data and knowledge of the shape and moments of the distributions (incertitude, see Chapter 2) adds another layer of uncertainty. Most risk assessments do not distinguish between these sources of uncertainty, and do not account for them separately when performing calculations (see Hilborn and Mangel 1997). For example, if

F(x)

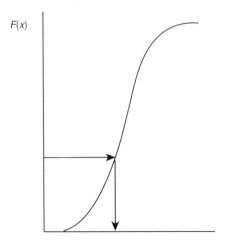

Figure 10.4. Representation of the inverse function, $Q(F(x)) = x$.

the task of the risk assessment is to make a prediction, then measurement error should be deducted from the uncertainty surrounding a parameter, before projections are made. The same issue arises for expert judgement. In practice, it is not easy to distinguish natural variation from incertitude, especially if the shapes of the tails of the distributions are important (Shlyakhter 1994, Hattis and Burmaster 1994, Haas 1997).

10.4 Generating answers

Once parameters have been defined and models have been selected to represent uncertainties, the (random) parameters have to be combined into a solution. Usually, the solutions are found using computers to iterate the problem over and over, thereby generating a distribution of answers that combines the uncertain parameters (and their dependencies, but more on that later).

Assume you have a cumulative probability distribution for a variable x which you call $F(x)$. Monte Carlo asks the question: 'What is the value of x for a given value of $F(x)$?' The function that outputs x values for given values of $F(x)$ is called the inverse function. Often it is written as $Q(F(x)) = x$, where Q is the inverse of F (sometimes written as $F-1$ or F^{-1}) (Figure 10.4).

This is a convoluted way of asking: 'What value of the variable will produce a risk less than or equal to the one specified?' Monte Carlo simulation provides an answer to that question by running repeated trials.

Table 10.3. *A table of random digits*[a]

86515	90795	66155	66434
69186	03393	42502	99224
41686	42163	85181	38967
86522	47171	88059	89342

[a] To generate a number with three significant figures, select the digits from the sequence (e.g. 0.865, 0.159 and so on).

10.4.1 Random numbers

A few simple examples illustrate the nuts and bolts of Monte Carlo simulation. How are random numbers drawn from the interval [0,1]? We could mark the digits 0, 1, 2, . . . , 9 on identical slips of paper, place them in a hat, mix them, take out one number, return it to the hat and repeat the process. If we wrote down the digits obtained in this way, it would form a table of random digits (e.g. Table 10.3).

We can simulate the toss of a coin using Table 10.3. Assign the numbers 0, 1, 2, 3, 4 to heads, and the numbers 5, 6, 7, 8, 9 to tails. Starting with the first row of numbers, the outcome would be T, T, T, H, T, T, H, T, T, T, T, T, H, T, T, T, T, H, H, H, giving just 6 heads out of 20.

Starting with the second row, the outcome is T, T, H, T, T, H, H, H, T, H, H, H, T, H, H, T, T, H, H, H, giving 12 heads.

This is a Monte Carlo simulation of a coin toss.

The long-run outcome of p(head) $= 0.5$ is trivial. But series like these can answer slightly more interesting questions than can be answered by providing the point estimate. For instance: 'What is the chance of more than 10 heads in a row (which, if I'm a gambler, may spell ruin)?' or, 'What is the chance of a sequence of 20 alternating heads and tails?'. The answer is obtained simply by going to the output and counting the number of times, out of the total, the event in question occurred.

Assume you manage a population of koalas and you know the average annual survival rate is 0.7. We could multiply the population size by the average, 0.7, to calculate the number of survivors expected in the following year. Or we could follow the fate of each individual.

At each time step, generate uniform random numbers between 0 and 1. If a random number is greater than the survival value (0.7 in this case), then the individual dies. Otherwise, the individual lives. We ask the question for each individual in the population, using a different random number each time. Thus, if there are 10 individuals in the population, there is no guarantee that 7 will survive, although it is the most likely outcome. There is some smaller chance that 6 or 8 will survive and some smaller chance still that 5 or 9 will survive.

This kind of uncertainty represents the chance events in the deaths of a real population, sometimes called demographic variation. We can then ask, what is the chance that all the koalas survive? Or none of them?

Binomial probabilities will also give the answers to these questions (the answer to the question, 'What is the chance that all the koalas survive?' is $0.7^{10} = 0.0282$). The beauty of Monte Carlo is that exactly the same logic may be applied to questions of (almost) arbitrary complexity, for which there is no convenient analytical answer (the majority of cases).

10.4.2 Pseudorandom numbers

A slightly more complex problem would be one in which there are three possible outcomes. We could assign a part of the uniform distribution to each outcome, dividing it into three buckets. Using Table 10.3 to select random numbers to generate a series of outcomes in a simulation, we could then take the first 3 digits from the table and divide by 1000. For the next random number from 0 to 1 select the next 3 digits and divide by 1000 and so on.

Suppose we have a discrete random variable X with the following distribution:

$$X \sim \begin{pmatrix} x_1 & x_2 & \cdots & x_n \\ p_1 & p_2 & \cdots & p_n \end{pmatrix}.$$

Consider partitioning the interval $[0,1]$ into n sub-intervals with lengths equal to $p_1, p_2, \ldots, p_n$.

$$y_1 = [0, p_1]$$
$$y_2 = [p_1, p_1 + p_2]$$
$$y_i = [p_1 + \cdots p_{i-1}, p_1 + \cdots + p_i]$$
$$y_n = [p_1 + \cdots p_{n-1}, p_1 + \cdots + p_n]$$

Each time we generate a random number between 0 and 1, we check to see which interval it falls into. If it falls into the interval labelled y_i, where $i = 1 \ldots n$, then the random variable X assumes the value x_i.

The trouble with this approach is that tables of random digits are cumbersome to use. We need a 'source' of variation, an engine that will generate arbitrarily long sequences of uniformly distributed random numbers, from which other distributions may be generated.

The most popular algorithm for generating pseudorandom numbers is called the congruential method. It provides a sequence of numbers from a two-step iterative process. The initial number (the 'seed') is specified by the user. The algorithm then produces numbers that satisfy a range of properties including that, after many iterations, pairs of random values densely fill the unit square. The period of the sequence is very large, so that repetition is unlikely (for details see Knuth 1981, Barry 1996).

The equations and algorithms that produce the answers are known as inverse probability distribution functions. Computational formulae exist for generating samples from almost any statistical distribution, using uniform random numbers as a source of variation (Box 10.3).

10.5   Dependencies

A correlation describes the extent to which two variables are associated (the degree to which they co-vary). A dependency implies that variation in one variable contributes to or causes the values in another variable. This kind of relationship is usually modelled with regression. Scatter plots are a useful way of establishing the form of a relationship between two variables. Typically, the dependent variable is plotted on the y-axis.

It may be argued that models that involve dependencies are not sufficiently resolved. If something is causing two variables to co-vary, then the cause should be included as an explicit process. In practice when variables are correlated, often there is no mechanistic or theoretical explanation for the pattern.

In general, positive correlations between variables in additive or multiplicative steps increase the spread of results and increase the risks of extreme events. The effects may be dramatic for strong correlations or for calculations involving several correlated variables. Dependent relationships demand careful consideration in risk assessments.

10.5.1   Rank correlations

Spearman's rank correlation measures the similarity in the rank order of objects (samples) in two lists. When objects have the same ordering, they have a rank correlation of 1. When the orders are reversed, the

Box 10.3 · *Algorithms for normal and lognormal random numbers*

One of the simplest algorithms for generating normal random deviates is (Knuth 1981):

1. Generate a list of 12 independent, uniform deviates from the interval [0,1].
2. Sum the 12 numbers.
3. Subtract 6.

This algorithm produces a single number from a normal distribution with a mean of 0 and a variance of 1 (a standard normal deviate). It may be translated to other locations and dispersions by:

$$x' = \overline{x} + sy$$

where s is the standard deviation of the distribution, $\overline{x}$ is its mean, and y is the standard normal deviate.

This algorithm produces an approximation with poor tails. The distribution is truncated between -6 and $+6$ standard deviations from the mean. Knuth (1981) described the alternative 'polar' method:

1. Sample two numbers from a uniform distribution, U_1 and U_2.
2. Let $V_1 = 2U_1 - 1$.
3. Let $V_2 = 2U_2 - 1$.
4. Compute $S = V_1^2 + V_2^2$.
5. If S is greater than or equal to 1, return to 1.
6. Compute $X_1 = V_1\sqrt{\frac{-2\ln S}{S}}$.
7. Compute $X_2 = V_2\sqrt{\frac{-2\ln S}{S}}$.
8. Compute $Y_1 = \mu + \sigma X_1$.
9. Compute $Y_2 = \mu + \sigma X_2$.

To sample a lognormal deviate, Knuth (1981) suggested:

1. Let $c = \sigma/\mu$.
2. Calculate $m = \ln(\mu) = 0.5\ln(c^2 + 1)$.
3. Calculate $s = \sqrt{\ln(c^2 + 1)}$.
4. Sample a random number, y, from a normal distribution with a mean m and a standard deviation, s.
5. The lognormal number, $l = e^y$.

The utility of a method depends on its purpose. The best is the fastest and easiest to implement that is 'sufficiently' accurate.

rank correlation is -1. When the orders are random with respect to one another, the rank correlation is 0. It is calculated by

$$r_s = 1 - \left(\frac{6 \sum (\Delta R)^2}{n(n^2 - 1)} \right),$$

where n is the number of objects (the number of pairs of samples) and ΔR is the difference in rank for each pair of objects. The statistic ignores the shape and dispersion of the distributions from which the lists are drawn.

Rank correlations are not intuitive and elicitation of rank correlations from experts is error prone (Vose 1996). However, rank correlations may be measured in some circumstances. They provide an alternative to unrealistic assumptions about the nature of dependencies when the processes causing dependencies are unknown.

In Monte Carlo simulation, it is possible to constrain the samples drawn from two distributions so that a specified rank correlation is generated. A method by Iman and Conover (1982) allows multiple rank correlations to be generated between many variables. It is not limited by the kinds of distributions involved.

10.5.2 Linear correlations

Dependencies are most often measured by linear correlations. Appropriate transformations may make the assumption of approximately linear relationships between variables reasonable. In complex models with many variables, we may wish to replicate the pairwise correlation coefficients that emerge from data. Pearson's correlation coefficient is given by:

$$r = \frac{\text{cov}(X, Y)}{s(X)s(Y)},$$

where $\text{cov}(X,Y)$ is the covariance between the two variables, and $s(.)$ is their standard deviations. Correlated random numbers may be generated for simulation using the algorithm in Box 10.4.

10.5.3 Fitted models of dependent relationships

If the form of the relationship between two variables is known from theory, or can be seen in scatter plots, then a function can be fitted to it (Box 10.5). Fitting a model removes the necessity of assuming linear relationships between the variables.

Box 10.4 · *An algorithm for correlated random numbers*

If correlations between pairs of variables have been estimated, linear relationships are assumed and unexplained variation is assumed to be normally distributed, then correlated random variables may be simulated by the following method:

1. Estimate the correlation, r, between the variables.
2. Generate two standard normal deviates, x_1 and x_2, with means, $m = 0$ and standard deviation, $s = 1$.
3. $y_1 = m_1 + s_1 x_1$.
4. $y_2 = m_2 + s_2(r x_1 + x_2 \sqrt{1 - r^2})$.

y_1 and y_2 will have appropriate means and standard deviations, with a correlation of r between them. Linear correlations may be generated between many variables simultaneously. These methods are outlined by Knuth (1981, p. 551) and Ripley (1987).

Box 10.5 · *A procedure for generating dependent variables (after Vose 1996)*

1. Plot the data with the dependent variable on the y-axis.
2. Find a line that best fits the relationship between the two variables, and that is consistent with whatever theory might suggest.
3. Use the equation of the line to find the expected value of the dependent variable for each value of the independent variable, $\hat{y} = f(x)$.
4. Decide on a model for the 'unexplained' portion of the variation. For example, you may choose a normal distribution because the residuals in a deviance plot appear to be normally distributed. Alternatively, you may use expert opinion of uncertainty together with some other distribution (such as the beta) to implement the judgement.
5. Estimate the parameters of the unexplained variation from the data. For instance, you may calculate the standard deviation, s, of the residuals in the deviance plot and use it as an estimate of the standard deviation of the source of variation.
6. For each value of the independent variable in each iteration of the simulation, calculate the corresponding value of the dependent variable [e.g. $y_t = \hat{y} + N(0, s)$].

If the magnitude of the variation is not constant, then the parameters of the model for the unexplained variation may themselves be a function of the independent variable (see Vose 1996 for some examples).

10.6 Extensions of Monte Carlo

10.6.1 Second-order Monte Carlo

If analysts are honest, in many circumstances it is difficult to judge the form of the relationship between two variables or the values of parameters. Two-dimensional Monte Carlo is used to separate variability ('true' natural variation) from incertitude (lack of knowledge; see Chapter 2). Variability may be represented by a set of statistical distributions for each naturally varying parameter. Incertitude may be represented by sets of alternative model scenarios, structures and parameter distributional shapes.

In ordinary, 'first-order' Monte Carlo, stochastic parameters are estimated to reflect true variability. The shape of the distribution, its parameters and other attributes of the model are fixed. In 'second-order' Monte Carlo, the stochastic parameters are themselves drawn from statistical distributions that reflect uncertainty about true values (Figure 10.5).

The results of these kinds of simulations generate families of risk curves, one set for each step along the axes representing incertitude and variability. Thus, the results may be presented as a set of risk curves, one set for each scenario and structural alternative. It produces a kind of qualitative sensitivity analysis in which the importance of structural and scenario assumptions may be explored and compared. For example, Cohen *et al.* (1996) used it to explore exposure and the effects of contaminant levels in the environment.

If the risk analyst judges that a lack of knowledge dominates some parameters, and if the risk assessment is sensitive to these parameters, then additional fieldwork may substantially improve a risk estimate. Despite its intuitive appeal, the method is rarely used because it involves many computations and dealing with dependencies is complex. Software supporting these analyses is improving and their complexity need not be debilitating.

Two-dimensional Monte Carlo cannot guarantee to include the correct answer. None of the models may be accurate. There are many possible choices for parameters, distributional shapes and kinds of dependence. While strategic choices might provide good coverage of possibilities, Monte Carlo cannot do them all (Ferson 1996b).

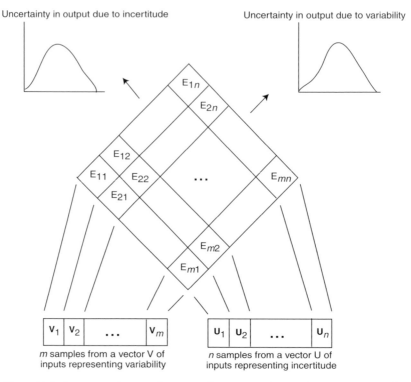

Uncertainty in output due to incertitude

Uncertainty in output due to variability

Figure 10.5. Structure of a two-dimensional Monte Carlo simulation to account for variability and incertitude (after Frey and Rhodes 1996).

Markov Chain Monte Carlo (known as MCMC) is useful in analysing large and complicated data sets. It is beyond the scope of this book to treat this method in detail, but it is especially useful in problems with some form of hierarchical structure such as chemical speciation and exposure relationships, and life history parameters in natural populations of plants and animals. The idea was introduced by Metropolis *et al.* (1953). It is used in conjunction with Bayesian prior distributions to estimate parameters when conventional methods are too complicated (see Link *et al.* 2002).

10.6.2   Models with incertitude and variability

Data usually include such things as measurements of input variables, parameters and output variables. The variation we observe in a time series for the size of a population (say) is composed of natural variation and measurement error (incertitude). To build a model, we need to decompose the variability in observations into its two components (Hilborn and Mangel 1997).

Taking the population example a step further, uncertainty in the number of animals we observe next year will be composed of variability,

$$N_{t+1} = s\,N_t + f_t + W_t,$$

and incertitude,

$$N_{\text{observed},t+1} = N_{t+1} + V_t,$$

where s and f are survival and fecundity, and W and V represent the process and observation error, respectively. Hilborn and Mangel (1997) provide guidance on how to estimate the two contributions to uncertainty and how to include them subsequently in making projections and inferences.

10.6.3 Model averaging

Analysts' lack of knowledge extends beyond the form of the relationship between two variables, the relative contributions of incertitude and variability, and the values of parameters. Hilborn and Mangel (1997) recommended that we retain different ideas and weight their credibility by how well they are supported by the data. This is an eminently sensible strategy, but in many circumstances it may not be helpful to provide decision-makers with a large number of assessments, one for each coherent set of ideas (a sample from the space of uncertainty), each with an associated likelihood. The alternative and most common strategy is to choose the single 'best' model, ignoring much uncertainty, resulting in overconfident predictions, particularly if plausible alternatives give different predictions (Wintle *et al.* 2003).

Model averaging combines the predictions of a set of plausible models into a single expectation in which individual weights based on information criteria reflect the degree to which each model is trusted (Draper 1995, Burnham and Anderson 2002). Prediction uncertainty is calculated from within- and between-model variance. More specifically (Raftery 1996, Wintle *et al.* 2003), the first step is to define a set of possible model structures (S_i) and a set of parameters specific to each structure. Data (D) are used together with the models to make predictions (O) such as the probability of an event (i.e. the presence of a species at a location, or the chance that a contaminant will exceed a threshold concentration),

$$P(O|D) = \sum_{i=1}^{I} P(O|S_i, D)\,P(S_i, D),$$

where S_i are the plausible models, $P(S_i | D)$ is the posterior model probability and $P(O | S_i, D)$ is the posterior prediction of O using model S_i and the data, D. Thus, the weight given to each model S_i is the degree of belief in that model, $P(S_i | D)$.

Model averaging has been used to incorporate model selection uncertainty in microbial risk assessment (e.g. Kang *et al.* 2000), nuclear waste disposal risk assessment (e.g. Draper *et al.* 1999), survival analysis (Raftery *et al.* 1996), and numerous health effects studies (e.g. Raftery 1996, Clyde 2000). Typically, the prediction intervals from model–averaging are much broader than those from any single model (e.g. Wintle *et al.* 2003) because the intervals incorporate uncertainty about the choice of models. The single best models often generate optimistic prediction intervals.

Wintle *et al.* (2003) give details and examples of how to estimate terms for average general linear models. Other guidance for averaging model predictions may be found in Draper (1995), Burnham and Anderson (2002), Hoeting *et al.* (1999) and other references in Wintle *et al.* (2003).

10.7 Sensitivity analyses

Models may be explored by examining how a model's output responds to changes in a variable or an assumption. Deterministic sensitivities may be ascertained analytically for many models, including those that use matrices to represent the structure of natural populations (see Caswell 2001). This is equivalent to asking, 'If a parameter is changed by a small amount in the region of the best estimate, what is the magnitude of change we should expect in model output, relative to the amount of change in the parameter?'.

Numerically, this is

$$s_P = \frac{\Delta V / V}{\Delta P / P},$$

where s_P is the sensitivity of the output variable V to a small change (Δ) in parameter P.

Values larger than 1 indicate the model is sensitive to P, and values close to 0 indicate that the parameter has little influence on V. Values for V and P may be summed over time in dynamic models.

The analytical approach examines the mathematical stability of a model in the region of the parameters. This perspective on sensitivity analysis can help us to understand which parameters in the model determine outcomes, or which might give the greatest return for management effort.

Risk analysts have broader interests than this. Different structural representations of an ecological system reflect different conceptual models. It is important to explore the sensitivity of model output to the uncertainties that arise because of different views of the way an ecological system works.

It is important to examine the sensitivities of risk estimates. Stochasticity in a model creates the opportunity to evaluate the sensitivity of model output to estimates of the shape and magnitude of uncertainty, and to estimates of dependencies among parameters.

Thus, sensitivity analysis should examine:

- *Parameter uncertainty*: rates, transitions, moments of distributions.
- *Structural uncertainty*: alternative representations of the model.
- *Shape uncertainty*: reflecting choices about statistical models for stochastic parameters.
- *Dependency uncertainty*: the existence, form and strength of dependencies.

The simplest approach is to estimate a risk and then calculate how much it changes for a small change in each parameter or structural element, relative to a 'standard' case. Sensitivities may be expressed as percentage changes in the output variable (say, the area under a cumulative probability distribution, or the area between two cumulative probability distributions).

Risk analysts may be interested in examining the importance of a variable over a range of realistic scenarios. They may limit their interest to those variables that are under some degree of management control. Changes in parameters may reflect management activities with different costs, so that the ΔP prescribed for each parameter may be standardized against the cost of an increment in P.

The analyst may explore sensitivities with a view to providing advice on further field studies, so the analysis may concentrate on those parameters that are amenable to further study. In general, the details of the analysis depend on the kind of model, the context of the problem, and the kinds of risks being considered.

10.8 Some examples

10.8.1 Monte Carlo for the dose equation

Each term in the equations used to estimate dose is uncertain. There are a number of alternatives to modelling this uncertainty.

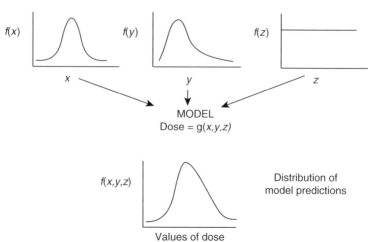

$f(x)$ $f(y)$ $f(z)$

x   y   z

MODEL
Dose = g(x,y,z)

$f(x,y,z)$ Distribution of
model predictions

Values of dose

Figure 10.6. Schematic representation of the combination of uncertain variables in the dose equation using Monte Carlo (after Suter 1993).

In Chapter 9 on interval arithmetic, data and expert judgement on the values in the equation were combined to generate bounds for the result. However, the bounds did not give any indication of the relative likelihood of values within them. Monte Carlo can be used to combine the terms, accounting for uncertainty and retaining information about relative likelihoods within prescribed bounds (Figure 10.6).

Data for crustaceans in a stream exposed to a chemical are:

$C = 0.63 \times 10^{-3}$ mg/l,
$IR = 5.0$ l/day,
$EF = 0.15$ (part of a year: unitless),
bw $= 25.11$ g,

$$\text{Dose} = \frac{C \times IR \times EF}{\text{bw}} = \frac{0.00063 \times 5 \times 0.15}{25.11}$$
$$= 1.88 \times 10^{-5} \text{(mg/day)/g.}$$

For the corresponding probability distribution functions for each of these parameters, assume that the following information is available describing inherent uncertainty for each parameter:

C = concentration of chemical in medium (mg/l), $C \sim N(0.00063, 0.000063)$,

IR = intake/contact rate (l/day), $IR \sim N(5, 0.5)$,
EF = exposure frequency, $EF \sim U(0.12, 0.18)$ and
bw = body weight (mg), bw $\sim N(25.11, 2.51)$.

The standard deviations for C, IR and bw represent coefficients of variation of 10% in each of the parameters. The Monte Carlo method combines the distributions for the random variables C, IR, EF and bw to form a probability distribution for the random variable dose (Box 10.6). The first 20 rows of random numbers sampled from the four parameters generated the first four columns in Table 10.4. These values were combined to give the dose (column 5). A total of 1000 replications of these operations generated the distribution of dose values in Figure 10.7.

This model has made a few assumptions. Each value in Table 10.4 was sampled independently of the other values. That is, there were no dependencies between the parameters. This may not be entirely reasonable. Intake rate is probably related to body weight. The greater the body weight, the higher the intake rate.

With this distribution in hand, it is possible to ask questions that were inaccessible when a deterministic (point) estimate for dose was calculated. Thus, even though the point estimate was less than 2×10^{-5} (mg/day)/g body weight, the distribution says there is more than a 2.5% chance that an animal will experience a dose in excess of 3×10^{-5} (mg/day)/g.

Box 10.6 · *Computation scheme for the dose equation*

1. Select a uniformly distributed pseudorandom number from the interval $[0,1]$ and calculate the corresponding C value using an appropriate inverse probability distribution function (e.g. Table 10.4, row 1, column 1).
2. Select another pseudorandom number and calculate IR from the appropriate inverse probability function (e.g. Table 10.4, row 1, column 2).
3. Continue the procedure for EF and bw (e.g. Table 10.4, row 1, columns 3 and 4).
4. Use the dose equation to obtain a single value for 'dose' for these parameter values (e.g. Table 10.4, row 1, column 5).
5. Repeat this procedure many times to obtain a distribution of 'dose' values.

Table 10.4. *First 20 of 1000 iterations of the process outlined in the text for sampling doses of a contaminant to freshwater crustaceans. Row 1 referred to in the text is shaded*

C	IR	EF	BW	Dose
0.000578	4.78	0.133	23.11	0.0000160
0.000651	5.07	0.165	25.83	0.0000212
0.000545	5.42	0.148	20.90	0.0000209
0.000617	4.25	0.124	22.00	0.0000148
0.000553	5.11	0.143	25.36	0.0000159
0.000575	5.25	0.151	26.95	0.0000169
0.000604	6.07	0.133	28.44	0.0000171
0.000637	4.13	0.155	24.01	0.0000169
0.000754	4.68	0.148	21.67	0.0000241
0.000560	4.70	0.133	25.80	0.0000136
0.000547	4.53	0.129	25.72	0.0000125
0.000534	4.77	0.171	26.51	0.0000164
0.000653	5.44	0.156	24.39	0.0000227
0.000670	4.97	0.151	20.21	0.0000248
0.000631	5.44	0.168	24.67	0.0000234
0.000704	4.69	0.159	27.23	0.0000193
0.000673	5.26	0.179	27.04	0.0000235
0.000645	5.55	0.159	25.43	0.0000224
0.000708	5.42	0.131	26.44	0.0000190
0.000617	5.42	0.176	25.64	0.0000230

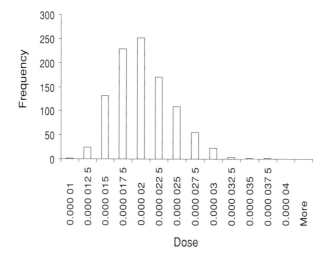

Figure 10.7. Frequency distribution of dose values resulting from 1000 samples taken from each of the four distributions defined above. The *x*-axis labels are the upper class marks for each interval.

The upper bound for the doses from interval arithmetic, using bounds from 95th percentage quantiles, is given by:

$$\text{Dose}_{upper} = \frac{0.000\,756 \times 5.84 \times 0.173}{20.09} = 3.80 \times 10^{-5}.$$

The distribution in Figure 10.7 indicates that this value has a probability of less than 0.1% of occurring (it is more extreme than any of the 1000 dose values generated in the simulation). Note that if C, IR and EF are perfectly positively correlated with each other, and they are perfectly negatively correlated with BW, the 95% confidence intervals would be the same as the interval arithmetic result.

The model could be modified to take dependencies into account. Other modifications are possible that would make the dose equation more realistic. We could consider the possibility that exposure frequency is also related to body weight, because different stages (sizes) reflect a dominance hierarchy, and larger animals inhabit the deeper waters, reducing their exposure. Thus, the dependence of intake rate on body size might be counteracted by a negative relationship between exposure and body size.

We may want to estimate doses received by this population using different parameter estimates for different seasons, because exposure to the contaminant depends strongly on flood conditions. The equation has embedded in it an implicit assumption about the long-term period of exposure, assuming some kind of equilibrium for the concentration of the contaminant in the stream.

10.8.2 Monte Carlo for algal blooms

A dynamic model for reducing phosphorus loads in a lake was developed by Lathrop *et al.* (1998). The model is:

$$[P]_t = [P]_{t-1} + \left(\frac{(1 - R)\, I_t + U}{V} \right) - \lambda[P]_{t-1},$$

where $[P]$ is the mid–April phosphorus concentration (mg/l), V is average lake volume (m^3), R is reduction (proportion) of P input loading after nonpoint pollution controls are implemented, I is the controllable input loading (kg/year), U is the uncontrollable input loading (kg/year) and λ is loss by sedimentation and outflow (unitless).

Uncertainty in the uncontrollable load and the rate of loss by sedimentation and outflow may be introduced by sampling U and λ from random distributions. Assume that, after accounting for measurement error, the empirical data suggest U is normally distributed with a mean of 100 and a coefficient of variation of 10%. λ is lognormally distributed, also with a coefficient of variation of 10%. The sources of random variation are independent. Thus, U is introduced into the system using:

$$\overline{U}_t = \overline{U} + s_R \, y_t$$

where $\overline{U}_t$ is the value taken by the variable x at time t, s_R is the standard deviation of x, and y_t is a deviate chosen from an appropriate distribution with mean $= 0$ and variance $= 1$ (see Box 10.3).

The values of λ and U at each time step were sampled using the algorithms described in Box 10.3. Simulations were run over 20 time steps. The resulting values of $[P]$ were used to generate a probability that $[P]$ will exceed a threshold.

Three scenarios were explored. In the first, management was unchanged ($R = 0$). In the second, controllable input was reduced by 30% ($R = 0.3$). In the third, the rate of sedimentation and outflow (λ) was doubled from 0.1 to 0.2. There is about a 22% chance that the phosphorus concentration will exceed 4.5 mg/l at least once in the next 20 years if management does not change. There is about a 10% chance that the threshold will be exceeded if controllable load is reduced by 30%. If the rate of loss from the lake can be doubled, there is almost no chance of exceeding the threshold (Figure 10.8).

The model could be developed to include some details of the sedimentation and outflow processes. Currently, they are lumped under the parameter λ. Sensitivity analysis could involve the construction of scenarios representing management options. These options would translate into different values for the parameters in the model. Each of the probability curves in Figure 10.8 could be bounded by an envelope of curves representing the extreme combinations of pessimistic and optimistic assumptions about model parameters and structures.

10.8.3 Population viability analysis

Model-based risk assessments for species are called population viability analyses (PVAs). In this context, risk is viewed as the magnitude of a decline of a population (or a set of populations) within some time frame, and the probability that a decline of that magnitude will occur.

Natural populations live in uncertain environments. The best that managers can do is to estimate the chances of particular outcomes based

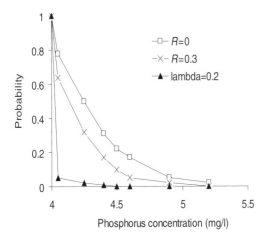

Figure 10.8. Cumulative probability distribution of [P] generated from the dynamic model for phosphorus and assuming uncertainty in R was normally distributed. The parameters were $\lambda = N(0.1, 0.01)$, $U = LN(100,10)$, $R = 0$, $I = 50$, $V = 500$, $[P]_{t=0} = 4$.

on the variations observed in the past and any mechanistic understanding of the processes that control the population. The objective of PVA is to provide insight into how resource managers can influence the probability of extinction (Boyce 1992, Possingham *et al.* 1993). PVA may be seen as any systematic attempt to understand the processes that make a population vulnerable to decline or extinction (Gilpin and Soulé 1986, Shaffer 1990). Much of what follows in this section is taken from Burgman *et al.* (1993), Burgman and Lindenmayer (1998), Akçakaya *et al.* (2000), Burgman (2000), Brook *et al.* (2002), Morris and Doak (2002), Ralls *et al.* (2002) and Reed *et al.* (2002).

The most appropriate model structure for a population depends on the availability of data, the essential features of the ecology of the species or population and the kinds of questions that the managers of the population need to answer. The model may include elements of age or stage structure, behavioural ecology, predation, competition, density dependence, or any other ecological mechanism that is important in determining the future of the population. The result of this formulation is a projection showing the expected future of the population, a single prediction made without any notion of uncertainty.

Once the deterministic form of the model is established, elements of stochasticity are added to represent specific kinds of uncertainty. Demographic uncertainty may be represented by sampling the number of offspring per pair from a binomial distribution. Environmental uncertainty may be represented by time-dependent survivorships and fecundities, often sampled from lognormal distributions (Burgman *et al.* 1993). The result is a cloud of possibilities for the future of the population. Each of these possibilities has a probability of occurrence.

The simplest way to represent population dynamics is as a process of births and deaths. The population size next year is the sum of:

1. The number of individuals that survive to the next time step (out of those that were already in the population for one time step).
2. The number of offspring produced by them that survive to the next time step.

This gives:

$$N(t + 1) = s\,N(t) + f\,N(t)$$
$$= (s + f)\,N(t)$$
$$= R N(t),$$

where s is the proportion of the population that survived from last year to this year, and f is the number of offspring raised by each individual and that survive to the next population census.

This model makes some implausible assumptions:

- There is no variability in model parameters due to the vagaries of the environment.
- Population abundance can be described by a real number. In other words, the model ignores that populations are composed of discrete numbers of individuals.
- Populations grow or decline exponentially for an undefined period.
- Births and deaths are independent of the ages, or other features of the individuals. Essentially, we assume that individuals are identical.
- The species exists as a single, closed population; there is no immigration or emigration.
- Within the population, the individuals are mixed.
- The processes of birth and death in the population can be approximated by pulses of reproduction and mortality; in other words, they happen in discrete time steps.

To add the potential for emigration from the population (say, a fixed proportion of the local population per year), and immigration into the population (a fixed number of individuals arrive each year, independent of the local population size), the model becomes:

$$N(t + 1) = s\,N(t) + f\,N(t) - e\,N(t) + I$$
$$= (s + f - e)N(t) + I.$$

Rates of dispersal between populations often are determined by their distances apart, usually represented by a (negative) exponential distribution. Things such as hunting, captive breeding, or supplementary feeding may be modelled by adding terms or adjusting parameter estimates to reflect the effect of the management actions.

It may be important to distinguish between different genders, ages or life stages because, for instance, different kinds of individuals respond in different ways to management activities. Age or stage structure is simply a matter of replicating the equations for birth and death for the different stages, and incrementing the composition of the stages appropriately. For example, to model adults, yearlings and juveniles separately,

$$J(t + 1) = A(t)f$$
$$Y(t + 1) = J(t)s_j$$
$$A(t + 1) = A(t)s_a + Y(t)s_y,$$

where A is the number of adults, Y is the number of yearlings, J is the number of juveniles, f is the average number of offspring born to adults alive at time t that survive to be counted at time $t + 1$. In this model, juveniles and yearlings do not breed, and juveniles, yearlings and adults have different survival rates (s). This may be represented by a simple conceptual model (Figure 10.9).

The inherent variability that results from random birth and death processes may be represented by sampling survival parameters from binomial distributions, and by sampling the number of births per adult from a Poisson distribution (Akçakaya 1990). Variation in the environment also will affect survival and fecundity, driving the parameters down in poor years. The biology of these processes suggests the variation would be well represented by a lognormal distribution. Thus, the equations for the number of juveniles and yearlings in the stage-structured model may be rewritten as:

$$J(t + 1) = \text{Poisson}[A(t).LN(\overline{f}, \sigma_f)]$$
$$Y(t + 1) = \text{Binomial}[J(t), LN(\overline{s}_j, \sigma_j)].$$

This equation says that the number of new juveniles born into the juvenile class between time t and time $t + 1$ is a Poisson sample of the number of

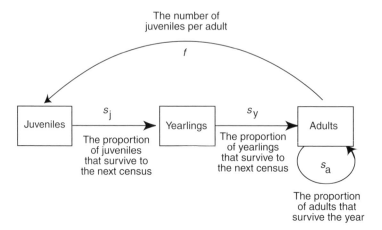

Figure 10.9. Conceptual model of a stage-structured population for a mammal.

adults alive at time *t* multiplied by their fecundity, in which the fecundity is drawn from a lognormal distribution with mean $\bar{f}$ and standard deviation σ_f. Similarly, the number of juveniles that survive from time *t* to time $t + 1$ to become yearlings is a binomial sample of the number alive at time *t*, in which the probability of survival is drawn from a lognormal distribution with mean s_j and standard deviation σ_j.

Wildlife managers implement plans to minimize risks of decline, and sometimes to maintain populations within specified limits, so that the chances of both increase and decline are managed. The probabilities generated by stochastic models allow different kinds of questions such as: 'What is the worst possible outcome for the population?' or 'Which parameter is most important if variability resulting from uncontrolled wildfires can be reduced?'.

Answers to such questions are a kind of sensitivity analysis. They provide guidance on where it would be best to spend resources in field measurements to estimate a parameter as accurately as possible, and to understand ecological processes.

10.8.4 Managing Sindh ibex

Population models sometimes include spatial structure. A metapopulation is defined as 'a set of local populations which interact via individuals moving between local populations' (Hanski and Gilpin 1991, p. 7). The concept is useful for describing the spatial arrangement of subpopulations of species in fragmented environments (Hanski and Gyllenberg 1993).

The ecological and behavioural mechanisms governing dispersal, the distances between patches and the strength of environmental correlation between patches are important considerations determining the persistence of species in fragmented habitats (Harrison 1991, Hanski 2002). A metapopulation may be in a state of dynamic equilibrium that depends sensitively on these parameters.

Ibex are found in most arid-zone mountain ranges in Pakistan. The Sindh ibex, *Capra aegagrus blythi*, is restricted to remote areas near Karachi in southern Pakistan. It is a large game animal and the most common native ungulate in Khirthar National Park. It has a high profile among overseas trophy hunters who prize the curved horns of adult males.

Ibex form groups that live around springs and other freestanding water within the mountain ranges. Their main requirements are crags, safe from direct disturbance because the terrain is inaccessible to domestic goats and shepherds. Ibex have survived here despite hunting, human encroachment on habitat and overgrazing of habitat by domestic livestock. Khirthar National Park is home to a large and growing human population, and there is speculation that oil and gas reserves may also exist within the Park boundaries. There may be more than 70 000 people living permanently within the boundaries. The total population of ibex decreased to about 200 before legal protection was introduced in 1967. Protection has resulted in significant increases in the population to about 12 500 animals (Yamada *et al.* 2004). There has been legal hunting within the adjacent Khirthar Game Reserve for over 50 years.

Park managers are faced with a range of potential pressures and alternative management options. Yamada *et al.* (2004) developed a population model for Sindh ibex to explore the importance of assumptions, to guide further research and to help make better management decisions for this species.

Yamada *et al.* (2004) began with a model for habitat based on terrain and elevation information (Figure 10.10). Each patch supported (or had supported) a more or less separate subpopulation. Yamada *et al.* (2004) modelled the species using a stage-structured population model within each patch, each with three stages (juveniles, yearlings and adults). They modelled both males and females. The model included occasional

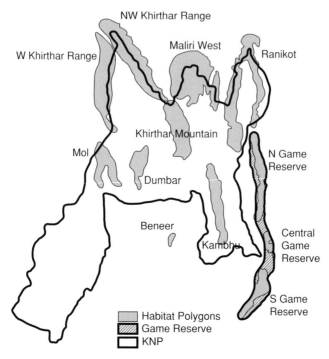

Figure 10.10. Habitat model for Sindh Ibex in Khirthar National Park, Pakistan (from Yamada *et al.* 2004).

dispersal between patches, spatial correlation of rainfall, droughts and limits to the carrying capacity of the environment.

Each of the curves in Figure 10.11 represents a different, plausible set of parameters and model structures. Each cumulative probability line is a different, plausible combination of parameter values and model assumptions. The result is that almost anything is possible.

Such variability is not debilitating. Rather, it is informative, not least because the uncertainty is transparent. Despite uncertainties in parameters and model details, sensitivity analyses proved it was easy to generate scenarios that led to substantial reductions or even the elimination of the ibex from the Park. For example, the species could be eliminated by a trend in habitat loss across the Park reflecting expansion of agricultural activities and unrestricted hunting. Simulations showed that the range of the species will almost certainly contract to the two central large populations if agriculture and other human activities expand in the other patches.

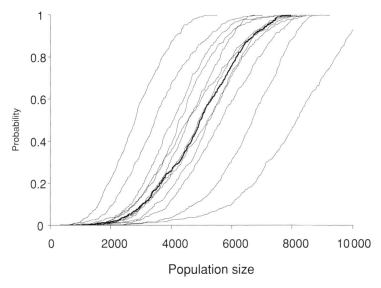

Figure 10.11. Probability curves for a range of parameter values and model structures. The probabilities represent the chance of the population of ibex in the park falling below the specified population size at least once in the next 20 years (from Yamada *et al.* 2004).

Given a broad range of plausible model assumptions, if current management prescriptions remain in place and the human population does not further encroach on habitat, the species is likely to persist. There is also some appreciable risk that the species may fall into the vulnerable category because of substantial population decline if vital rates are lower than expected, or if illegal harvest rates are higher than believed (Yamada *et al.* 2004).

Yamada *et al.* (2004) combined the model and field survey results to develop a monitoring strategy, closing the risk management cycle (Chapter 3). Monitoring effort was concentrated on the Khirthar and Dumbar outcrops, the two most important, central populations (Figure 10.10) with reconnaissance surveys in outlying patches to document persistence of the species and encroachment of agricultural activities.

10.8.5 Managing Baltic cod

It may be important to implement limitations to the size of the populations, reflecting changes in birth and death rates that occur because per capita resources decline with increasing population size. For example, if survival, s_j, or fecundity, f, is made a function of the difference between the total number of adults and the carrying capacity of the environment, it would create a feedback between population size and the rate at which the population grows. This is the standard definition of density dependence. The choice of the function that limits population growth should be determined by the biology and behaviour of the species (see Burgman *et al.* 1993 for some alternatives).

For example, Jonzen *et al.* (2002) used a model with density dependence for a commercially harvested fish population. Some management options may have no effect on mean population size but greatly affect variability, affecting the chances of population increase or decline. Their model may be written

$$N(t+1) = N(t)e^{((\bar{r}+sy)-N(t)/K)} - C(t),$$

where K is the total population size that the environment can support, $C(t)$ is the commercial catch, $\bar{r}$ is the average (long run) growth rate of the population in the absence of density dependence and harvesting, s is the standard deviation of the growth rate, and y is a standard normal deviate (the 'source' of variation in the Monte Carlo simulation).

Jonzen *et al.* (2002) used the stochastic version of their fish model to explore the consequences of different harvesting levels. They could not distinguish observational error from inherent variation in their data, so instead they performed a sensitivity analysis. They assumed first that the uncertainty in their observations was due to inherent variation. Then they assumed it was due exclusively to measurement error. The results were insensitive to the choice of the model for uncertainty. The exploration of management alternatives is one element in the sensitivity analysis of PVAs. There are numerous other options (Mills and Lindberg 2002).

10.8.6 Multispecies and food web risk assessments

Food webs involve multispecies interactions. They are important, for instance, in understanding the fate of contaminants in ecosystems. However, models for multiple species are rare, mainly because there is considerable uncertainty surrounding the ways in which species interact. Tools for building dynamic, population-based models for relatively complex ecological systems have only recently become available (see, for example, Spencer and Ferson 1998, Regan *et al.* 2002b).

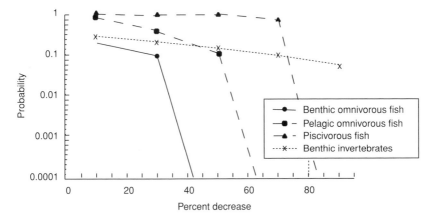

Figure 10.12. The probabilities of decline in the initial abundance of fish and benthic invertebrate populations in a lake in Quebec as a consequence of elevated concentrations of PCPs (after Bartell *et al.* 1999).

In an unstructured model, for instance, contaminant effects may be modelled simply as an additional mortality term:

$$N(t + 1) = (s + f - x) N(t).$$

Mortality, x, due to the contaminant may be a function of time or some other process. We may build a model for the toxicant itself, making it a function of flows (inputs and losses), internal and environmental concentrations, and rates of movement between environmental compartments.

When more than one species is involved, the relationships between the species must be specified. With such equations, it is possible to estimate toxicant concentrations and their consequences for predators, prey and the environment.

Bartell *et al.* (1999) used a detailed ecosystem model to predict the effects of a toxicant on a freshwater ecosystem. Their model included predators, prey, the physical environment and nutrient dynamics. They included both parameter and model uncertainty.

They assessed the consequences of elevated levels of contaminants including pentachlorophenol (PCP) for the biota of lakes and rivers in Quebec (Figure 10.12). The model suggested that substantial declines in fish and invertebrate populations were likely given plausible increases in PCP concentration.

10.9   How good are Monte Carlo predictions?

In Chapter 4, we asked the question, 'How good are subjective expert judgements?' It seems fair to ask the same question of more detailed, model-based predictions.

10.9.1   Predicting radioactive fallout

In the 1990s, 13 research groups in Europe participated in an exercise to predict (retrospectively) the consequences of the Chernobyl nuclear accident. Participants were provided with a detailed description of a site during the time of arrival of the Chernobyl plume of radioactive waste in central Bohemia and Finland.

They were given data on the amount of radionucleotides measured in air, soil and water in 1986, intake rates for various components of local human diets, and ^{137}Cs concentrations in foods. They were asked to predict the whole-body concentration of ^{137}Cs over time, including the variability in whole-body concentration among individuals. The participants used a total of 14 different models although only four groups provided a full set of answers. Some estimates were very good, some less so (Figure 10.13).

The geometric mean of the (approximately) lognormal distribution of concentrations predicted by Lindoz for 1987 matched the outcome quite well. However, the Lindoz group overestimated the geometric standard deviation resulting in underestimates at the low end and overestimates at the high end. In 1989, the predictions by the Lindoz group were nearly perfect. They calibrated their models using conditions in Romania, similar to the test site in central Bohemia.

The Ternirbu group overestimated individual ^{137}Cs concentrations in 1987 due to an overestimate of the amount of the contaminant in people's diets. The group underestimated ^{137}Cs concentrations in 1989 based on an incorrect assumption that soils would be tilled, which would have decreased bioavailability. Their agreement at the high end in 1989 was due to the fact that they also overestimated variability (a good example of compensating errors).

Most of the 13 groups confounded variability and incertitude, leading to estimates of standard deviations greater than reality and confidence intervals around predictions that failed to capture the true cumulative distribution. Only one participant used two-dimensional Monte Carlo.

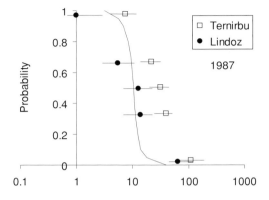

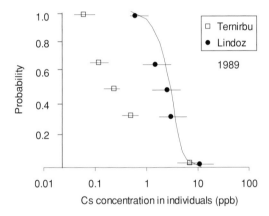

Figure 10.13. The predictions for Cs concentrations in individuals from Chernobyl fallout made by two research groups (Ternirbu and Lindoz) together with the actual outcomes for the site (continuous curve) in central Bohemia in 1987 and 1989. The curves represent the probability (the proportion of individuals) that the concentration exceeds the concentration indicated on the *x*-axis. The confidence intervals encapsulate incertitude and the slopes of the distributions (for the sets of points and the continuous curve) reflect modelled and actual variability among individuals (after Hoffman and Thiessen 1996).

10.9.2 Predicting extinction risk

McCarthy *et al.* (2004, see also Brook *et al.* 2000, Ellner *et al.* 2002) created four hypothetical species with different life histories (a frog, snail, small mammal and small plant). They created a scenario for each species. These data made up a hypothetical 'truth'. They then used simulation models to

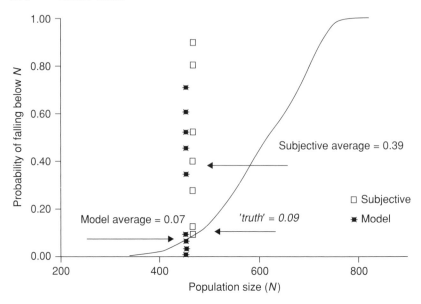

Figure 10.14. Models and subjective judgements were used to estimate the chance that a population of plants would fall below 50% of the initial population size (929 plants) at least once in the next 10 years. The results show the estimates from 12 independently constructed models and 10 independent subjective estimates.

generate survey data for each species, the kinds of data that are typically available in surveys of threatened species. They included observational bias, sampling error and several sources of natural variation.

They posted the survey results on the web and invited other researchers to predict the chances of decline of the species. One group of participants estimated a number of risk-based attributes for each population using Monte Carlo models. Another group used subjective judgement. Accuracy was assessed by comparing predictions with the 'true' outcomes from the models used to create the hypothetical scenarios.

Both the models and subjective judgements were more accurate than random numbers. Four of the five best predictions (as measured by absolute error) were from models, but this apparently superior performance of the models could have occurred by chance alone ($p = 0.218$, based on combinatorial probabilities). The five least-biased predictions were all from models, an unlikely chance occurrence ($p = 0.030$).

The results in Figure 10.14 are representative of those obtained by McCarthy *et al.* (2004) for a range of predictions (the chance of falling to zero, the chance of falling to 50% if reproductive rates fall, and so on) for the four hypothetical species. The accuracy (the spread) of estimates from models and subjective estimates were about the same. The average result over all models was close to the correct answer, whereas the average of the subjective estimates was consistently pessimistic, tending to overestimate risks (Figure 10.14). Changes in relative risks of decline were predicted more accurately than the absolute risks. Predictions over shorter time frames tended to be more accurate than over longer time frames.

Subjective judgements took between 1 and 2 hours for most entrants (most of that time was involved in synthesizing the information provided), whereas the predictions using models took about 1–2 days or more. Given the comparatively good performance of the subjective judgements in terms of overall error, the question remains: 'Is it worthwhile to develop models for species?'.

10.9.3 Limitations and strengths of Monte Carlo

Monte Carlo has a number of limitations, apart from sometimes getting it wrong. There are a number of things that Monte Carlo simulation cannot do (easily) (Ferson 1996b). It:

- cannot propagate nonstatistical uncertainty (ignorance versus statistical uncertainty),
- requires detailed knowledge of input distributions,
- cannot do back-calculations,
- cannot yield a realistic answer when:
  - dependencies are unknown (dependency uncertainty),
  - input distributions are unknown (parameter or shape uncertainty),
  - model structure is unknown (structural uncertainty).

Solutions such as two-dimensional Monte Carlo and trial and error back calculation are cumbersome and computationally costly.

It can be difficult to estimate the parameters, shapes and dependencies required for a simulation, and to accommodate the full spectrum of possibilities for parameters and alternative models, even when using two-dimensional Monte Carlo (Ferson 1996b, Ferson and Moore 2004). Most models confound variability with incertitude. To accomplish a successful analysis requires time and appropriate expertise.

Many analysts and managers find uncertainty dispiriting. Models for environmental risk assessment are hampered by lack of data and lack of validation. Results often are sensitive to uncertainty in the data (Taylor

1995, Ruckelshaus *et al.* 1997). It can be difficult to verify stochastic predictions. Models are sometimes misinterpreted and they impose overheads in terms of computational effort and technical skill that may not be warranted by the problem at hand (Beissinger and Westphal 1998, Burgman and Possingham 2000).

If all that is required is an estimate of risk, subjective judgements and other approaches may be cost-effective. However, model-based risk assessments have additional advantages (Brook *et al.* 2002, McCarthy *et al.* 2004). The rationale behind their predictions is explicit. The models are open to analysis, criticism and modification when new information becomes available. Their assumptions may be tested. The models can be used to help design data collection strategies. They help to resolve inconsistencies. Models may be more useful for their heuristic than for their predictive capacities.

Approaches that seem certain only submerge the breadth of uncertainty within assumptions. Despite their relative transparency and completeness, even Monte Carlo analyses make only a portion of the full range of uncertainty apparent. In fact, one could argue that the bounds are not broad enough because the suite of future possibilities is at the mercy and patience of the inventiveness of the analyst. They could easily have missed a combination of plausible parameters that further extended the set of possibilities. Other methods such as p-bounds make less restrictive assumptions.

10.10   p-bounds

The notion of bounds encapsulates a deeper philosophy about honesty in risk assessments. Risk assessments should be honest in the sense that they should not make unjustified assumptions, and should seek to convey the full extent of uncertainty about a forecast or decision. At the same time, they should not assume any greater uncertainty than is necessary. Thus, the machinery of risk assessment should strive to generate bounds on answers that are as narrow as possible, and that are faithful to what is known.

Monte Carlo techniques struggle to encompass the full breadth of uncertainty. For example, Cohen *et al.* (1996) developed a two-dimensional Monte Carlo for contaminant exposures and effects. At one point they noted, 'We know of no data quantifying the fraction of ingested soil from the site' (p. 950) and subsequently assumed the distribution was uniform over the interval [0,1]. However, this assumption is quite specific: a linear cumulative probability distribution with a mean of 0.5. It makes a number of unjustified assumptions. Methods need to be more honest about what is not known.

Robust Bayesian analyses estimate a set of posterior distributions for a quantity, based on prior distributions and likelihoods selected from classes believed to be plausible by the analyst. Standard Bayesian analyses are conducted for all combinations. Results are robust if the posterior distributions are not much affected by the choice of priors and likelihoods, at least with respect to a particular decision. This approach has a great deal in common with two-dimensional Monte Carlo conducted together with thorough sensitivity analyses. Like two-dimensional Monte Carlo, there is no comprehensive strategy for dealing with the full range of plausible alternatives for distributions, dependencies and model structures (Ferson and Moore 2004).

Instead of concentrating on estimating a quantity, bounding methods attempt to bound the value. Interval arithmetic introduced in Chapter 9 is the simplest approach to bounding. Unfortunately, interval arithmetic ignores whatever information may be available on values within the intervals such as distribution shapes and dependencies, and it does not quantify the likelihoods of extremes.

The 'p-bounds' method was developed by Frank *et al.* (1987) and implemented and extended by Williamson and Downs (1990) and Ferson *et al.* (1999). 'p-bounds' calculations bound arithmetic operations, making only those assumptions about dependencies, distribution shapes, moments of distributions, or logical operations that are justified by the data (Ferson 2002). 'p-boxes' are 'sure' bounds on cumulative distribution functions. It is easiest to think about p-boxes as though they were real numbers. They represent a quantity (in the form of bounds on a distribution) and the objective is to do calculations with them. The bounds can be any pair of lines that do not cross and that increase monotonically from 0 to 1 (e.g. Figure 10.15).

When bounds straddle a decision threshold, the results can be used to make it clear what more needs to be known before a robust decision is possible. The width of the bounds represents incertitude about the shape and other characteristics of a risk model. They provide an opportunity for a risk analyst to explore how sensitively the uncertainty about a risk estimate depends on lack of knowledge about parameters and other relationships (e.g. Regan *et al.* 2003).

If only range information is available, p-bounds provide the same answers as interval analysis. When information is sufficient to indicate

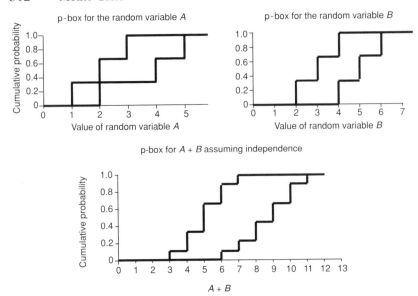

Figure 10.15. Two p-boxes corresponding to two random variables, A and B, and the p-box for the sum $A + B$ assuming independence (after Ferson and Moore 2004). The p-box for A is partitioned into the three interval-mass pairs: $A \varepsilon [1, 2]$, $p = 1/3$; $A \varepsilon [2, 4]$, $p = 1/3$ and $A \varepsilon [3, 5]$, $p = 1/3$. $A \varepsilon [1, 2]$ $p = 1/3$ means that the value of A is some value between 1 and 2, inclusive, with probability 0.33. The p-box for B is $B \varepsilon [2, 4]$, $p = 1/3$; $B \varepsilon [3, 5]$, $p = 1/3$ and $B \varepsilon [4, 6]$, $p = 1/3$ (from Ferson and Moore 2004).

precisely the distributions and dependencies for a problem, they give the same answers as Monte Carlo. Thus, they generalize both interval analysis and Monte Carlo (Ferson and Moore 2003). Possibilities within the bounds are not equally likely. Thus, bounding analyses do not replace the insight one might gain from a sensitivity analyses using a Monte Carlo model.

Dempster–Shafer structures (Shafer 1976) define sets of plausible values that the available evidence does not distinguish. The lower bound is termed a 'belief' function. The upper bound is termed a 'plausibility' function. The set defines the (epistemic) limits of the evidence. Cumulative plausibility and belief functions may define p-boxes (Ferson *et al.* 2003).

Alternative data-gathering options may be explored with separate Monte Carlo runs, generating distributions that lie within the envelope resulting from the bounding analysis. p-bounds may have technical difficulties with repeated variables in arithmetic expressions and all results must be expressed as cumulative probability distributions, which can be difficult to interpret (Ferson and Moore 2004). The choice of an analytical strategy will depend on the questions that need to be answered, and the technical skills available.

10.10.1 Monte Carlo and p-bounds: an exposure model example

Regan *et al.* (2002b) developed a detailed food web model of exposure of mink to a toxin, Aroclor-1254. They applied both two-dimensional Monte Carlo and p-bounds analysis. The p-bounds provided a useful way of summarizing the parameters that contributed most to the uncertainty in the risk estimates.

For example, the model included a parameter for the proportion of water in insect tissue (part of the exposure pathway of the chemical). The Monte Carlo analysis assumed natural variation resulted in a normal distribution. Regan *et al.* (2002b) estimated the mean and standard deviation from US EPA data. p-bounds used a minimum, maximum, mean and standard deviation but made no assumption about shape (Figure 10.16).

The envelope of cumulative density functions resulting from the p-bounds analysis captures the full extent of the uncertainty in knowledge of this parameter. Including both incertitude and natural variation gave a much broader picture of potential exposure than did Monte Carlo analysis of natural variation alone (Figure 10.17).

10.11 Discussion

Model-based risk assessments are transparent, relatively free from ambiguity and internally consistent. Explicit models can capture all available knowledge and be honest about uncertainty. Bounds can enclose the full scope of possibilities. So why is model-based risk assessment not universal?

One of the most serious impediments is regulatory inertia. Organizations and regulatory conditions can be slow to change. Appropriate numerical skills may be lacking, delaying the adoption of methods that would otherwise provide useful answers.

It can be costly and misleading if the quantitative tools drive decisions, rather than provide decision support. Often, the temptation to use a package as a 'black box' is too great and models are used without understanding their limitations. If stakeholders and managers are divorced from the model-building process, they will be unable to criticize its

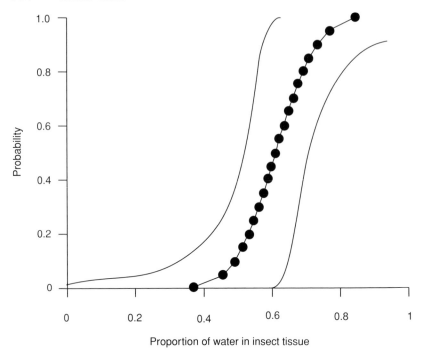

Figure 10.16. Normal cumulative density function and p–bounds for the proportion of water in insect tissue (after Regan *et al.* 2002b). The Monte Carlo simulation assumed a normal distribution (shown as the line marked by circles) whereas p–bounds (solid lines) made no assumption about shape.

assumptions. They will not develop an intuitive feel for its strengths and weaknesses. In many ways, this loses the most valuable aspect of the process.

There is also the perception that requirements for data are heavy. Monte Carlo analyses are sometimes criticized because they demand more data than are available. Both Monte Carlo and bounding methods are criticized because they provide such a wide range of potential risks. How is a manager to decide when anything is possible? Often, when confronted by images like Figure 10.11, managers and policy-makers despair of the risk assessment, calling it uninformative.

However, I see these outcomes as a consequence of being honest about uncertainty. The family of risk curves representing the range of possible outcomes is the strength of model-based risk assessments, rather than a reason for criticism or despair. The solution is not to dispense with model-based risk assessments. Rather, it is to learn from the model and change

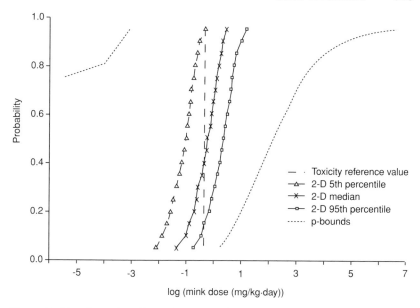

Figure 10.17. p-bounds and two-dimensional Monte Carlo estimates of Aroclor-1254 dose to mink, and the toxicity reference value for reproductive effects (after Regan *et al.* 2002b).

the questions so that they can be answered with available information, or to defer the decision until the critical information is available.

Each curve represents a particular set of assumptions. There is a single curve roughly in the centre of Figure 10.11 that reflects the best guess about parameters, dependencies and other assumptions. But the extent of our ignorance is such that a broad range of futures is possible. Futures are not all equally plausible. The extremes of the envelope will occur only if all the uncertainties turn out to be pessimistic (at one extreme) or optimistic (at the other extreme). It is unlikely that all judgements are biased in one direction.

Perhaps most importantly, it would be professionally negligent not to communicate the full extent of uncertainty to a manager. Such figures are an honest representation of what is not known. The decisions that flow from this representation of uncertainty will depend on the decision-making context and the attitude to risk of those involved in it.

The curves may also disaggregate variability and incertitude. The single risk curve reflecting the best guess at parameters and assumptions reflects the risks that result from inherent uncertainty. The spectrum of curves (the

Box 10.7 · *Steps in a Monte Carlo analysis designed to work in the context of the risk management cycle*

1. Build a conceptual model of important processes, on the basis of theory, data and expert opinion, linking human activities to critical endpoints.
2. Identify key parameters and relationships.
3. Write equations to represent the deterministic elements and processes.
4. Collect and analyse the data.
5. Quantify uncertainties in each term of the equations.
6. Decide on a development platform (choose a simulation environment with sufficient flexibility and computational power, with the kinds of features that suit the problem at hand).
7. Develop the model.
8. Design calibration, verification and validation protocols.
9. Make predictions for calibration.
10. Debug the model (eliminate logical inconsistencies, coding errors and related problems). The complexity of programs is related to the number of ways that their components interact. Debugging is hard and time consuming but various techniques exist to help minimize errors including using good design, good style, exploring boundary conditions, documenting assumptions and implementing sanity checks in the code.
11. Calibrate the model (iteratively improve the structure and parameter estimates of the model to agree with data and subjective judgement about plausible outcomes).
12. Verify the model's performance (reasonableness checks: confirm that the model predictions conform to common sense and to expert judgement of system behaviour, compare the results with other models, explore limiting cases).
13. Conduct sensitivity analyses.
14. Make predictions for validation.
15. Implement monitoring strategies to validate predictions and update model assumptions and parameter estimates.

envelope) encloses the extent of uncertainty, ignorance about processes and lack of knowledge about parameters.

Judgements as to what are and what are not acceptable risks involve social, political, economic and ethical reasoning. The relevance of models for environmental decision-making is in the mind of the policy-maker. An analyst who builds a model provides a service, a skill, and the end product is a set of recommendations that are bounded by assumptions and uncertainties. It is as important, if not more important, for the analyst to communicate those uncertainties and assumptions, as it is to communicate the set of predictions.

10.11.1 Monte Carlo for the risk management cycle

One of the reasons why managers become disenchanted with models is that different models often produce different expectations. This is because the bounds and structural details of the analysis are coloured by what the analyst believes to be important. The fact that models of the same natural system may generate different expectations is not surprising to modellers, but it is a source of frustration to decision-makers. The creation of a sense of frustration implies that the sensitivities, limitations and assumptions of the models have not been explained.

Much of the value comes from the process of building the model, rather than interpreting its output. The object of environmental models should be to improve communication and understanding. To achieve these ends, models must be carefully and thoroughly documented, and limitations, sensitivities and assumptions must be stated explicitly. The fact that model-based risk assessments encourage such thinking is one of their main strengths.

Numerous lists specify what makes a good Monte Carlo analysis. For instance, the US EPA (1997a, b) published policy and guiding principles for Monte Carlo analysis. Similar guidelines can be found for model building in other circumstances in Kammen and Hassenzahl (1999), Burgman and Possingham (2000) and Ralls et al. (2002).

11 · *Inference, decisions, monitoring and updating*

Risk assessments are prey to unacknowledged ambiguity and vagueness, as well as the psychological idiosyncracies and social contexts of those involved. Chapters 1 and 4 outlined the factors that lead people to colour their judgements, including such things as the level of personal control, as well as the visibility and dreadfulness of the outcome. These frailties lead to a number of identifiable symptoms including insensitivity to sample size and overconfidence (Fischhoff *et al.* 1981, 1982, Fischhoff 1995, Morgan *et al.* 1996).

Risk-based decisions should weight the probability of an incorrect decision by the consequences of an error. The preceding chapters outline techniques that can be used to build models that serve to protect stakeholders, risk analysts, experts and managers against some of the worst excesses of their own psychologies and contexts.

Once the analytical phase of the risk assessment is complete, the task remains to interpret the results, decide a course of action and design feedback mechanisms that will ensure that decision-making capability improves through time. This chapter outlines a number of methods that have particular utility for monitoring environmental systems, providing information to revise assumptions and models and to support decisions.

11.1   Monitoring and power

Monitoring is sampling and analysis to determine compliance with a standard or deviation from a target or prediction. It may be undertaken to gauge the effectiveness of policy or legislation, to test model assumptions, or to validate predictions.

Monitoring by focusing on the outputs of a system is like navigating 'by watching the white line in the rearview mirror' (Thompson and Koronacki 2002, p. 41). Monitoring should include both input variables ('pressures' such as nutrient loads, toxicant emissions, flows, hunting

Table 11.1. *The structure of inferences from a null hypothesis test*

		Conclusion of the study	
		Impact	No impact
Actual state of the environment	Impact has occurred	Correct	Type II error (β)
	No impact has occurred	Type I error (α)	Correct

levels) and output variables ('responses' such as fish kills, algal blooms, trends in species population sizes or epidemiological data).

It is the responsibility of the risk analyst to close the risk management cycle by developing monitoring strategies that will satisfy two primary goals: to measure the state and response of the system to management strategies, and to provide information about the components of the system that were both uncertain and important in influencing decisions. There are usually subsidiary goals, such as compliance with a set of regulatory thresholds and audits of environmental performance.

To achieve these goals, the monitoring system must be sufficiently sensitive to reliably detect changes that matter. The details will depend on the choice of assessment endpoints, the magnitudes of risks and the sizes of changes that are deemed unacceptable.

11.1.1 Null hypotheses

Most regulatory protocols assume that if no problem is observed, then none exists. Monitoring usually depends on traditional statistical inference. Significance tests evaluate the question, 'What is the probability that the deviations observed were caused by chance variation, assuming no underlying, true difference?'. In these circumstances, reliability depends on the ability of a method to detect real outcomes, usually against a background of natural environmental variation, measurement error, semantic ambiguity, vague concepts and ignorance of biological processes.

To protect the environment, society and the economy, monitoring systems should (a) tell us there *is* a serious problem when one exists (thus avoiding overconfidence, called 'false negatives' or type II errors) and (b) tell us there *is not* a serious problem when there isn't one (thus avoiding false alarms, called 'false positives' or type I errors) (Table 11.1).

The first is crucial for detecting serious damage to environmental and social values. The second is important to ensure that the economy is not damaged by unnecessary environmental regulations. The probability that a monitoring programme will detect important changes if they exist is known as 'statistical power'.

For example, in an experiment to test the effectiveness of a farming strategy designed to reduce the impact of an agricultural chemical on a population of fishes, one possibility is that the treatment has no effect while the alternative is that it is indeed effective. To distinguish between the two hypotheses, the possibility that the treatment has no effect is termed the null hypothesis (H_0). The other possibility, that it is effective, is the alternative hypothesis (H_1).

The example contains what is termed a 'negative' null hypothesis, because if it were true there would be no benefit from applying the treatment. Often environmental monitoring is designed to test a negative null hypothesis that a given human action has no impact on the environment (Fairweather 1991).

While type I and type II errors are two sides of a statistical coin, the attitude of the scientific community and of society in general to the two kinds of errors is different. The probability of a type II error is usually arbitrary and depends upon how a study was designed. Convention specifies that, generally, a type I error rate of 0.05 is acceptable. If scientific protocol, convention or regulatory guidelines were to specify a particular type II error rate, it would be necessary to plan for this before a study was undertaken.

Unfortunately, type II error rates are rarely calculated. Regulatory authorities and scientific conventions do not suggest acceptable thresholds. When reviewing more than 40 environmental impact statements, Fairweather (1991) did not find a single estimate of type II error rates. Mapstone (1995) noted that environmental assessment has inherited a preoccupation with type I error rates. It appears to be the result of nothing more than convention. The same focus on type I errors predominates in environmental management journals (Anderson *et al.* 2000).

11.1.2 Monitoring trends in a bird population

Consider the circumstance in which you are required to monitor the effects of tree death on a bird population (after Burgman and Lindenmayer 1998). The birds depend on large hollow-bearing trees for nest sites. The

Table 11.2. *Surveys of bird population size in an agricultural region conducted in two consecutive years*

Year	Bird population size	Standard error
2002	303	41
2003	291	37

number of hollows is known to limit the size of the population. While large hollow-bearing trees are retained on farms, no provision has been made for the recruitment of new hollow-bearing trees as the older trees decay and collapse. Thus, the number of large trees is thought to be decreasing exponentially.

Field observations and aerial photo interpretation establish that the rate of collapse of nest trees is about 2% per year. The conceptual model suggests that the bird population may be declining at the same rate, determined largely by the collapse of nest trees.

You collect some new data to test the hypothesis that the bird population is declining (Table 11.2). These data represent a change in population size of about 4%, somewhat higher than that predicted by the hypothesis that a decline in population size is due to nest tree collapse. However, a standard statistical test of the difference between the means in 2002 and 2003 shows no significant difference between these values. We cannot reject the null hypothesis. Convention stipulates that we accept that, so far, there is no 'significant' change in the size of the bird population.

Before reaching any final conclusions, we should ask how likely it is that the sampling programme could have detected an impact, if one actually exists. Given that the means of two populations must be about two standard deviation units apart before they could be considered different, we would have to observe a decline to about 220 individuals before the result would be statistically significant. This is a decline of about 25%, more than 10 times greater than the expected decline. Therefore, even without calculations, it is possible to see that the observations made in this test were insufficient to test the hypothesis of a 2% decline per year. The sampling design was inadequate.

The bottom line is that the absence of a statistically significant result should not be taken as evidence that there is no difference (no effect). Unfortunately, insensitivity to sample size and overconfidence would lead most people to conclude that there was no difference (no trend). This interpretation would be wrong.

The options are to increase the sample size, accept a higher type I error rate, reduce the measurement error by using more reliable counting methods, use a different method for inference, or temper the interpretation of results based on power calculations. The remainder of this chapter is devoted to outlining some of these alternatives.

11.2 Calculating power

Frequently, monitoring programmes are used when time and money are limited. For environmental monitoring to be useful, it must be sensitive to important changes. The notion of 'importance' is embodied in the effect size. The effect size is the magnitude of change that we wish to detect, if there is an impact. Low power for a particular test may mean insensitivity and inconclusive results (Fairweather 1991). High power may mean unnecessarily conservative protection of environmental values, or prohibitive costs. An effect size must be specified prior to an analysis of the power of a monitoring study (Cohen 1988, Fairweather 1991, Mapstone 1995). It is a decision that may involve considerations of biology, chemistry, physics, aesthetics, politics, ethics or economics. It is not simply a statistical or procedural decision, and involves a raft of judgements about the ecological and sociological importance of effects of different magnitudes (Mapstone 1995).

This section outlines the details of power calculations for three standard circumstances: comparing observations with a standard, comparing two samples and inferring the absence of an attribute. These are all relatively simple cases for which power calculations are easy. Luckily, they make up a large portion of the routine problems confronted by risk analysts.

11.2.1 How many samples?

This is one of the first questions to occur to people responsible for monitoring. Scientists and others have developed a bad habit of assuming that numbers between 5 and 30 are adequate, irrespective of the question, the decision context, or the within- or between-sample variability. This is at least in part due to psychological constraints such as insensitivity to sample size and judgement bias.

To answer the question, two value judgements are required. The first is, 'How far from the true mean can we afford to be, and not be seriously compromised?'. The second is, 'How reliably do we wish to know that our estimate is in fact within the critical range?'.

Put another way, we need to specify a tolerance for our estimate, and to specify a degree of reliability for our knowledge. An estimate of the number of samples comes from the formula for the confidence interval. The upper bound for a confidence interval is:

$$U = \mu + \frac{s}{\sqrt{n}} z_{(1-\alpha)}.$$

Rearranging the terms gives:

$$n = \frac{s^2 z_{(1-\alpha)}^2}{(U - \mu)^2} = \frac{s^2 z^2}{d^2}.$$

The 'tolerance', d, is a region around the true mean within which we would like our estimate to lie. The reliability is given by the value of z and reflects the chance $(1-\alpha)$ that the estimate lies within d of the true mean. To use the formula, we also need to come up with an estimate of s, which can be difficult in novel circumstances. Theory, experience or data from analogous circumstances can help.

The formula is used routinely to estimate the number of samples required to provide reliable measurements of timber volumes in forest resource inventories (Philip 1994). It was used by scientists from the United States, Canada and Japan to estimate the observer effort required to monitor high-seas drift-net fisheries (Hilborn and Mangel 1997). Despite these applications, it is rarely used to assist the design of monitoring programmes.

11.2.2 Comparing observations with a regulatory threshold

Toxicological analyses and environmental impact assessments routinely establish 'safe' thresholds. How many samples are needed to be reasonably certain of detecting noncompliance? Such cases include, for example, whether factory air emissions are safe, stream turbidity at a road crossing exceeds a limit set by the regulator, or the numbers of trees retained in harvested areas meet agreed standards.

What do we mean by noncompliance? If a single measurement on a single occasion exceeds the threshold, should we prosecute? If so, we may run a substantial risk of falsely accusing someone, especially if measurement error is large. Are we more concerned that the mean of the process should comply with the standard?

The appropriate choice for a decision rule (prosecute / don't prosecute) depends on the underlying conceptual model of cause and effect. If effects of the hazard are cumulative, the mean of the process matters. If effects are the result of spikes in exposure or single events, then single instances should trigger a regulatory response.

The conceptual model for uncertainty also plays a part. If we are concerned about single events that exceed a threshold and variation in measurements is due to process variation, all values that exceed a threshold should be acted on. If variation is due to measurement error, we may falsely accuse someone in up to 50% of instances, even if the true values are less than the threshold (or more, if the distribution of measurement uncertainty is skewed).

Take the case in which we want to be reasonably sure the mean of a process complies. Comparing a set of observations with a regulatory standard is equivalent to conducting a statistical test for the differences between two populations. Power is the probability of detecting a given true difference. It is the chance that you will catch someone who is, in fact, in breach of regulations.

Assume we set the significance level at 0.05. We will find the subject of the test in breach of regulations if the observed sample mean is significantly higher than the threshold, μ. The distance above the threshold is measured in terms of its standard deviation (Figure 11.1). For $\alpha = 0.05$, the critical value is $\mu + z/\sqrt{n}$, where $z = 1.645$ comes from the standard normal distribution.

Since we are only interested in values that exceed the threshold, the test is one-tailed. Essentially, we are relying on the fact that the sample mean will exceed μ by 1.645 or more of its own standard deviation about 5% of the time.

Expressed in algebraic form, the significance level α is equal to this probability, i.e.

$$\alpha = p\left(\bar{x} > \mu + z\frac{s}{\sqrt{n}}\right),$$

where the parameter x has been estimated from a sample of size n. Note that this formula is the expression for estimating confidence intervals. We use one-tailed tests when it is important to know if values exceed an expected value or a compliance threshold. For a one-tailed test, z is equal to 1.645. We use two-tailed tests when it is important to know if a process deviates from an expected value, but the direction of deviation (smaller or larger) is not important. For a two-tailed test, z is 1.96.

Consider an example in which natural levels for stream turbidity are around 6.3, based on long-term environmental monitoring. The biological implications of increases in stream turbidity are such that it is important

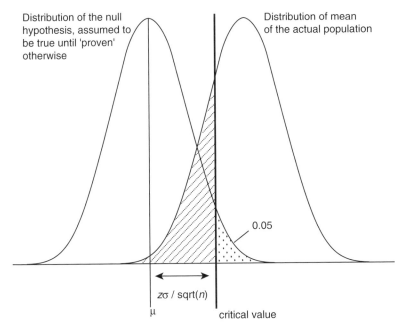

Distribution of the null hypothesis, assumed to be true until 'proven' otherwise

Distribution of mean of the actual population

0.05

$z\sigma / \sqrt{n}$

μ

critical value

Figure 11.1. Illustration of the terms and meaning of statistical power. Any population mean drawn from the hatched area may be (incorrectly) interpreted as confirming the null hypothesis. The proportion of the area under the curve of the alternative (true) hypothesis (the actual population) from which samples may be drawn that fail to lead to a rejection of the null hypothesis represents the chance of making a type II error. The power of the test is 1 − the type II error rate.

to detect any level greater than 8.8 (2.5 units above the threshold of 6.3). This is the effect size we wish to detect.

An environmental audit takes 20 samples of stream turbidity at a station. The standard deviation of this sample is 5.4. How likely it is that we can detect environmental changes that result in elevated turbidities that exceed the critical effect size of 8.8 units?

Assume initially that there has been no increase in turbidity. The probability that the mean of 20 measurements will exceed the background level $+z \times$ standard error of the mean is:

$$p\left(\overline{x} > 6.3 + 1.645 \frac{5.4}{\sqrt{20}}\right) = p(\overline{x} > 8.28) = 0.05.$$

Thus, there is a 5% chance that the mean of 20 measurements will exceed 8.28, even if there has been no change and the true mean is 6.3.

The value of 8.28 may become the critical value (the decision criterion) for a test. If we want to know if there has been any change in turbidity, the mean of any sample of 20 that exceeds 8.28 will be interpreted as significantly greater than the background of 6.3 when the standard deviation is 5.4.

The power of the test asks, 'What is the chance of observing a value greater than 8.28 (our decision criterion), given that the true state of the environment is that turbidity levels have a mean of 8.8?'. The power is then the chance of concluding, correctly, that the threshold of 6.3 has been exceeded:

$$\text{power} = \left(\bar{x} > 8.28 \,|\, \bar{x} \sim N\left(8.8, \frac{5.4}{\sqrt{20}}\right)\right)$$

$$= p\left(z > \frac{8.28 - 8.8}{5.4/\sqrt{20}}\right)$$

$$= p(z > -0.43), \text{ which from a table of areas}$$
$$\text{of the normal curve,}$$

$$= 1 - 0.3336$$
$$= 0.6664$$
$$= 67\%.$$

The result of this example says that, even in the case where the subject of the test exceeds the threshold by 2.5 units, the chance that we will catch them out, given 20 samples, is only about 67%.

With 10 samples, the chance is about 43%. The relationship between number of samples and the power of a test may be plotted on a graph (Figure 11.2). This kind of plot is known as a power curve, and it displays graphically the chance of successfully identifying a breach of regulations for the problem at hand. It can be used to decide if a particular monitoring strategy is effective, and how much more money and effort would be required to make the monitoring programme sufficiently reliable.

The three vertical lines on Figure 11.2 represent the number of samples needed if one wishes to be 80%, 90% and 95% certain of concluding correctly that there is a difference, when the true mean turbidity is 8.8 (i.e. the effect size is 2.5). Approximately 30, 40 and 50 samples, respectively, are required to achieve these goals. To be 99% certain of detecting a difference, under conditions in which the true mean is 8.8, more than 70 samples would be required.

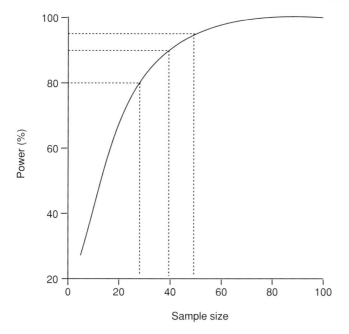

Figure 11.2. A power plot for the example in which we wish to detect deviations in water turbidity of 2.5 above the regulatory threshold, with a specified degree of reliability, given a standard deviation of the samples of 5.4.

We may use plots such as this in an iterative fashion, to explore the costs of detecting different effect sizes with different degrees of reliability. Budgetary constraints often are traded against the requirement for reliability. Power curves provide a means of making such trade-offs explicit and quantitative.

To avoid false prosecutions, we may conclude that a system is 'in compliance' if a proportion, β, of values are less than the threshold, T, with probability γ. That is, conformity is demonstrated if:

$$\overline{X} \le T - ks,$$

where s is the sample standard deviation and k depends on n (Table 11.3).

11.2.3 Comparing differences among means

The problem of comparing the mean of a set of observations with a standard may be generalized to the problem of finding the sample size required to detect a difference among two or more means. In circumstances

Table 11.3. *Values for* k *in calculating the compliance of a sample mean,* X, *given a requirement that a proportion of values,* β, *should be less than a threshold,* T, *with probability* γ *(after ANZECC / ARMCANZ 2000)*

	$\gamma = 0.95$		$\gamma = 0.90$		$\gamma = 0.50$	
n	$\beta = 0.95$	$\beta = 0.90$	$\beta = 0.95$	$\beta = 0.90$	$\beta = 0.95$	$\beta = 0.90$
2	22.26	20.58	13.09	10.25	2.34	1.78
3	7.66	6.16	5.31	4.26	1.94	1.50
4	5.14	4.16	3.96	3.19	1.83	1.42
5	4.20	3.41	3.40	2.74	1.78	1.38
6	3.71	3.01	3.09	2.49	1.75	1.36
7	3.40	2.76	2.89	2.33	1.73	1.35
8	3.19	2.58	2.75	2.22	1.72	1.34
9	3.03	2.45	2.65	2.13	1.71	1.33
10	2.91	2.36	2.57	2.07	1.70	1.32
15	2.57	2.07	2.33	1.87	1.68	1.31
20	2.40	1.93	2.21	1.77	1.67	1.30
30	2.22	1.78	2.08	1.67	1.66	1.29

in which there is a set of controls and a set of treatments, each of which may be replicated, it is important to know the chance of detecting an overall difference among the means of the samples.

One approach to this problem assumes the following things are known:

- The standard deviation (note that all populations are assumed to have the same underlying variability and to be approximately normally distributed).
- The number of groups in the planned experiment (>1).
- The smallest true difference that you desire to detect.

Sokal and Rohlf (1995) outlined a method for circumstances in which there are two groups. This method answers the question, 'How many samples are required such that there will be a probability, p, that the observed difference will be found to be statistically significant at the α level (the type I error rate)?'. The formula gives the number of replicates per sample such that the difference will be detected, with the reliability specified, between the means of any pair of samples. The formula is

$$n \geq 2 \left(\frac{\sigma}{\delta} \right)^2 \{ t_{\alpha[v]} + t_{2(1-\text{power})[v]} \}^2,$$

where:

n is the number of replications per sample.
σ is the true common standard deviation, estimated prior to the analysis.
δ is the smallest true difference it is desired to detect (note that it is necessary only to know the ratio of σ to δ).
υ is the degrees of freedom of the sample standard deviation; for example, with 2 samples and n replications per sample, $\upsilon = 2(n-1)$.
α is the significance level (the type I error rate).
Power is the desired probability that a difference will be found to be significant, if it is as small as δ.
$t_{\alpha[\upsilon]}$ and $t_{2(1-\text{power})[\upsilon]}$ are values from a 2-tailed t-table with υ degrees of freedom and corresponding probabilities α and (1-power).

These formulae answer the question, 'How large a sample is needed in each of the two groups so that the probability of a 'false alarm', an unnecessarily cautious approach, is no more than α, and the probability of detecting a real difference of magnitude δ among group means (the power) is at least $1-\beta$?'.

The following example is based on Sokal and Rohlf (1995). The co-efficient of variation of a variable is 6%. The plan is to conduct a t-test, comparing this variable in two populations. The question of power may be stated as, 'How many measurements are required from each population to be 80% certain of detecting a true 5% difference between the two means, at the 1% level of significance?'.

The equation above must be solved iteratively. This process begins by making a guess at n, the number of replicates required to achieve the required power. We try an initial value of 20. Then $\upsilon = 2(20-1) = 38$.

Since CV $= 6\%$, $s = 6\bar{x}/100$ (remembering that CV $= 100s/\bar{x}$). The problem, as stated above, specifies that we wish δ to be 5% of the mean, that is, $\delta = 5\bar{x}/100$. Thus, the ratio σ/δ becomes $(6\bar{x}/100)/(5\bar{x}/100) = 6/5$.

This example makes it clear that it is necessary to know only the ratio of σ to δ, not their actual values. Using these values, we have:

$$n \geq 2\left(\frac{6}{5}\right)^2 \{t_{0.01[38]} + t_{2(1-0.8)[38]}\}^2$$

$$n \geq 2\left(\frac{6}{5}\right)^2 \{2.713 + 0.853\}^2$$

$$= 36.6.$$

Our first guess for n was 20, which is not larger than 36.6. The inequality above is false, so we have to try again. This time, we use 37 as our initial guess. This gives $v = 2(37 - 1) = 72$, and:

$$n \geq 2 \left(\frac{6}{5}\right)^2 \{2.648 + 0.847\}^2 = 35.2.$$

The convergence between the left and right hand sides of the equation is close. We can assume that for the required power (to be 80% certain of detecting a true difference of 5%), we would require about 35 replicates per population.

In using this equation to explore the power of alternative designs, we could associate a monetary cost with the number of replicates and the number of populations sampled. We could attribute an environmental cost to various true differences between the populations.

Mapstone (1995) recommended using a ratio of α to $(1 - \text{power})$ that reflects the relative costs of the two kinds of errors. In the context of environmental risk assessment, this recommendation seems eminently sensible.

Marvier (2002) provided an interesting application of this formula. Some crops have enhanced insecticidal properties, a result of either conventional breeding or genetic modification. One of the concerns for environmental managers is the potential for harm to nontarget species.

Marvier (2002) focused on tests conducted on crops in which genes from *Bacillus thuringiensis* were inserted, causing the crops to produce *Bt* toxin. In the USA, the EPA is responsible for determining the environmental risks posed by a crop and for recommending a minimum number of replicates in trials. Petitions for deregulation of a genetically modified crop are accompanied by experimental evidence of the magnitude of the effect of the crops on nontarget insect species and, as of January 2001, the US Department of Agriculture had approved 15 such petitions.

Testing involves exposing nontarget organisms to high concentrations of *Bt* toxin, usually 10–100 times the concentration that is lethal to 50% of the target organisms. If no statistically significant effect is detected, it is assumed that the chance of effects at the more realistic lower doses is acceptably small.

The number of replicates specified by the EPA does not take into account the variation within samples. Marvier (2002) found for the five studies where sufficient data were available that variation and sample sizes were such that only one study had a 90% chance of detecting a

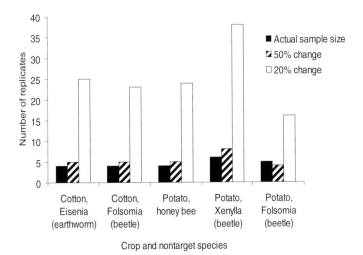

Figure 11.3. Actual and required sample sizes for studies of *Bt* toxicity of transgenic crops. The values give the sample sizes required to be 90% certain of detecting a 50% change and a 20% change in nontarget species (after Marvier 2002). Test 1, change in weight; Test 2, survival; Test 3, number emerged; Test 4, progeny production; Test 5, progeny production.

50% change in the nontarget species, and none of the studies had a 90% chance of detecting a 20% change (Figure 11.3).

Changes of this magnitude are likely to be ecologically important and yet the studies submitted in support of deregulation had a substantial chance of failing to detect them. The tests lacked power. The implicit conclusion that there was no effect was incorrect.

The only reliable way of establishing the ecological consequences would be to use a population model (such as the population viability models in Chapter 10) to establish the magnitude of change in these parameters that would result in important biological changes in the populations. Then, equations such as those above could be used to calculate the number of replicates necessary to detect important changes.

11.2.4 When is something absent?

Surveys are used to establish the presence or absence of important features within a prescribed area (such as the presence of rare or invasive species, or the detection of a disease or a contaminant). However, it is never possible to be absolutely sure that something is absent unless the sample takes in all of the statistical population.

For example, it was reported in *The Melbourne Times* (14 February 2001):

Mr Preston (the construction union health and safety boss) is at a loss to explain how the hazardous asbestos came to be at a site which was investigated by an environmental auditor and ... pronounced free of asbestos and a low contamination risk. But disturbingly, environmental scientist David Raymond says audits are simply 'indicative' and give 'an approximation of what's on a site'.

Environmental scientists in such circumstances need to communicate how reliably they can assert that a site is free of a contaminant. The reliability of methods to detect the presence or absence of an attribute depends on the spatial and temporal allocation of sampling units, the sampling effort at a site and the detectability of the event or attribute.

McArdle (1990) defined f to be the probability that an attribute (a species, a disease, an indicator) would appear in a single, randomly selected sampling unit. This number includes the probability that the attribute will be present in a sample, and the probability that it will be detected, given that it is present. The probability with which the attribute will be detected in a sample of size n is:

$$p = 1 - (1 - f)^n.$$

For example, if an attribute on average turns up in 1 sample in 50 ($f = 0.02$), the chance that it would be found at least once in 20 samples is only 0.33. If it is necessary to determine what level of scarcity could be detected with a given likelihood, then we may use:

$$f = 1 - (1 - p)^{1/n}.$$

This gives the upper $100 \times p$ confidence interval (one-sided) for the proportion, if n samples are taken without detecting the species or attribute of interest. For example, a total of 20 samples will detect, with a probability α of 0.9, an attribute of scarcity, f, of 0.1.

For the design of surveys, it is often useful to know the number of samples required to achieve a given level of effectiveness. To estimate the number of samples to detect an attribute with a level of scarcity f, at least once with a probability p,

$$n = \frac{\log(1 - p)}{\log(1 - f)}.$$

Thus, to detect species of scarcity $f = 0.02$ with a probability of 0.9 would require more than 100 samples.

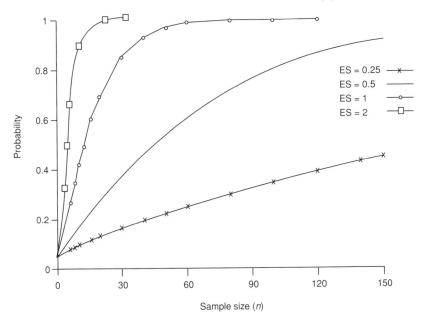

Figure 11.4. The probability of detecting a species (or other attribute) in a sample of size n. The parameter f is the probability that the attribute will appear in a single, randomly selected sampling unit.

For example, Figure 11.4 answers the question, 'How much survey effort is required to be reasonably certain that a particular area does not contain a given species?'. The approach is independent of the sampling strategy used to establish the presence or absence of a species at a location. For species that are relatively rare in samples (in fewer than about 1 in 50 samples), considerably more than 100 samples are required to be reasonably certain (with a probability of more than 95%) of detecting these species at least once, in places where they occur.

Jones (1997) calculated the probability that habitat patches suitable for the lizard *Aprasia parapulchella* were inhabited. The species is cryptic and fossorial in sites dominated by native grasses (particularly *Themeda* spp.). Its diet consists only of the eggs and larvae of ants and it may be located only by lifting rocks. Jones (1997) used individual rocks as her sample unit and found the species to be present at a density of 1 in 250 rocks at two known locations. How many rocks need to be lifted before we are more than 95% certain that the species is absent from a patch of potential habitat?

$$n = \ln(1 - 0.95)/\ln(1 - 0.004)$$
$$\sim 750.$$

The study involved sampling 855 and 8063 rocks at two sites of potential habitat. In both cases, no animals were found, and the study concluded that the species was absent from these locations or, if present, then at densities significantly less than 1 in 250 rocks. The same equation has been used to estimate the disease rate in feral animal populations (Hone and Pech 1990), the number of patches that should be sampled to be certain of detecting species of forest birds (Robbins et al. 1989), the number of times it is necessary to visit a site to be sure that a particular frog species is absent (Parris et al. 1999) and the likelihood that species have become extinct (see Solow 1993, Solow and Roberts 2003, Reed 1996).

McArdle (1990) pointed out that these formulae assume that the value of f is constant in all sampling units (true for any random sample) and that the sampling units are independent. In the case of the lizard, for instance, the probability of detection may be assumed to be independent of the search, and samples may be assumed to be independent of each other because the animal is sedentary and not territorial. Erdfelder et al. (1996) provide convenient software for power calculations.

11.3   Flawed inference and the precautionary principle

Psychology interacts with the structure of null hypothesis significance tests, leading people to make irrational inferences. There is a yawning gap between the ways in which statistical tests are described in textbooks and the ways in which the tests are routinely used and interpreted. As noted above, an important and quite common error in environmental risk assessment is to take the lack of a significant result as evidence that the null hypothesis is true. For instance, p values are often interpreted as a measure of confirmation of the null hypothesis, something that neither Neyman and Pearson nor Fisher would have agreed with (Mayo 1996, Johnson 1999b).

For example, Crawley et al. (2001) conducted a 10-year-long study, planting four different genetically modified, herbicide-resistant or pest-resistant crops (oilseed rape, corn, sugar beet and potato) and their conventional counterparts in 12 different habitats. Within four years, all plots had died out naturally. One plot of potatoes survived the tenth year, but that was a conventional potato plot.

Goklany (2001, p. 44) interpreted this result as follows, 'In other words, GM plants were no more invasive or persistent in the wild than their conventional counterparts . . . The study confirms that such GM plants do not have a competitive advantage in a natural system unless that system is treated with the herbicide in question.' Goklany (2001) was convinced that there is no risk because no difference was observed between the treatment and the control.

The same flawed logic extends beyond null hypothesis tests. In another example, the US Department of Energy evaluated the suitability of the Yucca Mountains in Nevada to store high-level nuclear wastes for 10 000 years. They argued that, '. . . no mechanisms have been identified whereby the expected tectonic processes or events could lead to unacceptable radionucleotide releases. Therefore . . . the evidence does not support a finding that the site is not likely to meet the qualifying condition for postclosure tectonics.' (DOE 1986, in Shrader-Frechette 1996b). Shrader-Frechette (1996b) criticized the inadequacies of the decision framework, which implied that if a site cannot be proved unsuitable, scientists assume it is suitable.

It is hard to make intuitive estimates of adequate sample size reliably. Consequently, the power of most tests is too low. Because power typically is low, null hypothesis tests applied to environmental monitoring imply it is important to avoid declaring an impact when there is none, but less important to avoid declaring that an activity is benign when there is an impact. Thresholds for statistical significance usually are unrelated to biologically important thresholds. Poor survey, monitoring and testing procedures reduce apparent impacts.

These observations are not new (see Johnson 1999b, Anderson *et al.* 2000). Some summary judgements on null hypothesis significance tests include (all made by eminent statisticians, in Johnson 1999b):

'. . . no longer a sound or fruitful basis for statistical investigation'
'. . . essential mindlessness in the conduct of research'
'. . . the reason students have problems understanding hypothesis tests is that they may be trying to think'
'. . . significance testing should be eliminated; it is not only useless, it is also harmful . . . '
'In practice, of course, tests of significance are not taken seriously'
'. . . hypothesis testing does not tell us what we want to know . . . out of desperation, we nevertheless believe that it does'

Power calculations can be considerably more complicated than those above. Many standard statistical packages do not offer them although Monte Carlo simulation can be used to calculate the power of arbitrarily complex null hypothesis tests, so there is really no excuse. The omission is simply the result of the preoccupation of mainstream science with type I error rates. Type II error rates are forgotten. However, they matter to

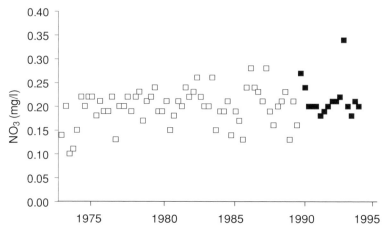

Figure 11.5. Time series plot of nitrate concentration at USGS benchmark station on Upper Three Runs Creek, Aiken South Carolina (after Dixon 1998). Closed squares represent measurements taken after the station was moved in October 1989.

environmental risk analysts because errors are costly. It is the responsibility of the analyst to be aware of the chance of false inference, even if the machinery is unavailable.

Monitoring is usually the first element of the risk control cycle to be jettisoned when resources are constrained. Yet it plays two essential roles. It provides a means of validating model predictions and sensitivities. And it provides new knowledge that may be used to revise assumptions, ideas, decisions, model structures and parameter estimates. Monitoring programmes should be designed with both objectives in mind.

11.3.1 The precautionary principle

In applications such as the examples in the previous section, scientists have lost sight of type II errors, blinkered by the constraints of null hypothesis testing and urged on by pathological overconfidence. Such mistakes are common (Johnson 1999b).

For example, a standard approach to interpreting the data in Figure 11.5 would be to fit a trend line and examine the statistical significance of the coefficient. A simple linear regression fitted to the data for the period from 1973 (when measurements commenced) to October 1989 results in a line with a slope of 0.000417 mg/l (per time step between measurements). However, the null hypothesis for the statistical test of the slope is that the true slope is zero. The analysis results in a p value of 0.124, a nonsignificant result. The classical inference is that we have insufficient evidence (for a threshold of $\alpha = 0.05$) to reject the null hypothesis that the slope is zero.

Common sense suggests that there might be a true trend but the test was not powerful enough to detect it. How long do we have to wait before enough evidence accumulates that we can confidently conclude there is a trend in nitrates in the creek? Waiting until 1995 would have been enough. The full data set, including the solid squares, gives a statistically significant slope coefficient, although we might worry about the effect of moving the station. If nitrates are increasing, the cost is that we would have waited four or five years before acting.

But even if scientists have become confused, the general public has been savvy. Stakeholders have argued that even if 'statistically significant' changes in the environment have not been detected, environmental degradation may still occur. We may need to act before 'scientific certainty' (a statistically significant result) is available.

The term 'precautionary principle' was coined by German bureaucrats in 1965 (see Cross 1996), although it almost certainly has earlier roots (Goklany 2001). It appears in various forms in numerous national and international agreements and treaties. The most common definition is from Principle 15 of the Rio Declaration (UN 1992:10):

Where there are threats of serious or irreversible damage, lack of full scientific certainty shall not be used as a reason for postponing cost-effective measures to prevent environmental degradation.

The principle has been used as a basis for broad-reaching policy goals by international nongovernmental organizations (Myers 1993, Harremoes et al. 2001). Irreversibility should lead to more conservative and flexible decisions, a rationale that is long established in economics (Arrow and Fischer 1974, Gollier et al. 2000). Yet the principle has been repudiated as an unnecessary evil by others. It is viewed by some as a tool of technological scepticism, casting a too-pessimistic light on the inventiveness and adaptability of human technology to solve problems (Brunton 1995, Cross 1996, Goklany 2001; see Deville and Harding 1997).

For instance, Anderson (1998a) suggested, in a talk entitled 'Caution: precautionary principle at work' that

... recently, a sinister and irrational variation of this principle has also evolved, which effectively reverses the burden of proof... This has particular impact on the mining industry, which is by definition a risky business – given geological, political and market uncertainties, the role of a mining executive has more to do with risk management than digging holes.

Brunton (1995) argued that the principle would be used to justify nonexistent links between human actions and environmental effects based on spurious correlations. It would jeopardize the development of resilient management systems, ignoring the potential benefits of development on the environment.

The precautionary principle is an antidote to the scientific myopia of null hypothesis significance testing that results in failures to detect important environmental impacts. Goklany (2001) argued that propositions examined in the light of the precautionary principle should weigh the risks they may reduce against those they may create. Routine, effective application of statistical power tests and consensus on effects sizes is one way to operationalize it.

However, there is a long and unsuccessful record of people trying to correct the slanted vision of scientists (Anderson *et al.* 2000). Few environmental studies consider statistical power in the design of sampling and monitoring programmes or report confidence intervals (Peterman 1990, Fairweather 1991, Taylor and Gerrodette 1993, Underwood 1997). Statistical power can be a difficult concept to teach or to explain to an audience of stakeholders.

Below, we explore other tools that avoid some of the problems of null hypothesis tests, that provide information to update models and test assumptions, and to estimate and trade risks of different kinds.

11.4 Overcoming cognitive fallacies: confidence intervals and detectable effect sizes

Routine measurements are used to determine whether air emissions are safe, stream turbidities at road crossings are acceptable, the numbers of retained trees in harvested areas meet agreed standards, or a species should be classified as vulnerable. In all these circumstances, inferences are made from samples.

The discussion above describes a few well-documented fallacies associated with null hypothesis testing. Empirical studies in cognitive psychology have demonstrated that these fallacies are widespread and deeply engrained (e.g. Tversky and Kahneman 1971, Oakes 1986; see Chapters 1 and 4). As outlined above, a common and serious error is to interpret a statistically nonsignificant result as 'no impact', that is to treat the p value as a measure of *confirmation* of the null hypothesis. In fact, probabilistic statements of this kind (e.g. the p value of a traditional null hypothesis test) belong to tests, not hypotheses. They tell us how often, when a particular hypothesis is true, our test will lead to an error. This section outlines some tools that help overcome the cognitive fallacies.

There are two potential outcomes of a traditional null hypothesis significance test: 'reject H_0' or 'do not reject H_0' (Table 11.1). If the statistical power of the test is low, the chance of committing a type II error may be unacceptably large. This is frequently the case when environmental risks are evaluated. In fact, there are several possible outcomes:

a. There is evidence of an important trend or effect ('reject H_0').
b. There is evidence of an effect but it is unimportant ('reject H_0').
c. The evidence is equivocal; there is no evidence of an effect but the study was unlikely to find an effect even if there was one ('do not reject H_0').
d. There is no sufficient evidence of an effect ('do not reject H_0').
e. There is evidence of no effect or, rather, if there is an effect, it is small ('do not reject H_0').

Conventional interpretations of statistical data confuse these possibilities ('fallacies of acceptance and fallacies of rejection', Mayo 1985; see also Shrader-Frechette and McCoy 1992). These interpretations do not distinguish a statistically powerful test from a test that is not powerful enough to detect a real, potentially important effect.

Confidence intervals make the possibilities clearer by presenting them graphically (Tukey 1991, Cumming and Finch 2004). They provide information about the size of the effect, not just its presence or absence. Furthermore, the width of a confidence interval is a measure of precision; a wide interval indicates a lot of uncertainty.

Consider an example in which treatments to control invasive weeds are expected to increase the numbers of a threatened plant. Managers are aware that the treatments might also be detrimental. According to planning guidelines, a 20% increase in abundance would be considered a success.

Confidence intervals and tests are based on a given statistic (such as the mean), with assumptions about its distribution and the independence of samples. If the 95% confidence interval for the estimate of the change in the population of the threatened plant includes the null value (no change in population size) and also the biologically important threshold

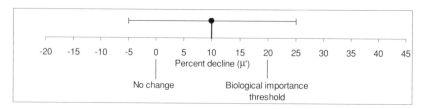

Figure 11.6. Percentage change in the abundance of threatened species. This provides an example of an uninformative test of the null hypothesis (H_0: that there was no change in population size). There is evidence that the true increase does not exceed 25% and that there was less than a 5% decrease in population size.

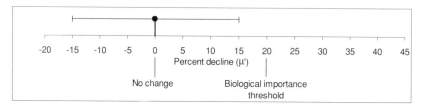

Figure 11.7. There is evidence that the population did not decline at a rate greater than 15%, or increase at a rate more than 15%. There is evidence that the rate of change was not acceptable.

(a 20% increase), more data are needed before an informed decision can be made (Figure 11.6 illustrates such a result). The confidence interval draws our attention to the fact that the data are consistent with there being a biologically important effect. However, a traditional hypothesis testing interpretation would lead us to accept the null hypothesis, because the null value is also within the interval; it would clearly be erroneous to regard the data as evidence that there has been no increase.

In many instances, a one-sided test for positive discrepancies from 0 is appropriate – the threshold of interest only concerns increases (in decline rate or chemical concentration, for instance). A one-sided interval corresponding to a one-sided test would give a lower bound. A supplementary principle such as power (or severity, see Mayo 1996, 2003) is necessary to justify looking at upper bounds to avoid fallacies of acceptance.

Consider a different outcome (Figure 11.7) in which the 95% confidence interval included the null hypothesis but *excluded* the values reflecting important change. Here, there is evidence that the rate of increase in the size of the threatened population is unacceptably small. In fact, the two-sided test would reject the null hypothesis (at $\alpha = 0.05$) only if the observed increase was about 15%.

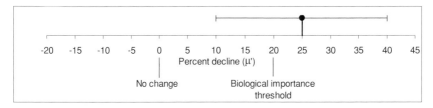

Figure 11.8. There is evidence that the true rate of increase was at least 10%, although less than 40%. Given a target rate of increase of 20% or greater, there is evidence the treatment was successful.

The third possibility (Figure 11.8) is that the 95% confidence interval includes the target value and excludes the null hypothesis. In this case, we may infer that the treatment was successful.

Confidence intervals provide a visual interpretation of precision. Three-valued frames have been used for some time together with confidence intervals to try to improve the interpretation of evidence in medicine (e.g. Berry 1986, Tukey 1991). They should assist scientists and others to distinguish the decision of setting the criteria for 'important' effects from the scientific data gathered from the study. For example, Dixon (1998) recommended an alternative approach to the definition of no observed effect concentrations (NOECs, Chapter 7). The analyst specifies an 'equivalence' region, the discrepancy between the null case and a response that may be considered unimportant or tolerable. The analyst should conclude that two groups (such as a treatment and a control) are equivalent if there is a sufficiently high probability that the true difference lies within the equivalence region.

However, one of the drawbacks of confidence intervals is that they display only central tendency and limits. Often, in practice, all values within the intervals are assigned equal plausibility. A confidence interval licenses us to exclude values greater than the upper bound and less than the lower bound. However, it does not alert us that, for example, we have reasonable evidence that μ is less than values slightly below the upper bound.

However, if a testing perspective is adopted, a $(1 - \alpha)$ confidence interval can be regarded as the set of null hypotheses that would not be rejected with an α level test. Mayo and Spanos (2004) described the 'detectable effect size' principle: if data x are not statistically significantly different from H_0, and the power to detect effect d is high, then x constitutes good evidence that the actual effect is no greater than d. Conversely, if data x are not statistically significantly different from H_0, and the power to detect effect d is low, then x constitutes poor evidence that the actual effect is no greater than d.

Looking at our previous example, the most we can say with confidence intervals is that we are not sure (Figure 11.6), the effect is insufficient (Figure 11.7) or the treatment is successful (Figure 11.8). But, in the case of Figure 11.7, what is the evidence for the proposition that the decline is less than 10%? What is the evidence that it is greater than 15% or greater than 20%? The confidence interval alone cannot give us answers to these questions.

Mayo (1996) introduced the idea of making post-test inferences that make fuller use of the data than is possible with confidence intervals. The idea is to distinguish effects (or effect sizes) that are and are not warranted from the results of tests. In the case of 'accept H_0', i.e. a statistically nonsignificant result, this combats the fallacy of acceptance by indicating the upper bounds that are (and are not) licensed; in the case of reject H_0, it helps to combat fallacies of rejection by indicating which discrepancies from the null are and are not indicated (Mayo 1996, 2003).

11.5   Control charts and statistical process control

Statistical process control was created to improve the stability and operation of production systems. It was invented by Shewhart (1931) in Bell National Laboratories in the 1920s and 1930s. It was forgotten for a while, and then picked up by an engineer, W. E. Deming, who applied it in Japanese automobile and audio/video production (Thompson and Koronacki 2002). There are few applications in environmental monitoring although it has potential to be useful.

Systems influenced only by chance variation are said to be in 'statistical control'. If properties of the system can be attributed to 'assignable causes', the convention is that patterns should be explored and the explanations verified and perhaps corrected by intervention.

Statistical process control has a few essential components (Thompson and Koronacki 2002):

- Flowcharting of the production process (constructing a conceptual model).
- Random sampling over time at numerous stages of the process.
- Identification of nonrandom characteristics in the measurements.
- Identification and removal of causes of unwanted change.

It acknowledges that stochasticity is unavoidable. However, if samples from the system display nonrandom behaviour, then the system's manager should explain them. The kind of management response may be calibrated against the importance of different kinds of patterns. Thus, statistical process control differentiates between the deterministic processes that may push a system towards an unwanted state, and the stochastic processes that result from natural variation. Both kinds of processes may be managed.

11.5.1   Planning

Statistical process plans are developed with the intention of improving management of the system. The first step is to identify what can be measured that will provide useful information about the system, a process analogous to the identification of measurement endpoints and indicators (Chapter 3). The plans include specification of:

- The aspects of the process it is intended to improve.
- What can be measured and how it is to be measured.
- Details of data collection: time of sampling, by whom, what qualifications and experience are necessary in the data collection team, how data collection protocols will be standardized, how measurement error will be calibrated, the number of replications, use of composite samples, observations, and so on.

In industrial applications, the overriding motivations are to 'delight the customer', manage by fact, manage both people and processes, and create circumstances that lead to continuous improvement (Kanji 2000). Industry motivations are the same as those of environmental management. Customers are stakeholders. The statistical process control 'cycle' has all of the important attributes of the risk management cycle: explicit recognition of uncertainty and of the importance of stakeholders, planning, modelling, implementation, monitoring, acting on monitoring outcomes to improve management and iteration for further improvement.

A process is 'in control' when most of the endpoints in a monitoring programme fall within specified bounds or 'control limits'. 'Shewhart' control charts use multiples of the standard deviation to set control limits. The threshold values for triggering a test failure may be adjusted to suit circumstances. The method assumes the process under observation generates values that are independent, normally distributed and with a

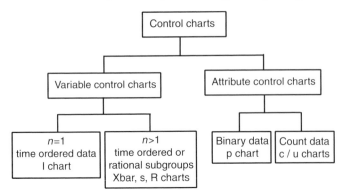

Figure 11.9. A subset of control chart applications. If the effect size is small relative to background variation, then most authors recommend using CUSUM or EWMA charts (after Montgomery 2001). Variable control charts are applied to continuous variables. Attribute control charts are applied to binary or discrete variables.

constant mean and variance,

$$x_i = m + \varepsilon,$$

where ε is drawn from a normal distribution with mean zero and standard deviation s. Control limits may be specified as a product of the magnitude of the natural variation in the system,

$$\text{control limits} = \mu \pm L\sigma,$$

where L is the distance of the limits from the centre line (the mean) in standard deviation units. The approach is flexible because it provides a huge range of potential numerical devices for detecting changes in the behaviour of a system, and for conditioning management responses tailored for specific kinds of processes.

Control charts are useful for tracking environmental parameters over time and detecting the presence of abnormal events or trends. Variable control charts plot statistics from measurement data, such as salinity. Attribute control charts plot count data, such as the number of times a criterion was exceeded or the number of noncompliances in a set of samples (Figure 11.9).

ANZECC/ARMCANZ (2000) highlighted the potential for application of these methods in environmental risk assessment when it commented:

They are particularly relevant to water quality monitoring and assessment. Regulatory agencies are . . . recognising that, in monitoring, the data generated from

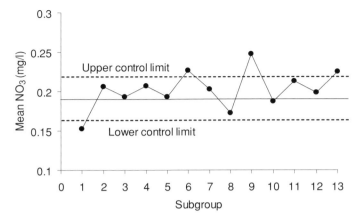

Figure 11.10. Essential features of an Xbar control chart (after Gilbert 1987, Thompson and Koronacki 2002). The statistic is the average nitrate concentration in Upper Three Runs Creek (Dixon 1998). For the sake of the example, each group of six points is assumed to be sampled simultaneously rather than consecutively, forming 'rational subgroups' (see below). Each point is the mean of six samples shown in Figure 11.5. The control limits are two times the standard error (0.012 mg/l) of the mean (0.19 mg/l) based on the first 24 individual samples.

environmental sampling are inherently 'noisy'. The data's occasional excursion beyond a notional guideline value may be a chance occurrence or may indicate a potential problem. This is precisely the situation that control charts target. They not only provide a visual display of an evolving process, but also offer 'early warning' of a shift in the process level (mean) or dispersion (variability).

11.5.2 Xbar charts

An Xbar chart is a control chart of subgroup means. Figure 11.10 shows the essential features of a control chart for means. It tracks the mean response of a system and detects the presence of unusual events.

The Xbar chart requires an estimate of process variation, usually the pooled standard deviation s, from a set of samples. Historical data or theoretical understanding of the system may also provide a basis for an estimate. The context may suggest that values that exceed ±2 standard deviations are important and should be investigated. The control limits in Figure 11.10 were drawn ±2 standard errors either side of a long-term average. Each sample (subgroup) was composed of six individual samples.

Subgroup means are plotted. This strategy takes advantage of the central limit theorem, so that, irrespective of the kind of distribution

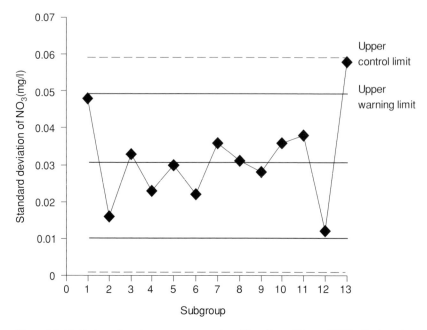

Figure 11.11. S chart for nitrate concentration (data from Dixon 1998). Each point is the subgroup standard deviation for six consecutive measurements (Figure 11.5). The centre line was constructed from the average of the first four sets of measurements. The control limits were set using scaling factors in Table 11.5.

from which the samples are drawn, the subgroup averages will have approximately normal distributions. The approximation to the normal will be determined by the size of the subgroups, and how different the underlying distributions are from the normal.

Warning limits may be placed inside the control limits initiating different kinds of actions. Actions may be tiered, reflecting the seriousness of deviations.

The control limits may be updated as new data are added to historical data, generating a revised estimate of the standard errors for the subgroup samples. They may vary seasonally, with sample size or with location.

11.5.3 R and S charts

An R chart is a control chart of subgroup ranges. An S chart is a control chart of subgroup standard deviations (Figure 11.11). They are used to monitor process variation. If processes become more variable, subgroup standard deviations and ranges will increase. They may also detect

Table 11.4. *Form of data for Xbar, R and S charts*

Subgroup	Number of measurements in the subgroup	Average	Range	Standard deviation
1	n_1	$\overline{x}_1$	R_1	s_1
2	n_2	$\overline{x}_2$	R_2	s_2
. . .	. . .	. . .	. . .	. . .
k	n_k	$\overline{x}_k$	R_k	s_k

abnormal events that affect process variation. Usually, R charts are used to track process variation for samples of size 5 or less, while S charts are used for larger samples. In the example in Figure 11.11, managers are not concerned about downward trends. Only the upper control and warning limits are labelled.

11.5.4 Rational subgroups

Points may lie outside control limits because the mean of the process drifts, or because it becomes more variable. Shewhart (1931) termed the subgroups selected for display 'rational subgroups'. The term is intended to emphasize that the choice should generate averages that are sensitive to the kinds of processes the monitoring system is intended to reflect. Subgroups should minimize measurement error and the effects of other sources of variation that might cloud the detection of trends or changes in the nature or magnitude of variation.

Montgomery (2001) defined rational subgroups as those that maximize the effects of 'assignable causes' between groups and minimize their effects within groups. The selection of subgroups, allocation of sampling effort, setting response thresholds and specification of management actions should be sensitive to deviations from a simple random process when unacceptable 'assignable' causes operate.

An example of a rational subgroup for air quality monitoring might be five air quality measurements collected once per week at a particular point in a city, by a given instrument, over a given period, at a given time of the day and week. For the water samples above, subgroups may be made up of samples collected at locations in a single reach of the river, in late summer when water levels are at their lowest ebb. If collection conditions are made as uniform as possible, extraneous variation within subgroups is minimized, maximizing the potential to detect trends and changes.

Table 11.5. *Factors for computation of control limits in Xbar and S charts.*
These limits generate thresholds equivalent to 3 standard deviations (3 sigma
limits), accounting for subgroup sample size. These values are appropriate when
about 25 subgroups are plotted (see Montgomery 2001, p. 761, for formulae
and additional values). The factor labels (A, B and D) are conventions in process
control literature

n	A_2	A_3	B_3	B_4	D_3	D_4
2	1.880	2.659	0	3.267	0	3.267
3	1.023	1.954	0	2.568	0	2.575
4	0.729	1.628	0	2.266	0	2.282
5	0.577	1.427	0	2.089	0	2.115
6	0.483	1.287	0.030	1.970	0	2.004
7	0.419	1.182	0.118	1.882	0.076	1.924
8	0.373	1.099	0.185	1.815	0.136	1.864
9	0.337	1.032	0.239	1.761	0.184	1.816
10	0.308	0.975	0.284	1.716	0.223	1.777
12	0.266	0.886	0.354	1.646	0.283	1.717
15	0.223	0.789	0.428	1.572	0.347	1.653
25	0.153	0.606	0.565	1.435	0.459	1.541

11.5.5 Control chart parameters

The data for the control charts outlined above generally have a form like
that in Table 11.4.

The Xbar centre line is the weighted mean of the observations,

$$\bar{\bar{x}} = \frac{\sum n_i \bar{x}_i}{\sum n_i}.$$

Control limits for Xbar charts and the centre lines and control limits for S
charts and R charts are given by various formulae that account for sample
sizes in the subgroups (see Gilbert 1987, Montgomery 2001, Hart and
Hart 2002, for details).

The upper and lower control limits for Xbar charts are

$$\bar{\bar{x}} \pm A_3 \bar{s},$$

where A_3 is from Table 11.5 and s is the standard deviation. An alternative
construction is

$$\bar{\bar{x}} \pm A_2 \bar{R},$$

based on the notion that, in small samples, the range provides a good
approximation of the sample standard deviation (Montgomery 2001).

The upper and lower control limits for S charts are $\bar{s}\,B_4$ and $\bar{s}\,B_3$. The upper and lower control limits for R charts are $\overline{R}D_4$ and $\overline{R}D_3$. When substantially fewer or more than about 25 groups are plotted, control limits need to be further adjusted to control the experiment-wise error rate (see Montgomery 2001, Hart and Hart 2002, Thompson and Koronacki 2002).

Often Xbar, S and R charts are plotted underneath one another so that the analyst can look for covariation in fluctuations. For instance, if Xbar and S charts vary in phase, it is likely that the underlying data are skewed to the right, generating correlations between the mean and the variance. Transformation may be in order. Or it may mean that attributes of the system explain the covariation and may be monitored independently or eliminated.

11.5.6   p-charts

p-charts monitor the relative frequency of binary events (such as alive / dead, flawed / not flawed, present / absent). The centre line, p, is the long-run mean relative frequency (the frequency of one of the conditions divided by the frequency of both conditions in the reference set). The standard deviation is simply the standard deviation of the binomial distribution. Control chart limits are created by adding and subtracting the standard deviation from the mean:

$$\overline{p} \pm 3s_p = \overline{p} \pm 3\sqrt{\frac{\overline{p}(1-\overline{p})}{n}},$$

where $\overline{p}$ is the mean relative frequency of the event and n is the subgroup size. A disadvantage is that points may be left or right skewed, depending on whether p is greater or less than 0.5, unless subgroup size exceeds $4/p$ (Hart and Hart 2002; see Montgomery 2001).

11.5.7   u- and c-charts

u- and c-charts monitor count data in time-ordered or subgroup data. Count data are assumed to be Poisson distributed and, therefore, right skewed. c-charts show raw count data and assume the population from which they are drawn does not vary in size from time to time, or from subgroup to subgroup.

u-charts are used when subgroup size varies, and counts are standardized by dividing each count by the relevant subgroup size. The centre line, $\overline{u}$, is the long-run mean (standardized) count. To create a control chart, usually a multiple of the standard deviation is added to and subtracted from the mean:

$$\bar{u} \pm 3s_u = \bar{u} \pm 3\sqrt{\frac{\bar{u}}{n}}.$$

11.5.8 CUSUM and EWMA Charts

CUSUM and EWMA charts make use of sequential information and are more sensitive to correlated processes and small changes than are the other kinds of charts described so far.

Cumulative sum (CUSUM) charts accumulate deviations above the target in one statistic (C^+) and deviations below the target in another (C^-). There are numerous forms of CUSUM charts (Thompson and Koronacki 2002) including kinds that apply when the data are binary or Poisson (Reynolds and Stoumbos 2000). One of the most common forms for continuous variables is:

$$C_i^+ = \max[0, x_i - (\mu + K) + C_{i-1}^+]$$
$$C_i^- = \max[0, (\mu - K) - x_i + C_{i-1}^-]$$

where μ is a target value for the process mean (Montgomery 2001). Starting values for C^+ and C^- usually are 0. K is usually selected to lie halfway between the target μ and a value for the process that would be considered out of control (a value we are interested in detecting quickly).

Thresholds are chosen and if either C^+ or C^- exceeds them, the process is considered to be out of control. Limits often are selected to be five times the process standard deviation (Box and Luceno 1997, Montgomery 2001), although they may be adjusted to be sensitive to the costs of different kinds of errors.

An EWMA chart is a chart of exponentially weighted moving averages, defined as:

$$z_i = \lambda x_i + (1 - \lambda)z_{i-1},$$

where λ is a value between 0 and 1 and the starting value z_0 is the process target or the average of preliminary data (μ). Thus, each point is simply a weighted average of all previous sample means and λ is the discount rate applied to earlier observations. The rate may be adjusted so that (squared) deviations between fitted values of z and observed values of z are minimized (Box and Luceno 1997).

EWMA charts can be tailored to detect any size shift in a process. Because of this, they are often used to monitor processes to detect small shifts

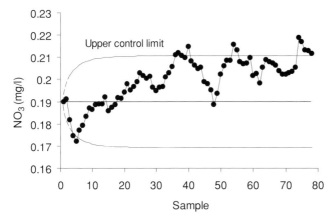

Figure 11.12. EWMA chart for nitrates in Upper Three Runs Creek (Dixon 1998). The plot used the data in Figure 11.5. The reference value (the centre line) is the mean (0.19 mg/l) based on the first 24 samples. The calculations used a value for L of 3, a value for λ of 0.1, and the standard deviation of the first 24 samples (0.03 mg/l).

away from a target. They are insensitive to assumptions about normality and are therefore ideal to use with individual observations (Montgomery 2001). The usual way to set control limits is:

$$\mu \pm Ls \sqrt{\frac{\lambda}{(2-\lambda)}[1-(1-\lambda)^{2i}]}.$$

The plot points can be based on either subgroup means or individual observations, where L is a weight defined by the user. Exponentially weighted moving averages are formed from subgroup means. By default, the process standard deviation, s, is estimated from a pooled standard deviation. In a broad range of applications, values of λ between about 0.05 and 0.4, and a value of L of 3 provide useful starting points (Box and Luceno 1997, Montgomery 2001; e.g. Figure 11.12).

The trend in the data is clear, and there are warnings that action may be warranted as early as halfway through the sampling period. The choice of parameters (L and λ) should be adjusted to account for the relative costs of false-positive and false-negative warnings. The value of s may be updated at each time step.

Like CUSUM charts, EWMA charts can be applied when data are Poisson or binary. Like CUSUM charts, they react sensitively to small shifts in process behaviour but do not react to large shifts as quickly as the

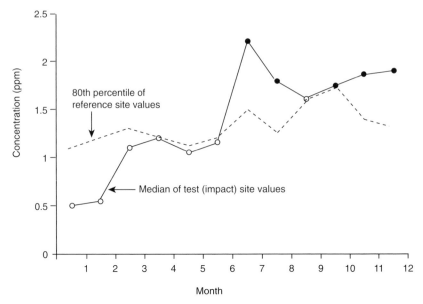

Figure 11.13. Test/reference (control/impact) chart for concentration of heavy metals in an estuary. The 80th percentile of the reference site provides an upper control limit for the monitoring programme. Values from the test site that exceed the control limit (filled circles) trigger detailed investigations of the sources of the pollutant. At the reference site, the 24 most recent monthly observations are ranked from lowest to highest. The reference value is the 80th percentile, calculated as the mean of the 19th and 20th observations (from ANZECC/ARMCANZ 2000, Figure 6.7).

more traditional Shewhart control charts. If subgroups are used instead of individual observations, then x_i is replaced by $\overline{x}_i$, and $s_x = s/\sqrt{n}$. Montgomery (2001) gives formulae for EWMA charts for Poisson data and other applications.

11.5.9 Test/reference (control/impact) charts

In many circumstances, ecological systems are not stationary, but change in response to such things as disturbance and successional processes. It is nevertheless important to gauge the response of a system to management actions. ANZECC/ARMCANZ (2000) devised a control chart for triggering action through comparison with a control area (Figure 11.13).

Conceptually, the method makes use of the control/impact scenarios used routinely in environmental impact assessment studies. The system also makes use of medians and quantiles, to avoid the difficulties that may

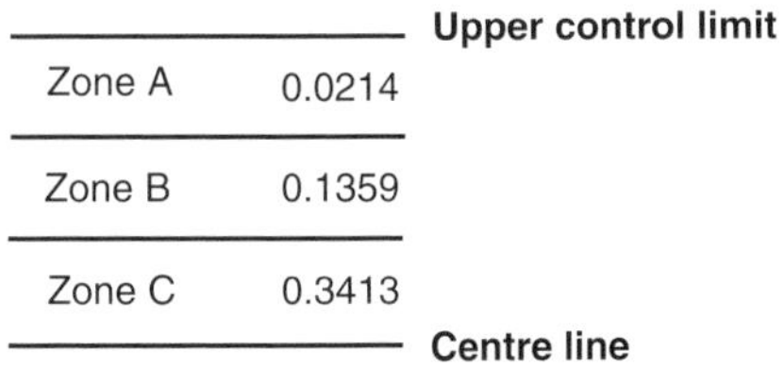

		Upper control limit
Zone A	0.0214	
Zone B	0.1359	
Zone C	0.3413	Centre line

Figure 11.14. Zones and the probability of a point being in the zone, given normally distributed data (after Hart and Hart 2002).

arise from non-normal data. Action is required when the median of the test site exceeds the 80th percentile of the reference site. Values close to the threshold (from months 4, 5 and 6, for instance) may trigger a tiered response, so that action in the form of investigations about the source would have begun before the concentrations substantially exceeded the control limits.

11.5.10 Pattern response and decision thresholds

Setting thresholds for decisions is a balancing act between type I and type II errors. Too frequent, unnecessary action is expensive and counterproductive. Too infrequent action leads to unacceptable environmental damage. To some extent, the choice of thresholds involves social decisions about the merits and costs of action and inaction.

Experience will also shape decisions about thresholds and modes of response. We may elect to use a 90th quantile, or a 70th, depending on how past events have been dealt with and are perceived by the broader community. Other tiers of action may be invoked to better balance the tradeoffs between intervention and inaction. The strength of control charts lies in their flexibility and adaptability to changing circumstances.

Triggering management actions by exceeding a threshold is just one of many options for interpreting control charts. More sensitive decision rules may be based on the frequency with which points have particular attributes. For normally distributed Xbar charts, Nelson (1984) suggested dividing the region around the centre line into three zones (A, B and C; Figure 11.14) so that interpretation could follow general rules such as:

a. One point beyond zone A.
b. Nine points in a row in zone C or beyond.
c. Six points in a row increasing or decreasing.
d. Fourteen points in a row alternating up and down.

e. Two out of three points in a row in zone A or beyond.
f. Four out of five points in zone B or beyond.
g. Fifteen points in a row in zone C (above and below the centre line).
h. Eight points in a row on both sides of the centre line with none in zone C.

Of course, these decision rules may be generalized with:

- K points in a row on the same side of the centre line.
- K points in a row, all increasing or all decreasing,
- K out of K − one point in a row more than one standard deviation from the centre line,
- K out of $K + n$ points in a row decreasing or increasing, and so on.

The number of possibilities is almost limitless. This flexibility allows the monitoring programme to be designed to be sensitive to specific attributes. If anticipated changes generate unique signals that affect either the trend of the process or its stochastic properties, the monitoring strategy could anticipate them. These expectations may be derived from theories or from the model of the system.

Given a control limit or trigger value, the probability of getting r values worse than the limit from a total of n samples is:

$$p(r) = [n!/(n - r)!r!] p^r (1 - p)^{n-r}$$

To compute a p value, sum the probability of this event and those that are more extreme. The probability of at least one exceedence is then $1 - p$(zero exceedences) (ANZECC/ARMCANZ 2000).

11.5.11 Dependencies

Most control charts assume that samples or sets of samples in rational subgroups are independent. It may be that the characteristics of the system and the sampling strategy result in samples that are not independent. If the value of a variable, x, at time t depends in some way on the value of x at time $t - 1$, the variables will be autocorrelated. If x at time $t - 1$ is relatively high, the chances are that x at time t will be high.

Many ecological processes generate time-dependent correlations. For instance, take the case of a lake with volume V being fed by a stream with a flow rate f. Samples taken from the outflowing water may be correlated, even if the concentration flowing into the lake is a random variable. Assuming the lake is mixed, the correlation between samples from the outflowing stream will be:

$$r = e^{-\Delta t / T},$$

where Δt is the interval between samples and $T = V/f$ (Montgomery and Mastrangelo 2000). The correlations will generate patterns that will elicit a response in normal Shewhart control charts.

When data are autocorrelated, the alternatives are to use an appropriate time series model and apply control charts to the residuals, or use an EWMA chart with an appropriate choice of λ. Box and Luceno (1997) and Montgomery and Mastrangelo (2000) provided details and examples. However, care must be taken to establish that the trend eliminated from the data is an artefact of sampling or uninteresting physical characteristics of the system, and is not an important trend from an assignable cause that warrants action.

11.5.12   Power and operating characteristic (OC) curves

The cost of employing a large number of triggers is that each one contributes to the chance of a type II error. If there are k rules, each with a type I error rate α, then the false alarm probability is:

$$\alpha = 1 - \prod_{i=1}^{k} (1 - \alpha_i).$$

If false alarms are expensive or damaging to corporate reputation or public confidence compared to failures to act, a regulator may set limits so that a response is elicited only when the evidence for nonrandom behaviour is relatively strong. The values for 'K' above may be adjusted upwards, and the limits may be extended further from the mean. Alternatively, if failures to act are expensive or unacceptably damaging, compared to false alarms, then 'K' may be reduced or the control limits contracted.

Montgomery (2001) recommended adaptive (variable) sampling to combat the problem of type I and type II errors. In adaptive sampling, the sample size or the interval between samples depends on the position of the current sample. The relative costs of 'false alarms' versus 'failure to act' should contribute to the allocation of sampling effort such that warning thresholds equalize the relative costs of the two kinds of errors.

The average run length (ARL) for in-control processes is defined as the average number of samples between points that results in false alarms. The average interval is:

$$ARL_0 = \frac{1}{\alpha}.$$

Characteristics such as the average number of samples to a signal (following a change in the process) are used to design limits and other triggers (see Reynolds and Stoumbos 2000). The ARL for out-of-control processes is defined as the average number of samples between true alarms, the average delay before an alarm is raised. It is:

$$ARL_1 = \frac{1}{1 - \beta}.$$

Average run lengths may not be the best measure of the performance capabilities of a control chart because they are geometrically distributed and therefore skewed. Errors in estimates of process parameters lead to overestimates of ARLs (Montgomery 2001). Despite these drawbacks, ARLs illustrate the potential for manipulating decision thresholds to give acceptable decision errors.

The ability of Xbar and R charts to detect shifts in process quality is described by their operating characteristic curves (OC curves). In the simplest case of an Xbar chart with a known standard deviation, s, the probability of failing to detect a shift of ks in the mean of the first sample following the shift is (Montgomery 2001):

$$\beta = \Phi(L - k\sqrt{n}) - \Phi(-L - k\sqrt{n}),$$

where Φ is the cumulative standard normal distribution, k is the size of the shift in standard deviation units, and n is the subsample size. If the chart has $3 - \sigma$ limits ($L = 3$) and the sample size if 5, then the chance of failing to detect a shift equal to $2s$ is:

$$\begin{aligned} \beta &= \Phi(3 - 2\sqrt{5}) - \Phi(-3 - 2\sqrt{5}) \\ &= \Phi(-1.47) - \Phi(-7.47) \\ &= 0.0708. \end{aligned}$$

The chance of failing to detect a $2s$ shift in the mean of the process is about 7%. Plots of β versus k are called operating characteristic curves (Figure 11.15).

Many of the difficulties that stem from hypothesis tests and the interpretation of numerical thresholds such as $\alpha = 0.05$ in monitoring

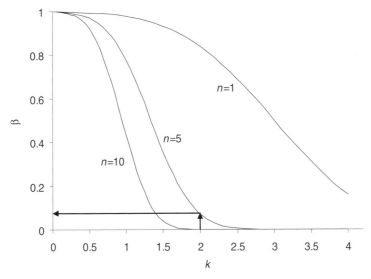

Figure 11.15. Operating characteristic curves for the Xbar control chart with 3*s* control limits, and three sample sizes (after Montgomery 2001). The value of β for $k = 2$ and $n = 5$ given in the text is shown by arrows on the figure.

contexts may be circumvented by adopting control charts for interpreting evidence and making decisions.

A decision-maker may condition control limits to make them sensitive to the relative magnitudes of the costs of type I and type II errors (for methods, see Montgomery 2001). The response parameters may be adjusted as more is learned about the system, creating opportunities to develop more sensitive and better-conditioned rules, so that false–positive and false–negative error rates are reduced over time.

11.6 Receiver operating characteristic (ROC) curves

Receiver operating characteristic (ROC) curves are used to judge the effectiveness of predictions for repeated binary decisions (act / don't act, present / absent, diseased / healthy, impact / no impact). They are a special case of operating characteristic curves in which a binary decision is judged against a threshold for a continuous variable (an indicator). They are built around confusion matrices that summarize the frequencies of false and true positive and negative predictions, for various values of a prediction threshold.

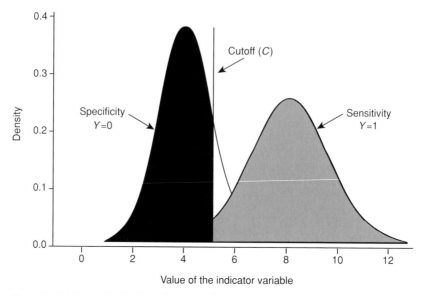

Figure 11.16. Hypothetical distributions of indicator values, *X*, when $Y = 0$ (left) and when $Y = 1$ (right). Where the value of the indicator is less than *C*, we conclude there is no impact (for instance, that the contaminant is absent). Where the value is greater, we conclude there is an impact (after Murtaugh 1996).

11.6.1 Confusion matrices

When we select a continuous indicator variable, *X*, its value may depend on a binary response variable, *Y*. For instance, Murtaugh (1996) used counts of diatoms (a continuous variable) to indicate the presence of a contaminant in water because, in the conceptual model, diatom abundance depended on the contaminant. Diatoms are sometimes present, usually in lower numbers, even when the contaminant is absent. Therefore, we define a critical threshold, *c*, above which we assume the contaminant is present, and below which we assume it is absent.

Diatom abundance has two statistical distributions, one in the absence of the contaminant and one in its presence. Figure 11.16 shows statistical distributions of an indicator variable (*X*) in unstressed and stressed environments (denoted by $Y = 0$ and $Y = 1$, respectively) and *C* is the threshold (cutoff) for the variable.

Binary predictions may be either true or false, giving the two-by-two classification table called a confusion matrix (Table 11.6). Its structure is identical to the structure of inferences from null hypothesis tests (Table 11.1).

Table 11.6. *Confusion matrix: classification of true and false prediction rates*

True situation	Indicator (prediction)	
	Positive	Negative
Positive	True positive (TP)	False negative (FN)
Negative	False positive (FP)	True negative (TN)

Table 11.7. *Predicted and actual ecological status (weed and nonweed) for 980 plant species introduced onto the Australian continent (after Smith et al. 1999b)*

True outcome	Predicted outcome	
	Weed	Nonweed
Weed	17	3
Nonweed	147	833

Figure 11.16 illustrates that error rates depend on the choice of the threshold. If it is more costly to mistakenly predict that there is an impact than to mistakenly predict that there is none, then we might move the cutoff, C, to the right, increasing the frequency with which we conclude there is no impact. Thus, the decision threshold should be conditioned by the cost of a wrong decision.

The following statistics summarize the performance of the indicator:

- Number correctly classified $= TP + TN$.
- True positive fraction (Sensitivity) $= TP/(TP + FN) =$ proportion of outcomes correctly predicted positive.
- True negative fraction (Specificity) $= TN/(FP + TN) =$ proportion of outcomes correctly predicted negative.
- False positive fraction $= FP/(FP + TN)$.
- False negative fraction $= FN/(TP + FN)$.

Smith *et al.* (1999b; Lonsdale and Smith 2001) gave an example in which introduced plant species were assessed (retrospectively) as being weeds or nonweeds. They defined a weed as an invasive species that causes significant damage to agricultural or natural ecosystems (Table 11.7).

Table 11.8. *Raw data for algal bloom occurrence and phosphorus concentration in a lake*

Indicator [P]	Response (algal bloom)
0.11	Absent
0.15	Absent
0.33	Absent
0.41	Absent
0.44	Absent
0.45	Present
0.48	Present
0.52	Absent
0.61	Present
0.86	Present

Their forecast was based on a point-scoring system that took into account each species' life history, reproduction, dispersal potential, toxicity and environmental tolerance in its home range and in other places where it had been introduced (Pheloung 1995).

Smith *et al.* (1999b) defined the 'accuracy' of the method to be the total number of weeds rejected divided by the total number of weeds assessed ($17/20 = 85\%$). They estimated the average cost of introducing a new weed species to be 3×10^6, and the worth of a useful species (discounted by the uncertainty that a purposefully introduced species will be useful) to be 2×10^5, making the cost of allowing in a weed 15 times greater than the opportunity cost of disallowing a useful plant (see Lindblad 2001 for another example).

11.6.2 ROC curves

The values in a confusion matrix and the associated measures of sensitivity and specificity depend on the cutoff, C. We could move the value of C from small values to large ones, and watch how specificity and sensitivity change. The plot of these values is an ROC curve.

Consider an example in which we attempt to predict the occurrence of an algal bloom from the value of a related indicator, the phosphorus concentration in the water. We have data from previous observations (Table 11.8).

A threshold of 0.50 gives:

		Indicator		
		+	−	
Response	+	2	2	True positive fraction = 2/4
	−	1	5	True negative fraction = 5/6

If we use a decision threshold of 0.45, then the classification table would be:

		Indicator		
		+	−	
Response	+	4	0	True positive fraction = 4/4
	−	1	5	True negative fraction = 5/6

The lower decision threshold improves our ability to predict positive outcomes (a bloom), without compromising our ability to predict negative outcomes (no bloom).

A receiver operating characteristic (ROC) curve is a plot of sensitivity versus specificity or (more traditionally) sensitivity versus 1 − specificity, for all values of C. The area under the ROC curve summarizes the overall accuracy of the indicator. It is expected to be 0.5 for a noninformative indicator, and 1 for a perfect indicator.

Figure 11.17 is a map of the potential habitat of a plant species. The model for its distribution was generated by linking the probability of occurrence of the species (presence / absence) to a set of predictor variables using logistic regression. The predictor variables are spatially distributed attributes such as aspect, radiation and temperature. Predictions of presence can be made at all points in the landscape. A prediction of occurrence was made for each point on Figure 11.17 and, subsequently, data were collected to validate the model's predictions (Elith 2000).

The rate of true positive and true negative predictions of the species' presence depends on the threshold chosen for making a judgement. As the value of the threshold increases from 0 to 1, the frequency with which presence is predicted falls. Figure 11.18 shows the ROC curve for Figure 11.17.

ROC curves can help to assess the utility of a continuous variable for predicting a binary outcome. The area under the curve summarizes performance over all decision thresholds. The shape of the curve can be

Figure 11.17. Predictions of the presence of a plant (*Leptospermum*) in a southern region of eastern Australia. The model was a logistic regression giving the probability of occurrence of the species as a function of terrain, soil and climate variables (after Elith 2000). Darker areas are higher probabilities of occurrence.

used to select a threshold that minimizes the relative costs of false-positive and false-negative errors (Zweig and Campbell 1992).

Shine *et al.* (2003) explored the utility of different approaches to assessing toxicity of metals in marine sediments. Acute toxicity was defined as a concentration that killed more than 24% of test organisms in toxicity tests. The standard compliance threshold used by managers is a ratio of SEM/AVS of 1 (see Figure 11.19).

These bounds divide the response figure into four regions. The top right quadrant represents correct predictions of toxic effects (true positive predictions). The bottom left quadrant represents correct predictions of nontoxic effects (assuming less than 24% mortality is nontoxic; true negative predictions). The off-diagonal quadrants represent false predictions.

The ROC curve for these data (Figure 11.20) illustrates the performance of the indicator (SEM/AVS) in predicting toxicity. Shine *et al.*

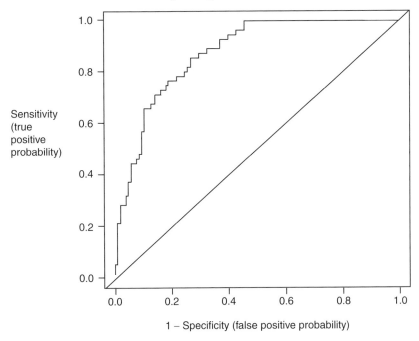

Figure 11.18. ROC curve for the predictive map of *Leptospermum* habitat in Figure 11.17 (after Elith 2000).

(2003) argued that thresholds commonly used by environmental managers such as SEM/AVS = 1, although yielding high sensitivity, come at the expense of low specificity. That is, the threshold does a good job of avoiding false negative predictions of toxic effects, but at the cost of a large number of false-positive predictions. They suggested that a threshold providing a more 'desirable' tradeoff between sensitivity and specificity would be higher than the commonly used threshold.

This interpretation relies on the definition of 'acute toxicity', a concentration that results in more than 24% mortality in test organisms. A threshold set at, say, 10% mortality would generate a different ROC curve. The desirability of a change in a threshold should take into account the ecological importance of an effect, and should make explicit account of the costs of decisions that lead to false-positive and false-negative outcomes.

Most countries import exotic plant species for ornamental purposes or for agricultural trials. Some of these species become weeds detrimental to agriculture or the environment. Various federal agencies worldwide

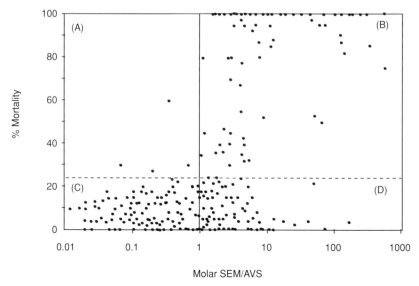

Figure 11.19. Percentage mortality of test organisms as a function of SEM/AVS (heavy metals in sediments, measured as the ratio of simultaneously extracted metals to acid-volatile sulphides; from Shine *et al.* 2003). Acute toxicity was defined as >24% mortality (horizontal dashed line). An SEM/AVS value of 1 (vertical line) is used commonly as a threshold to separate toxic from nontoxic samples. The four quadrants are areas where toxicity is correctly predicted (B), toxicity is incorrectly predicted (D), nontoxicity is correctly predicted (C), and nontoxicity is incorrectly predicted (A).

are responsible for decisions about which plants to allow and which to exclude. Allowing weedy species may result in substantial economic and environmental costs. Excluding valuable species may carry substantial economic opportunity costs. Table 11.7 summarized the success and failures recorded by a quarantine service that applied a scoring system to assess the risks posed by new plant species based on their ecological attributes (Smith *et al.* 1999b).

Hughes and Madden (2003) examined the data used to predict if 370 plant species would become weeds. Attributes were scored and the scores added into a weed risk assessment (WRA) score (Pheloung 1995, Pheloung *et al.* 1999, Lonsdale and Smith 2001) (Figure 11.21).

Outcomes were evaluated after introduction so that true-positive, false positive, true-negative and false-negative predictions could be compiled (Pheloung *et al.* 1999). Hughes and Madden (2003) used logistic regression to link the predictor variable (WRA score) to the outcome (weed

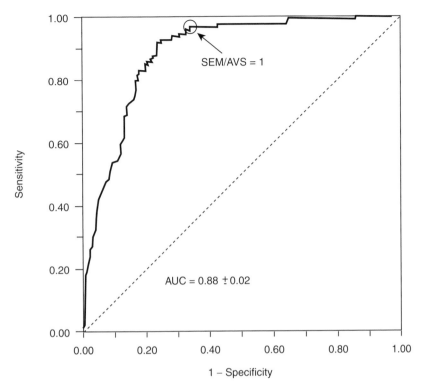

Figure 11.20. ROC curve for percentage mortality of test organisms as a function of heavy metal concentration in marine sediments (from Shine *et al.* 2003). The common threshold for distinguishing toxic from nontoxic sediments (SEM/AVS = 1) is shown. The area under the curve (AUC) measures the performance of the measure over a range of thresholds.

or nonweed) (Figure 11.22). A probability of 0.5 corresponds to a WRA score of about −1.

The people responsible for deciding whether to admit a new plant species use a value of the WRA score as a decision threshold. The use of a threshold is analogous to applying a diagnostic test and is guided by the need to balance the costs of false-positive and false-negative predictions.

Hughes and Madden (2003) used the logistic regression relationship to construct an ROC curve for predictions about the future weed status of a candidate species (Figure 11.23).

A threshold that falls towards the top right-hand corner of Figure 11.23 would reflect a regulatory policy that avoids false-positive predictions. Plant species will be prohibited unless there is strong evidence that they

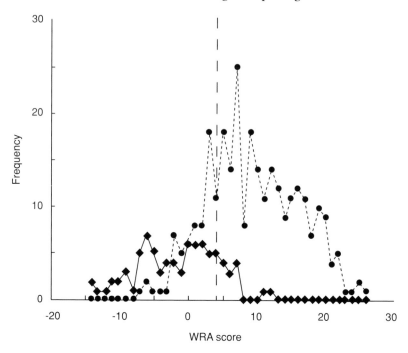

Figure 11.21. Frequency distribution of weed risk assessment (WRA) scores plotted separately for weeds (circles) and nonweeds (diamonds) (after Hughes and Madden 2003). The status of species (weeds or nonweeds) was determined retrospectively from data on invasive behaviour after introduction. The vertical dashed line represents a threshold score of 4 used to discriminate weeds from nonweeds before introduction.

are safe. A threshold in the extreme top right-hand corner of the plot would reflect a total embargo on the import of all new exotic plant species. A threshold towards the bottom left-hand corner of the plot would reflect a regulatory policy that avoids false-negative predictions. Plants will be allowed unless there is strong evidence that they will become weeds. A threshold in the extreme bottom left-hand corner would represent no control over imports (Hughes and Madden 2003).

As in the case of deciding a threshold for heavy metals, the choice should be conditioned by the relative costs of false positives and false negatives. Decision thresholds can be interpreted in terms of benefits and costs using cost-weighted probabilities. The utility (C_T) of a decision threshold is given by the costs of the four possible outcomes (Table 11.6),

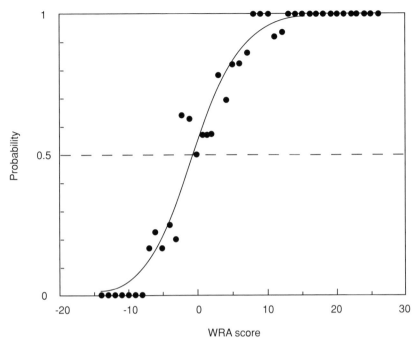

Figure 11.22. Logistic regression analysis for data linking ecological attributes to weed status, for 286 weed species and 84 nonweed species (after Hughes and Madden 2003). The curve shows the fitted relationship between probability (of being a weed, equal to the proportion of species in a class that were weeds) and the explanatory variable, their WRA score, based on a scoring system summarizing ecological attributes.

weighted by their probabilities of occurrence (Metz 1978):

$$C_T = (p_{TP} \times C_{TP}) + (p_{TN} \times C_{TN}) + (p_{FP} \times C_{FP}) + (p_{FN} \times C_{FN}).$$

where, for instance, p_{TP} is the true positive fraction, p_{FP} is the false positive fraction, p_{FN} is the false negative fraction and p_{TN} is the true negative fraction. The threshold may be adjusted and the costs recomputed to find the threshold that minimizes overall cost. This approach can be generalized to find the optimal decision point such that the overall costs of wrong decisions are minimized (see Mathews 1997, Smith *et al.* 1999b, Hughes and Madden 2003).

While most formulations of these methods require that the costs and benefits be accounted using the same currency (usually money), there

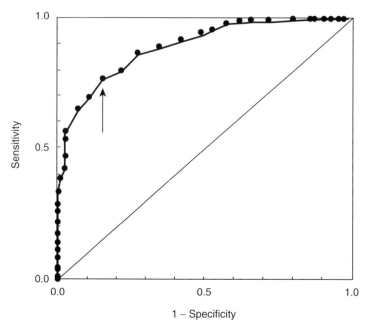

Figure 11.23. ROC curve for WRA scores (after Hughes and Madden 2003). The arrow indicates a threshold WRA score of 4.

is ample opportunity for decision-makers to weigh the relative worth of nonmonetary outcomes. Chapter 12 outlines some mechanisms for estimating and comparing nonmonetary values.

11.7 Discussion

Monitoring detects unacceptable ecological impacts and provides information that may improve conceptual models, forecasts and decisions.

The notion of effect size is embedded in all monitoring programmes and all decisions. How big an effect is it important to detect? Who cares if the population declines by X? At what point should we react, prosecute or invoke option X? How bad would it be to ban a harmless pesticide? How bad would it be to license a dangerous pesticide?

The methods above illustrate that it is possible to control the twin statistical errors: false positives and false negatives. For instance, ROC curves oblige the user to consider type I and type II errors simultaneously (Tables 11.1 and 11.6; Murtaugh 1996, Swets *et al.* 2000). However, the question remains of how to specify the sizes of the environmental mistakes that society is willing to tolerate, and therefore how to set the standards of proof by which the statistical tests are to be evaluated.

Setting a standard of proof is a device for both controlling error and assuring fairness. It requires an assessment of the priority of the risk in question. For example, people are generally more willing to risk environmental error that causes the deaths or damage of natural populations than to risk human lives (Fischhoff *et al.* 1982). A standard of proof cannot reasonably be set without an appreciation of its consequences for the perceived fairness of the decision-making process. Such judgements require legal, technical, social and ethical considerations.

Ideally, the relative costs and benefits of different actions could be converted into a common currency. Risk assessments could then provide a platform to summarize ecological processes, hazards, human activities and management options in an internally consistent form. If a change in an ecological parameter produced a change in risk of an unwanted event of a given magnitude, it would be possible to make a social judgement about the acceptability of that change. All that would then be required would be to specify the reliability with which a change of that magnitude should be detected (a question of power).

Benefit cost analysis suggests that policy debates may be translated into analytical, value-free, technical questions. The approach assumes rights can be balanced with expected utilities (Brown 1996). There are other positions (e.g. MacLean 1996) in which treating 'public trust' resources as commodities is viewed as subverting the responsibility of stewardship. For instance, Adams (1995) argued that 'Decisions about risk are essentially decisions about social priorities. . . . The degree and mode of compensation and behavioural adaptation exhibited by people can be explained, in large part, by their beliefs.' Chapter 12 explores methods that accommodate different ways of viewing the world.

Monitoring produces observations of trends and deviations from expectations. The expectations embody the best judgement about ecological processes and their sensitivities to human activities. When the monitored processes drift from expectations, or when the model's prediction fails to occur, the monitoring data provide the basis for revising conceptual models, re-estimating parameters, re-evaluating sensitivities, generating new predictions and re-designing the monitoring programme. In this fashion, the risk management cycle can be completed, resulting in an iterative process that will produce environmental decisions that are sensitive to the costs and benefits of false alarm and unjustified security.

12 · Decisions and risk management

Risk management makes use of the results and insights from risk assessment to manage the environment.

Morgan *et al.* (1996) argued that risk management requires negotiating human perception and evaluation processes of the kind outlined in Chapters 1 and 4 (Figure 12.1). It is easy to misinterpret Figure 12.1. It emphasizes that human perceptions overlie the physical interactions between human actions and environmental responses. It does not intend to suggest a kind of linear system in which risks are identified and analysed and in which, subsequently, the range of options are passed to an evaluative box in which social and political consequences of decisions are assessed.

In Morgan *et al.*'s (1996) view, risk analysis includes issues such as accountability and trustworthiness. Those bearing the risk should enter the picture before the risks have been identified and analysed.

This view of risk assessment accords with the risk management cycle (Chapter 3) that suggests the people bearing the risk should be involved from the outset in all stages, including hazard identification and model building. It is important to recognize that human perceptions and values affect experts and analysts as strongly as other participants.

12.1 Policy and risk

Policy makers are concerned with ensemble risks. They are obliged to consider all potential sources, including those not examined formally by risk analysts. In many circumstances, risks compete with one another for attention. Some may be political or intangible. Some may affect policy makers personally. For example, a consequence of risk aversion is that resources are allocated disproportionately to preventing any concentrated loss of human life. The newsworthiness of an event and the desire of officials to avoid being held to blame become important (Graham and Hammitt 1996).

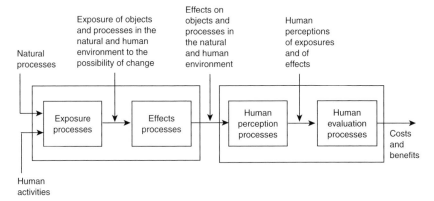

Figure 12.1. The interaction between physical and ecological processes and the way they are perceived and evaluated (from Morgan *et al.* 1996).

Policy makers decide among risk-reduction options. Ideally, options may be ranked in terms of net risk reduction, accounting for both target risks and competing risks. It may then be possible to compute the expected amount of risk reduction for each option, and rank decision options according to the 'mean' estimate of risk reduction. Expected value and 'rational choice' arguments suggest that options ranked by expected risk reduction will result in the maximum long-run risk reduction (Graham and Hammitt 1996).

If feedback from monitoring is timely and informative, then the ranks of decision options could reflect the degree of scientific certainty about net risk reduction and resource costs. Monitoring may provide useful information about the effects of management on risk. If there is an option to learn more, then actions about which there is greater uncertainty may become relatively more attractive.

Despite such seemingly fundamental tenets, policy decisions do not attempt to maximize net benefits. Risk, costs and benefits are highly uncertain. Often, communication is vague and ambiguous, even when measurable risks are thought to exist. Decision options that reduce catastrophic possibilities are sometimes upgraded because of public perceptions and outrage. Rational aversion to seemingly attractive gambles occurs when the consequences of the loss are not acceptable. Risks that are correlated among individuals or populations (spatially or temporally) may be viewed as more serious than similar, but uncorrelated risks. In circumstances such as these, policy makers appeal to the concepts of comparative risks and acceptable risks.

12.1.1 Comparative risks

Before decisions are made, hazards may be ranked by their relative risks, a process known as comparative risk ranking. It is used to compare risks from two similar sources (e.g. cancer from two pesticides), from dissimilar sources (e.g. accidents and benzene exposure) and the risk reduction from two options (e.g. use of chlorine or ozone to purify water). Comparative risk ranking is used to set regulatory and budgetary priorities. It may involve comparisons among large numbers of hazards before deciding strategies.

However, the task is complicated by the fact that different social segments and demographic groups are exposed to different risks (Silbergeld 1994). Contributing factors usually interact and there are limited data, further limiting the ability of analysts to evaluate risk distributions completely and equitably. For example, Freudenburg and Rursch (1994) documented an example in which people became less tolerant of a proposed hazardous waste incinerator after its risks were compared to the risks of smoking. The objective of the comparison was to show that the risks of the incinerator were relatively small. The additional information was overwhelmed by the change of context, which affected trust in the proponents and the government.

UDMH, a breakdown product of Alar used on apple crops, was considered to be less dangerous to humans than eating Aflatoxin found in peanut butter (see Chapter 6, Section 6.4.8). Finkel (1995) analysed the uncertainty in the judgement and revealed that one could be only 90% certain that the relative risk of Aflatoxin to UDMH lay somewhere between 300 : 1 in favour of Aflatoxin and 35 : 1 in favour of UDMH. According to Finkel, this meant that decision makers must balance 'the 5% chance that ignoring or delaying action on [UDMH] would erroneously leave unaddressed a problem 34 times greater than aflatoxin, against an equal chance that the opposite decision would focus attention on a problem 376 times smaller than aflatoxin...' (p. 383). Finkel concluded that either substance could be considered riskier, depending on what outcome the decision maker wanted to avoid: 'it is entirely a question of policy and values, not of science' whether this analysis, or even an analysis that showed a smaller probability of Alar being riskier than aflatoxin, 'could legitimately be reduced to the overconfident pronouncement that "peanut butter is riskier"' (Finkel 1995, p. 381).

People use a variety of techniques to estimate risks, making risk comparisons difficult. For instance, cost per life-year saved is sometimes used to help set risk reduction priorities (Table 12.1).

Table 12.1. *Cost effectiveness of life-saving interventions in the USA (after Paté-Cornell 1998)*

Intervention	Cost per life-year	Risk assessment method
Mandatory seat belt use and child restraint laws	$1 300	Current risk, subjective estimate of benefit
Chlorination of drinking water	$4 200	Before/after measures (existing statistics)
Smoke detectors in airplane lavatories	$30 000	Probabilistic analysis
Asbestos ban in pipeline wraps	$65 000	Plausible worst-case, dose–response
Vinyl chloride emission standards	$1 700 000	Plausible worst-case, dose–response
Sickle cell screening for newborns	$65 000 000	Current risk, subjective estimate of benefit

Some interventions are unlikely to provide the benefits indicated. For instance, risks from exposure to asbestos and vinyl chloride are conservative estimates by regulators resulting from worst-case bounds. In many such cases, the estimates are made in response to public perceptions of threat. Analysts have a vested interest in being conservative. There is a large, personal and direct cost to them if they underestimate the risk (the cost of a false negative is high). There are few costs to the analysts from overestimating the risk (to the analyst, the relative cost of a false positive is low).

12.1.2 'Real' and perceived risks

Policy makers have to deal with different advice from different risk analysts about the nature of risk and how best to deal with it. To some analysts, there is a sharp dichotomy between real risks (objective, analytical, evidenced based) and perceived risks (subjective, emotional, irrational). To other analysts, no such dichotomy exists.

The dichotomy is maintained mainly by people with technical and scientific training. It was reflected in the observation made by P. Sandman (in Watts, 1998),

the essence of environmental risk for companies is the hazard itself, which they define in technical terms, where you multiply the magnitude of the hazard by the frequency of the problem . . . For the public, risk is not a technical phenomenon at all. It is influenced by factors like fairness, trust and who has control.

For example, apparent disparities in spending to save lives can be traced to public perceptions of risk (Tengs *et al.* 1995). Epidemiologists and occupational safety analysts are often frustrated in their attempts to be 'efficient' by media reports that increase the visibility of particular hazards. In some cases, the political and social imperatives to act are insensitive to the 'real' risks.

Salvation from 'irrational' responses to perceived risk was sought through risk communication. However, experience over the last few decades shows that communication has been only partly successful in resolving disputes between scientists and engineers on the one hand, and the public on the other (Freudenburg 1996, Slovic 1999).

An alternative view is that risk assessments are inherently subjective and value laden because subjective judgements are used at every step (Adams 1995, Slovic 1999). Subjectivity enters through problem formulation, adoption of a definition of risk, selecting experts, creating conceptual models, deciding endpoints, estimating and extrapolating exposures, setting thresholds for risk classification, and so on.

Scientists, too, are subjective. They are encouraged to indulge their biases behind a mask of scientific authority, protected by pervasive linguistic and epistemic uncertainty and flawed conventions for inference.

For example, Meer (2001) documented the subjective judgements required of people involved in comparative risk ranking of toxic chemicals (Table 12.2). The table underscored that the central judgements in comparative risk assessment are to decide what types of evidence will be considered, and what weight will be given to different types of evidence. The key choice was between overprotection (accepting less certain evidence) and underprotection (accepting only more certain evidence). There are no professional standards to guide the analyst to the right position. It is a matter of personal values and personal exposure to wrong decisions.

Naïve interpretations of context may be misleading. Slovic (1999) gave the example that society may be best served by minimizing the number of worker deaths per tonne of coal produced, whereas a union representative is obliged to minimize the number of deaths per worker per year. All lives may be equal under the law, but the death of a young person elicits a stronger social reaction than the inadvertent death of an old person, suggesting people are sensitive to loss of life expectancy,

Table 12.2. *Some of the value judgements associated with the New York Comparative Risk Project, a comparative risk assessment of toxic chemicals (after Meer 2001)*

Source of uncertainty	Nature of value judgement required
Only a limited number of stressors can be screened and a small number selected for further evaluation	Choose what type and weight of evidence will qualify a stressor for screening; choose what type and weight of evidence will qualify a stressor for further evaluation (e.g. decide whether or not to consider evidence beyond quantitative data)
Data are lacking for the majority of chemicals used	Determine whether and how to consider the potential risk from unstudied chemicals
Data are lacking for noncancer risks	Determine what type of evidence regarding noncancer risk will be considered
Data are lacking on additive, synergistic or antagonistic effects	Determine whether the potential for such effects should be considered; determine what type of evidence will be considered
Data are lacking on low-dose and chronic effects	Determine whether the potential for low-dose and chronic effects should be considered; determine what type of evidence will be considered
Data are lacking on variations in human sensitivity, including the sensitivity of children	Determine whether to consider the potential for increased sensitivity in certain population groups, including children; determine what type of evidence of sensitivity will be considered
Direct human evidence of toxicity (i.e. clinical or epidemiological data) is not available for most chemicals	Determine whether animal data will be considered; determine the weight that will be given to different kinds of evidence for ranking purposes; for example, determine how to compare evidence based on human exposure to animal studies, and how to compare quantitative evidence to qualitative evidence
Data are lacking on cumulative exposures and pathways	Determine whether the potential for such exposures should be considered; determine what type of evidence will be considered and the weight evidence will be given in ranking
Uncertainty is magnified when toxicity data are compared to exposure data	Determine how to combine toxicity and exposure data into one overall rank for each stressor

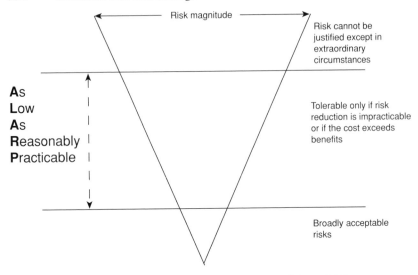

Risk magnitude

Risk cannot be justified except in extraordinary circumstances

As **Low A**s **Reasonably Practicable**

Tolerable only if risk reduction is impracticable or if the cost exceeds benefits

Broadly acceptable risks

Figure 12.2. The ALARP principle (after Stewart and Melchers 1997).

rather than simple loss of life (Johansson 2002). Thus, the importance of different kinds of fatalities depends on your point of view.

12.1.3 'As low as reasonably practicable': defining acceptable risks

Many standard treatments of risk assessment take the traditional view that real risks are measurable. Risk acceptance is achieved through trust and appeal to the authority of scientific principle. This notion is embodied in the idea that risks should be reduced to a level called 'as low as reasonably practicable', sometimes termed the ALARP principle (Figure 12.2).

There's nothing wrong with this principle as it stands. However, equitability depends on who defines terms such as 'tolerable' and 'reasonable'.

Acceptable risks have been defined operationally in a range of circumstances. For instance, the US Nuclear Regulatory Commission supported risks to people in the vicinity of nuclear power plants from nuclear accidents of < 0.1% of the sum of fatalities from all other accidents, and a risk of cancer fatality that was < 0.1% of the sum of cancer risks from all other sources (Morgan and Henrion 1990).

If a risk of a given magnitude is deemed acceptable in one domain, often the acceptability is extrapolated to another. However, there is no general standard for acceptable risk. If time and experience lead to public acceptance of a standard, it may become embedded in regulations and operating guidelines. But it would be unreasonable to assume that such standards can be identified in all technical and social circumstances (see Freudenburg 1996).

Acceptability is a social quality (Fischhoff 1994). It will always be influenced by context, visibility, trust and equitability. Slovic (1999) argued that because risks are socially constructed (see Adams 1995), whoever controls the definition of risk controls the 'rational' solution to the problem. The antidote he proposed is to increase public participation to make risk assessments more democratic. As a result, he argued, the quality of technical analysis would improve and the results would have increased legitimacy and public acceptance.

In a more inclusive paradigm for risk assessment, likelihood and consequence would be two of several factors. The others would include voluntariness, equity, catastrophic potential, novelty and control. None would be essential. Rather, the rules governing the conduct and outcomes of the risk assessment would be agreed by participants. The risk management cycle recommended in Chapter 3 puts these suggestions into operation.

12.2   Strategic decisions

Strategic risk management involves using risk assessment to determine organizational activities, the process of deciding what actions to take in response to a risk. It involves forecasting, setting priorities, formal decision making and reconciling viewpoints (Beer and Ziolkowski 1995). Typically, judgements consider environmental, social, economic and political factors. They involve determining the acceptability of damage that could result from an event, and what action, if any, should be taken (Suter 1995).

12.2.1   Decision criteria

The problem formulation stage of a risk assessment carries an implicit value judgement regarding the criteria to be used to discriminate good from bad decisions (Morgan and Henrion 1990). Utility-based criteria involve decisions based on the valuation of outcomes (e.g. Table 12.3). Rights-based criteria involve consideration of process, allowed actions and equity independent of benefits and costs.

When the values under consideration can be simplified to one or a few criteria, valuation and hybrid techniques are useful. For example,

Table 12.3. *Examples of decision criteria that might be employed in risk management (after Morgan and Henrion 1990, p. 26)*

Criterion	Description
Utility-based criteria	
Probabilistic benefit–cost	Estimate benefits and costs of alternatives in economic terms, and use expected value (weighted by risk) to find the option with the greatest net benefit
Maximize multi-attribute utility	In place of economic value, use a utility function that incorporates the outcomes in terms of all important attributes
Maximize / minimize chances of extreme outcomes	Minimize the chance of the worst outcome, or maximize the chance of the best outcome, usually dictated by the political or social context.
Rights-based criteria	
Zero risk	Eliminate risks entirely, irrespective of costs or benefits
Constrained risk	Constrain risk so that it does not exceed a specified threshold
Approval / compensation	Impose risks on only those parts of the population that have given consent, perhaps after compensation
Hybrid	Maximize probabilistic cost-benefit within the constraint of an upper bound on risk to an element of the system

Haight (1995) sought to maximize cost-benefit estimates for a species adversely affected by timber harvesting within a constraint of an upper bound on risk. He calculated optimal strategies for harvesting, given that the option should result in a 99% chance of achieving a population goal that gave a required standard of safety for resident species (an upper bound on acceptable extinction risk, Figure 12.3).

Haight (1995) measured population risk by the size of the population at the end of the harvesting period. The uncertainty in the projection of future populations meant that decisions on acceptable risk could be based on the mean expected population size (with a 50% chance that the outcome would be above or below this target), or on an assessment that gave a 99% chance of providing at least the degree of security wanted. The economic costs of these different attitudes to risk are plotted.

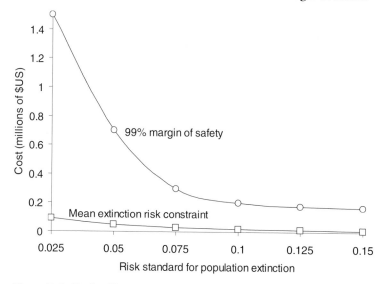

Figure 12.3. Tradeoff between risk to a threatened species and foregone timber revenue in a model with three patches of forest habitat (after Haight 1995).

In the cases above, the choices of decision criteria are social and ethical, rather than scientific. Safety costs money. There is no simple solution to the problem of whether to accept a 50% or a 1% chance of an unwanted outcome from timber harvesting (Figure 12.3). The acceptability of an institutional strategy dealing with risk will be affected by the same factors that affect personal attitudes to risk; namely, trust, control, visibility, equitability and so on.

12.2.2 Risk regulation

Environmental managers regulate the environmental effects of industry with two basic approaches: regulatory rules and agreements (applied before the impacts occur) and liability and damage penalties (applied after effects have occurred). Shavell (1987) pointed out that rules and agreements are preferable to penalties when industry is unable to pay for damage, if it is difficult to attribute responsibility, or industry does not anticipate the risks.

Bier (2004) argued that environmental regulation may be better couched as a game than a set of decisions. If monitoring is regular (so that a breach of compliance is likely to be detected) and penalties are stringently applied, the firm should weigh the costs of compliance against the penalties of noncompliance. If a firm's liability is limited by what it actually knows, the private value of information may be negative. It may be better to be ignorant of consequences, a rationale that Bier (2004) argued supported failures in the tobacco and asbestos industries.

Bier noted that regulators often establish cooperative relationships with the people they regulate. Penalties may be diluted by the probability of nondetection, especially if the frequency of monitoring is low and penalties are rare. But a firm may have an incentive to comply, even though the costs of compliance exceed the penalties of noncompliance. A more lenient, cooperative system saves the regulator the cost of monitoring. The firm saves in the long run through cost-of-compliance tradeoffs that allow it to use lower cost methods to achieve goals. Both sides avoid the cost of prosecution. Such systems work best if the regulator demonstrates a willingness to prosecute noncooperative firms and if the penalty for failing to disclose exceeds the penalty for violation (Bier 2004).

Viscusi (2000) provided an example of the negative private value of information. He surveyed the attitudes of almost 500 jury-eligible citizens towards cases involving risk and compensation. He found that if the group responsible for a hazard had completed a systematic analysis of risks and costs, and had decided the benefits of proceeding outweighed the risks, it triggered a bias against the defendant. Juries were likely to penalize a defendant for having undertaken a risk analysis, substantially increasing the chance of punitive damages. Furthermore, juries used the value placed on safety by a firm as an anchor when awarding punitive damages. Firms that valued safety more highly were more likely to experience higher damages.

The risk regulation environment will determine, to a large extent, the strategies available to individuals, firms and industries. Those who set the contexts for risk assessments should be aware that the criteria used to compare risks and evaluate the success of management strategies will mould outcomes, and are sensitive to personal values.

12.2.3 Where model-based assessments fit in

The results of risk assessments do not translate easily to policy and management decisions. I have argued that model-based risk assessments may provide results that are internally consistent, relatively transparent and free of linguistic ambiguity. They make an attempt to incorporate uncertainty plainly and, in doing so, oblige those who interpret the results to consider the possibility and consequences of wrong decisions.

The impediments to the use of model-based risk assessments include regulatory inertia, a lack of requisite skills and the perception that data requirements are heavy. There is also a perception that the model will be a black box that will tend to drive decisions. It can be difficult to dissent from the predictions of a model, particularly if the modeller is part of a kind of scientific priesthood and the model's assumptions are inaccessible.

Many models assume decision makers are 'rational' in the sense that they will act to maximize expected net social benefits, even though there is ample evidence to the contrary. Power and McCarty (1998) examined the ecological risk protocols adopted or recommended by agencies in the Netherlands, the US EPA, Australia, UK Department of the Environment, US National Research Council, Canadian Standards Association, and the US Risk Commission. They concluded:

Trends toward greater stakeholder involvement, decreased emphasis on quantitative characterization of risk and uncertainty, and development of iterative decision-based analysis . . . suggest a move toward embedding risk assessment within risk management . . . This is because technical analysis and command-and-control regulation have either failed to deal satisfactorily with environmental problems or, in suggesting solutions, have created conflict with other valued social objectives.'

This apparent dichotomy of ideas ('rationality' on one hand, 'subjective judgement' on the other) is a consequence of the level at which some risk assessments are conducted and their results viewed. Separating analysis from the communication phase erodes the credence they are given in a decision-making framework. Tools exist that may assist decision makers to deal with different values and preferences. Some of these are outlined below.

12.2.4   The advantages of deciding under uncertainty

When people are presented with a point estimate for the future, they are obliged to be risk-neutral. People rarely are. The consequence of dealing in point estimates is that the people involved in a decision will be more often dissatisfied with the outcome than they otherwise would have been.

Providing bounds on judgements creates new opportunities for making good decisions. Risk seeking or risk averse behaviour may be warranted by context (Morgan and Henrion 1990, Bernstein 1996).

Burgman et al. (2001) provided an example of a risk-based environmental tradeoff. Eight patches of habitat were proposed for development

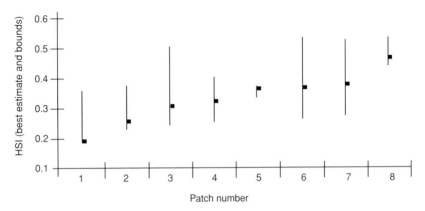

Figure 12.4. Habitat suitability index (HSI) estimates and bounds for eight patches of habitat for the Florida Scrub Jay (after Burgman *et al.* 2001).

within the range of a threatened species (the Florida Scrub Jay). The decision problem was, which patch? The developer was indifferent to the choice on other grounds, and so supported the selection of the patch that minimized environmental harm. Burgman *et al.* (2001) constructed a model based on expert judgement of the relationship between the bird and a range of environmental variables. These variables were mapped. The subjective uncertainties about relationships were included in the model. The suitability of each patch was calculated using interval arithmetic, resulting in a best estimate and bounds for each patch (Figure 12.4).

Attitude to the decision problem should affect the way the sites are ranked. If we protect one site, we may decide to choose the most valuable location. Patch 8 is the best, although it is possible that patches 3, 6 and 7 have higher suitability.

If the consequences of being wrong are catastrophic, our strategy should change. Assume, for instance, that these are the last remaining patches for the bird. If we get it wrong, we lose the species. Then the best strategy may be to rank the sites on the basis of the lower bounds of the intervals. We would select a site for which we are certain the habitat suitability should be at least this good. This is termed 'minimizing maximum regret' (French 1986, Morgan and Henrion 1990). Again, patch 8 has the highest lower bound, making it the best choice from the perspective of expected value and of avoiding unacceptable risk.

Now consider the strategy if we are asked to eliminate one site. Patch 1 has the lowest best estimate. But we may consider it to be more important to avoid losing high-quality habitat. We would then rank on the basis of the upper bounds of the intervals, and then choose the smallest upper bound. Under this view, patches 1 and 2 are about equal.

If we are risk averse, a patch that is known with greater certainty may be more valuable than another with higher expected value. For instance, if the decision problem involved choosing between patches 5, 6 and 7, we may choose patch 5, even though it has a slightly lower expected value. Its value is much more reliably estimated than the other two, both of which carry a possibility of being much worse (and much better) than we expect.

The full spectrum of decision possibilities is only available to us because we have taken the trouble to carry uncertainties through the chain of calculations that produce a habitat suitability index, and to present them in an accessible form.

12.3   Stochastic analyses and decisions

Decisions involve choices among options. Because the expected utilities of various options are uncertain, choices between options often are not self-evident. This section describes some grounds for making choices.

12.3.1   Stochastic dominance

Stochastic dominance describes the extent to which a given decision is 'best', depending on the source or magnitude of uncertainty. Management options based on forecasted cumulative probabilities of benefit or loss oblige the decision maker to be explicit about risk aversion or risk tolerance at different levels of likelihood of outcomes.

In the simplest case, strategies are clearly distinguished because cumulative probability plots do not cross (Figure 12.5). Choosing the least risky option requires the assumption that more of the 'benefit' is always better.

To simplify the decision context, decisions that are always inferior may be eliminated from the list of candidate actions, irrespective of the status of the uncertainty surrounding parameters (Morgan and Henrion 1990, Clemen 1996).

In some instances, scenarios are broadly distinguished but the cumulative probability plots cross at one extreme or the other. An option may be *almost* always better than the others. The choice of a strategy requires an additional assumption of at least some risk aversion.

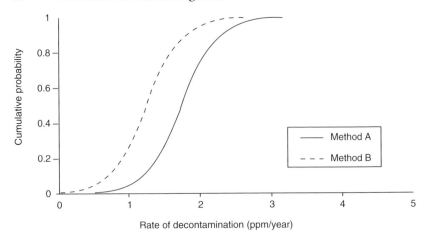

Figure 12.5. Scenarios in which one option is always better than another.

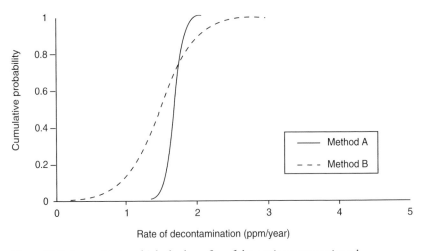

Figure 12.6. Scenarios in which the benefits of the options are equivocal.

Management options become more equivocal when the cumulative probability distributions cross (Figure 12.6). It is not clear what the best decision should be without making additional assumptions about the decision maker's attitude to risk. The probability that the rate of decontamination will be below 0.5 ppm/year is higher under Method B. The probability that the rate will be below 2 ppm/year is higher under Method A.

The decision should be weighted by the decision maker's attitude to either outcome. If one is much more costly than the other, it may be best to be risk averse and choose a decontamination method that is more likely to achieve at least a specified rate. For instance, if it is socially, politically, or ecologically important to achieve a rate of at least 1 ppm/year, and a rate lower than 1 would be damaging, then the best option is to choose Method A because it is unlikely to deliver a rate less than 1. When we choose it, we forgo the prospect that Method B could result in rates above 2 ppm/year. This thinking is embedded in the benefit-cost analyses used to set decision thresholds in receiver operating characteristic (ROC) curves (Chapter 11).

In most instances, choices are equivocal. Because circumstances are often politically and emotionally charged, the risk analyst is obliged to manage the social context as much as the analytical one, using tools such as those outlined in Chapters 4 and Sections 12.4 and 12.5, together with results such as those above.

12.3.2 Benefit-cost analysis

Usually, benefit-cost analysis takes into consideration the consequences of options in monetary terms. Valuations are based on well-developed theories for willingness-to-pay or to accept compensation.

The discussion and methods in preceding sections outline several approaches to benefit-cost analysis for environmental decisions. Risk treatment, remediation and education cost money and there is never sufficient to reduce risks from all sources, to the satisfaction of everyone. A traditional view of the optimal strategy is the one that reduces risk most efficiently (Figure 12.7).

In hypothesis tests (Chapter 11), the error rates α and β define acceptance or rejection of the null hypothesis. They may be adjusted so that their ratio equals the ratio of the 'costs' of making a type I error versus a type II error (Mapstone 1995). Costs may be measured in any currency (dollars, jobs, quality of life, habitat loss). The acceptable ratio is a social choice, not a scientific one.

When using control charts (Chapter 11), warning and action thresholds combine with the number of different triggers to determine the effective error rates ('false alarm' and 'failure to act'). Operating curves show these rates explicitly. The rates may be adjusted by manipulating control limits and the number and kinds of tests. In the end, the problem reduces to the same one as is confronted in traditional hypothesis testing. The costs and benefits need to be expressed in a form so that people can make judgements about acceptable tradeoffs (see Chapters 8 and 11).

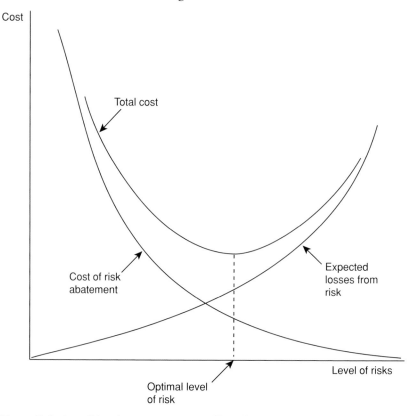

Figure 12.7. A traditional, economic view of benefit–cost analysis (from Morgan 1990) in which the optimal level of risk minimizes the sum of abatement cost and losses from risk.

Finding equivalent currencies can be difficult. Among economists, contingent valuation is perhaps the most popular method. It has been applied to a wide range of environmental issues including wilderness protection, water quality and soil erosion. It uses questionnaires and/or interviews to elicit preferences and demand functions for environmental goods and services (Freeman 1993, Garrod and Willis 1999). Variations take into account ownership, access, social context and perceptual biases (Slovic 1999).

The method has unresolved technical and theoretical problems (Chee 2004) including the influences of context and framing (Bingham *et al.* 1995), and free-riding (Garrod and Willis 1999). There are alternatives including 'hedonic' pricing (using market valuations such as house prices), and stated preference techniques that use direct consumer valuations of environmental values. Most rely on converting environmental preferences to monetary preferences and each has its own peculiarities (Chee 2004). Haimes *et al.* (2000) recommended multi-objective tradeoff and surrogate worth tradeoff to evaluate options. There is no easy solution when the problem involves values that have inherently different scales.

But these are not reasons for dispensing with benefit-cost analysis. Rather, it is important to be aware of assumptions and to interpret results accordingly. They are practical in situations in which the values of alternatives are expressed naturally in the same currency. These data can be important in a broader multi-criteria analysis (Section 12.5.2).

For example, Akçakaya and Atwood (1997) used benefit-cost analysis to evaluate management options resulting from a Monte Carlo model for the California gnatcatcher (*Polioptila c. californica*). It is a threatened subspecies inhabiting coastal sage scrub in southern California. Its habitat has declined due to agricultural and urban development.

Akçakaya and Atwood (1997) first developed a habitat model using logistic regression to link GIS data to records of species occurrences. Variables included percentage of coastal sage scrub, elevation, distance from grasslands, distance from trees and various interactions among these variables. They validated the habitat model by building it on data from the northern half of its range and predicting observations in the southern half. They then used it to define patches of habitat as a basis for building a metapopulation model. This model included demographic data such as fecundity, survival, as well as variability in these demographic rates.

The model predicted a high risk of decline in the next 50 years with most combinations of parameters. However, there was a considerable range of outcomes due to uncertainties in parameters. Results were most sensitive to assumptions about density-dependent effects, the probability of weather-related catastrophes, adult survival and adult fecundity.

Akçakaya and Atwood (1997) then explored management and conservation alternatives. They examined the scenario in which three of the habitat patches were potential candidates for restoration. If these patches vary in size, then there would be a total of seven alternatives (ranging from restoring only the smallest patch to restoring all three).

These, plus the 'no action' alternative, were evaluated by running a series of simulations. They incorporated the expected improvements in the carrying capacity and other parameters of the patches where habitat would be restored (Figure 12.8).

They ranked the eight options in order of increasing effectiveness (measured by the reduction in the risk of extinction). The obvious choice

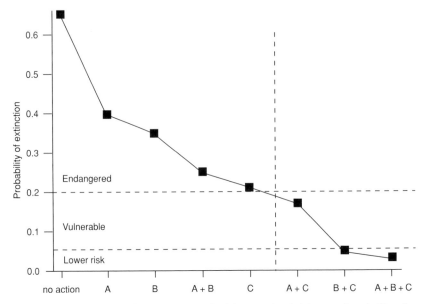

Figure 12.8. The risks and costs associated with restoring habitat patches A, B and C for the California gnatcatcher (from Akçakaya and Atwood 1997).

is to improve the habitat in all three patches. In reality, improving all three patches may exceed the budget for conservation of the species. The ranked options allow managers and stakeholders to evaluate each action in terms of costs and benefits (Akçakaya and Atwood 1997).

In another example, Reckhow (1994) explored the economic costs associated with the control of biological oxygen demand (BOD) in industrial waste water. Deviations from a target BOD resulted in costs of additional treatment and the potential for fines if permit conditions were violated. Designs to reduce BOD were uncertain because of natural variation in BOD. The cost function was asymmetric. As a result, Reckow (1994) concluded that the greater the cost of fines, the larger the safety margin should be.

Weitzman (1998) proposed a method to set priorities for species recovery that accounted for opportunity, cost and outcome (gain). Taxon i is given a rank, R_i, with,

$$R_i = [(I_i + D_i)\Delta P_i]/c_i$$

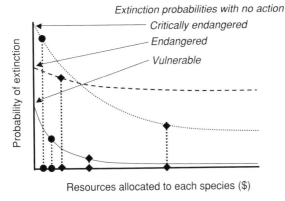

Figure 12.9. Allocating resources to recover threatened species. The lines represent the way in which risk of extinction declines for each species as money is spent on that species. One species is listed as critically endangered (dots), one as endangered (dashes) and another as vulnerable (continuous line). To minimize loss of species, allocate funds so that the marginal rate of gain for each species is the same. With a small allocation of funds (●) the optimal strategy allocates most funding to the vulnerable species, less to the critically endangered and none to the endangered. With a larger budget (◆) most of the money should be spent on the critically endangered species, some on the vulnerable species and little on the endangered species (after Possingham *et al.* 2002a).

where I_i is 'intrinsic' utility yielded by taxon i, D_i is genetic distinctiveness, ΔP_i is reduction in extinction risk and c_i is conservation effort (expenditure). The term $(I + D)$ allows taxa to be differently weighted (from whatever perspective). Possingham *et al.* (2002a) used a similar idea when they argued that resource allocation would be globally more efficient, thereby minimizing species loss, if funds are allocated to recovery to maximize extinction risk reduction (Figure 12.9).

Such ideas have been extended to the valuation of ecosystem services. The probabilities of different environmental outcomes may be evaluated against cost under alternative management options. For instance, Carpenter *et al.* (1998) used expected utility to evaluate environmental remediation versus 'business as usual' for a catchment in which phosphorus loads and algal blooms were linked in an ecological model.

Benefit-cost analyses are prey to the full range of uncertainties that affect other kinds of analysis. The points on Figure 12.8, for instance, are themselves uncertain. They do not illustrate the extent of uncertainty. All such analyses should be subjected to sensitivity analyses to evaluate the sensitivity of the best course of action to the full range of knowledge limitations.

If minor changes in probabilities, utilities or model structure give rise to significant changes in the best course of action, the options are to gamble on what appears to be the best strategy (Bier *et al.* 1999), or to choose a strategy that delivers an acceptable outcome, no matter what the true state of the world (Ben–Haim 2001; see below). The latter course may deliver suboptimal outcomes but it will avoid catastrophes. The best strategies usually will involve decisions that leave options open and gather useful information.

12.3.3 Stochastic dynamic programming

When the objective is to control a stochastic process, stochastic dynamic programming (SDP) finds the optimal decision at each time step in the control period (Possingham *et al.* 2002b).

The first step is to define 'payoffs' (benefits) for achieving each state of a system. For example, when managing a threatened species, the simplest interpretation is that the benefit is measured by population size. In an environment with several patches, one could account the payoff as '1' if the population persists, or '0' if it becomes extinct in a patch. In a contaminated stream, benefit may be '1' if water quality measurements fall within safe bounds and '0' if regulatory thresholds are exceeded.

Starting at the end of a specified time period, the optimal strategy is found by stepping back through time, choosing the optimal decision each time step (the one that maximizes payoffs), assuming that later decisions are made optimally.

Possingham *et al.* (2002b) provided the following example. A threatened species occurs in two patches. Suitability of the patches changes with time since the last fire. The environmental fluctuations experienced by the two patches are correlated. Managers may make one of four possible decisions at each time step: burn neither patch, burn patch 1, burn patch 2, or burn both patches.

To determine the best decision for every state of the system, we begin at some time horizon, T. The payoff is 0 if both patches are empty, and is 1 otherwise. This may be written

$$\text{Payoff}_T(x_1, x_2, F_1, F_2) = 0 \quad \text{if } x_1 = x_2 = 0$$
$$= 1 \quad \text{otherwise,}$$

where x_1 and x_2 are the population sizes in patches 1 and 2, and F_1 and F_2 are the times since the last fire in patches 1 and 2.

The best decision at time $T - 1$ optimizes the expected payoff at time T, in state (x_1, x_2, F_1, F_2). The expected payoff is a function of transition probabilities between states. These transition probabilities are derived from the population model, depending on the current state of the system and the decision chosen from the strategy set.

This generates the best strategy for decisions one year ahead. The best long-term strategy is found by back stepping repeatedly until an equilibrium strategy is found. It turned out in their example that the optimal strategy depended on whether organisms moved between patches.

Stochastic dynamic programming depends on a Markov chain that gives the transition probabilities among the possible states. Unfortunately, most environmental applications are plagued by gaps in knowledge, including complete ignorance of processes, parameters and uncertainties. The following section explores methods that can be applied to generate assessments when information is severely limited.

12.4   Info-gaps

Simon (1959) introduced the idea of 'bounded rationality'. He argued that perception and uncertainty limit the ability of people to achieve goals. People construct simple models to deal with these difficulties. The key idea is that people 'satisfice' rather than optimize. Satisficing means taking decisions that do well enough and that are robust to a wide range of uncertainty. Ideas began to emerge in the 1950s about how to make good decisions when critical information is missing (Box 12.1).

For example, Cooper (1961, in Ignizio 1985) suggested that, instead of trying to maximize some objective, y, it may be better to aspire to 'at least x units of y'. Techniques such as goal programming (Ignizio 1985) may be used to minimize unwanted deviations from an objective, one way of implementing the idea of satisficing.

Info-gap theory (a term borrowed from economics) was invented by Ben-Haim (2001) to perform model-based decision analysis in circumstances in which reliable probability models are unavailable. It can function sensibly when there are 'severe' knowledge gaps.

Info-gap analysis requires three elements:

1. A mathematical process model.
2. A measure of performance.
3. A model for the uncertainty.

Box 12.1 · *Minimizing maximum regret*

Savage (1951, in French 1986) defined 'regret' as the difference between the result of the best action and the result of the action taken. Given n actions, a_i, and m states of the world with probabilities, p_j, regret of action a_i is

$$r_{ij} = \max_{l=1}^{m}\{v_{lj}\} - v_{ij}.$$

The worst regret possible from action a_i is given by:

$$\max_{j=1}^{n}\{r_{ij}\}.$$

To minimize regret, choose the option with the smallest regret. That is, choose a_i such that:

$$\min_{i=1}^{m}\{\rho_i\} = \min_{i=1}^{m}\left\{\max_{i=1}^{n}\{r_{ij}\}\right\}.$$

This was called the 'minimax regret' criterion (see French 1986, Morgan and Henrion 1990).

Info-gap asks, 'What do we need to know to make a decision?'. It entertains two dimensions in making a decision: robustness (immunity from unacceptably bad outcomes) and opportunity (chances of windfall that exceed our expectations). Thus, it recognizes implicitly that uncertainty can be good or bad.

The process model is a mathematical representation of a conceptual model. It summarizes what the analyst (or the expert) believes to be true and important in the system. It could be a population model, an expert system, a logic tree or any other quantitative model.

The quality of the outcome of a decision is assessed by the measure of performance. It may be the risk of population decline, the concentration of a contaminant, the density of algal cells in a freshwater stream, or the size of a managed fish population. The objective may be to reduce the first three, or to enhance the latter. Performance measures may include multiple attributes.

The model for uncertainty describes what is unknown about the parameters in the process model. An info-gap model is an unbounded family of nested sets of possibilities. Typically it is a set that encloses all possible values for a parameter or function. For instance, if the process model is a function for the growth of a population, the uncertainty in the growth rate may be bounded by an interval. But it is not an interval in the usual

Box 12.2 · *Some axioms of info-gap theory*

This section is taken from Regan *et al.* (2004). Info-gap models have an uncertainty parameter, α, and a centre, $\tilde{v}$. They have two basic properties, nesting and contraction. Nesting means that α is a horizon of uncertainty such that $\alpha \leq \alpha'$ implies

$$U(\alpha, \tilde{v}) \subseteq U(\alpha', \tilde{v}).$$

For any given value of α, $U(\alpha, \tilde{v})$ is a set of possible values of v. What 'nesting' means is that as α gets larger, the set $U(\alpha, \tilde{v})$ gets more inclusive. This imbues α with its meaning as a 'horizon of uncertainty'. An info-gap model of uncertainty is a family of nested sets of possible values of the uncertainty entity, $U(\alpha, \tilde{v})$, $\alpha \geq 0$, rather than a single

a.

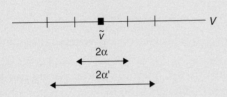

b.

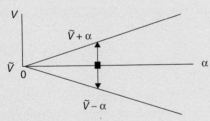

Figure 12.10. Nesting and contraction properties of the interval uncertainty model (after Regan *et al.* 2004). Contraction is illustrated in b. v is a scalar; its value is shown on the vertical axis. The horizon of uncertainty, α, increases to the right. At any given value of α the range of variation of v is shown by the growing cone of uncertainty. For instance, the vertical double arrow in b corresponds to the upper of the two horizontal double arrows in a. As the horizon of uncertainty is reduced, the range of variation of v diminishes, until v precisely equals the nominal value when there is no uncertainty: at $\alpha = 0$ the info-gap model 'contracts' to the singleton set containing only the centrepoint $\tilde{v}$. Note that the centrepoint is a possible value of v at all horizons of uncertainty. In contrast, $\tilde{v}$ is the only possibility in the absence of uncertainty.

set (Ben–Haim 2001). Contraction means the nominal value, $\tilde{\nu}$, is the only possible value in the absence of uncertainty,

$$U(0, \tilde{\nu}) = \{\tilde{\nu}\}.$$

The elements of the set $U(\alpha, \tilde{\nu})$ are often a scalar, but they could be virtually anything including functions or vectors.

The properties of nesting and contraction are illustrated in Figure 12.10 where the uncertain entity, ν, is a scalar. In Figure 12.10a the value of ν is displayed along an axis. The nominal, or best-estimate, of ν is $\tilde{\nu}$. At the horizon of uncertainty α, ν can vary in an interval of size 2α, from $\tilde{\nu} - \alpha$ up to $\tilde{\nu} + \alpha$, as indicated by the double arrow labelled '2α'. At a greater horizon of uncertainty, α', the range of possible variation of ν is wider as shown by the longer double arrow. Since the horizon of uncertainty is unknown, the info-gap model is an unbounded family of nested intervals like those shown in Figure 12.10a that contract to zero uncertainty at the nominal value (Figure 12.10b).

sense. The uncertainty parameter, α, may take on any value so the interval is infinitely wide.

The robustness function answers the question, 'How wrong can the model be without causing failure?'. The decision maker can trade robustness for performance. The opportunity function answers the question, 'How much should the model be changed to allow 'windfall' performance?'. It implies that we may make decisions that facilitate the possibility of better-than-expected outcomes. This provides a formal framework to explore the kinds of speculations that occur intuitively when examining decision options.

The process model, performance measure and uncertainty model provide a system of equations that may be solved for estimates of robustness and opportunity. The decision maker can then play 'what-if' games, usually focusing on a critical parameter that represents failure (a threshold reflecting the limit of unacceptable performance).

12.4.1 A process model and measure of performance

The orange-bellied parrot is listed as 'critically endangered' by the IUCN (2001). The total population size is less than 200. Birds breed during summer in the coastal areas of Tasmania. In winter, they migrate north to mainland Australia, sheltering and feeding in coastal salt-marsh areas

Table 12.4. *Decision table (utilities and probabilities) for three management actions and four states. Utilities are the minimum expected population sizes resulting from management aimed at alleviating the effects of the potential cause of decline (Chapter 8). The example is hypothetical but is based on a plausible set of scenarios (after Drechsler* et al. *1998)*

System state (i) (cause of decline)	Likelihood of each state (subjective probability that this is the primary factor inhibiting recovery)	Action 1 (predator control)	Action 2 (habitat rehabilitation)	Action 3 (reducing exposure to toxins)
		(a_1)	(a_2)	(a_3)
	p	v_1	v_2	v_3
Feral predators	0.2	30	5	0
Grazing impacts	0.3	5	5	0
Loss of habitat area	0.4	5	10	0
Ecotoxicological effects	0.1	0	5	30
Expected utilities		$\sum p_i v_{j1} = 9.5$	$\sum p_i v_{j2} = 7$	$\sum p_i v_{j3} = 3$

of south-eastern Australia. High mortality during winter seems to be responsible for its persistent small population size. The population was many thousands of individuals around 1900, but declined steadily until reaching its current size about 20 years ago. The options for managing winter habitat include control of grazing, expansion of suitable habitat, minimizing exposure to potentially contaminated waste disposal ponds, and control of predators and competitors (Drechsler *et al.* 1998).

Expected utilities have been used to assist environmental decision making (e.g. Maguire 1986, Ralls and Starfield 1995, Possingham 1996). Decision tables or decision trees involve identifying three main components – acts, states and outcomes (Table 12.4; see Chapter 8). The acts refer to the decision alternatives, the states refer to the relevant possible states of the system, and the outcomes refer to what will occur if an act is implemented in a given state. Usually, it is assumed the state of the system does not change substantially through time, so that the chosen alternative remains relevant or applicable.

For decision making under uncertainty the usual procedure is to assign probabilities to each of the relevant states and utilities to each of the outcomes. The task then reduces to a choice that maximizes expected utility. Probabilities assigned to the states represent the likelihood that the system is in that state. So while the particular state in which a system exists is uncertain, the probability that the system is in that state is assumed to be known with certainty.

The main objective of the management team is to minimize the probability of population decline. Drechsler *et al.* (1998) wrote a population viability model for the dynamics of orange-bellied parrots. We used this to generate forecasts of population dynamics under different management options. McCarthy and Thompson (2001) proposed measuring the utility of management decisions using minimum expected population size. It summarizes the chances of a population falling below a lower threshold within a specified time period. Table 12.4 provides estimates of the response of the parrot population to a range of management options, in terms of the minimum expected population size of adult females.

Under ordinary utility theory, the expected utility of an action is the sum of the products of individual utilities (assuming the system state is true) times the probability that the state is true (Chapter 8). The best action is the one that maximizes expected utility, in this case, action 1.

12.4.2 A model for uncertainty

A problem with using decision tables is that they are composed of utilities v and probabilities p that are uncertain. It is difficult to quantify these uncertainties. Utilities and probabilities of the states of the world are assessed subjectively. Intervals can be assigned to probabilities and utilities to incorporate the range of values these parameters might take (Walley 2000b, Walley and De Cooman 2001; see Chapter 9). In this example, we represent uncertainty in the probabilities and utilities with unbounded families of nested intervals (Box 12.3).

Table 12.4 was evaluated using the info-gap interval uncertainty and process models outlined above (see Regan *et al.* 2004). $\hat{\alpha} = 0$ in Figure 12.11 corresponds to the usual maximum expected utility solution (Table 12.4). Using an expected utility strategy is risky because it ignores uncertainty in utilities and probabilities.

Action 1 is best only when there is relatively little uncertainty in the states of the world and in the expected utilities. For wider ranges of uncertainty, action 2 is more robust. Action 3 is suboptimal, irrespective of the magnitude of uncertainty.

Box 12.3 · *Info-gap uncertainty model for the orange-bellied parrot decision table*

The info-gap model for utility uncertainty, is the family of nested intervals

$$U_v(\alpha, \tilde{v}) = \left\{ v : \left| \frac{v_{ij} - \tilde{v}_{ij}}{\tilde{v}_{ij}} \right| \leq \alpha, i = 1, \ldots, I, j = 1, \ldots, J \right\}, \alpha \geq 0.$$

This specifies a rectangle with dimensions IJ, oriented along the co-ordinate axes, in which v_{ij} varies from its nominal value, $\tilde{v}_{ij}$, by no more than α. The horizon of uncertainty, α, is unknown and unbounded. In the application below we restrict ourselves to positive utilities although under the formulation here they can take any value. The model for uncertainty in the probabilities, p, is similar. There are the additional constraints that the p's must be positive and normalized to sum to 1 (because they are probabilities). To bound the p values, we may express them as fractions of the nominal value in a manner similar to the bounds on the utilities,

$$\frac{|p_j - \tilde{p}_j|}{\tilde{p}_j} \leq \alpha,$$

which implies that, at the horizon of uncertainty α, the jth probability is in the interval,

$$(1 - \alpha)\tilde{p}_j \leq p_j \leq (1 + \alpha)\tilde{p}_j.$$

Uncertainties for both utilities and probabilities are defined to have identical relative uncertainties. To keep the p values non-negative and their sum normalized, we define

$$U_p(\alpha, \tilde{p}) = \left\{ p : 1 = \sum_{j=1}^{J} p_j; \max[0, (1 - \alpha)\tilde{p}_j] \right.$$

$$\left. \leq p_j \leq \min[1, (1 + \alpha)\tilde{p}_j] j = 1, \ldots, J \right\}, \alpha \geq 0.$$

Info-gap theory takes the position that the best strategy is the one that provides an outcome that is both 'good enough' and that keeps us immune from unacceptable outcomes given some level of uncertainty. That is, we choose a strategy that maximizes the reliability of an adequate outcome. This is termed 'robust satisficing' (Ben–Haim 2001).

We let C be the smallest value of a utility that is acceptable. In our case, a value of zero is unacceptable because it implies the population is expected to become extinct. We would like the value of minimum population size to be as large as possible. C thus represents an aspiration. There is no need to choose it a priori. We will return to it below. Since p_j and v_{ij} are imperfectly known, and not effectively bounded, we cannot determine whether or not the outcome of any action will be adequate. Instead, we determine the greatest horizon of relative uncertainty, $\hat{\alpha}$, within which all of the outcomes of a given action result in an adequate performance (that is, they result in expected utilities greater than C). We express the robustness of an action in terms of the maximum uncertainty that allows us to reach the performance aspiration. The robustness of action a_i, given performance aspiration C, is

$$\hat{\alpha}(a_i, C) = \max \left\{ \alpha : \min_{\substack{v \in U_v(\alpha, \tilde{v}) \\ p \in U_p(\alpha, \tilde{p})}} \sum_{j=1}^{J} v_{ij} p_j \geq C \right\}.$$

The robustness $\hat{\alpha}(a_i, C)$ of action a_i, with aspiration C, is the greatest horizon of uncertainty α up to which all utilities v [in $U_v(\alpha, \tilde{v})$] and all probabilities p [in $U_p(\alpha, \tilde{p})$] result in expected utilities no worse than C. Given a lower bound for an acceptable outcome, C, more robustness [$\hat{\alpha}(a_i, C)$] is better than less. Hence the robustness function determines a preference ranking for the management alternatives. The action that maximizes robustness, for a given threshold of tolerance, is the best.

Robustness functions have the important property that they always decrease monotonically as the aspiration for utility becomes higher. If our aspiration is low (say, a minimum of four adult females), we can tolerate a broad range of uncertainty in utilities and probabilities. If our aspirations are higher (no less than eight females is acceptable), we tolerate a narrower range of uncertainty and the best management option may change. Another way of looking at these results is that if fewer than nine pairs are unacceptable, we have little tolerance for uncertainty. We should choose action 1. It's the only option that has any chance of delivering an outcome we can live with.

This analysis reflects the ever-present tradeoff between immunity to uncertainty and aspiration for performance. Very demanding aspirations

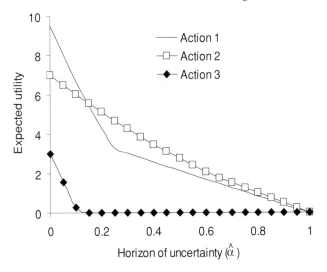

Figure 12.11. Maximal robustness curve $\hat{\alpha}(a_i, Ec)$ versus aspiration for expected utility, Ec, for the three management actions in Table 12.4. The expected utility is measured by the minimum expected number of adult female birds in the population.

become more vulnerable to uncertainty. In the extreme, an expectation of maximum utility (the default in most present applications) is maximally vulnerable to uncertainty.

What does an $\hat{\alpha}$ of 0.5 mean? It means that all the parameters – utilities v_{ij} and probabilities p_j – can vary from their nominal values fractionally by as much as 0.5, without causing the expected utility to fall below the critical value C. For example, it implies that the probabilities may be as large as 0.6 or as small as 0.2 for loss of habitat area (which has a nominal value of 0.4). The value of predator control, if predators are important, may be anywhere between 15 and 45 (given a nominal value of 30). The interpretation of uncertainties in each case will depend on context and the attitudes of stakeholders and decision makers to expected outcomes and risk.

12.5 Evaluating attitudes to decisions

12.5.1 Scenario analysis

Risk management requires a conceptual model of the future. To serve this need, scenario analysis began to evolve at about the same time as the

Delphi method (Chapter 4). Scenarios are shared, agreed mental models. They are internally consistent descriptions of possible futures created in structured brainstorming exercises (Van der Heijden 1996 describes the methods in detail). A scenario is a hypothetical sequence of events that focuses thinking on chains of causal relationships and their interaction with the decisions a person or a corporation might make.

Scenario analysis shares common roots with the Delphi method, developing from methods created to deal with defence decision making in the USA (Kahn and Wiener 1967; see Chapter 4). Over the last few decades, it grew largely in the realm of corporate decision making.

Scenarios are constructed and communicated with a story line in which events unfold through time through a series of imagined causes and effects. The idea is to engage a group in the value-free exploration of options. It distinguishes between the organization itself, over which the strategist has control, the transitional environment made up of people and things the strategist can influence, and the contextual environment over which the strategist has no influence.

Scenario planning, as envisaged by Kahn and Wiener (1967), does not rely on probabilistic forecasts although quantitative projections may support the creation of scenarios. Rather, it uses qualitative causal thinking to generate decisions that are robust to as wide a range of plausible, alternative futures as possible (Van der Heijden 1996). In that sense, it shares a philosophical position with info-gap theory (above).

Scenario planning depends on the cooperation of participants. It requires that they are forthcoming about outcomes, dependencies and uncertainties. Open dialogue may be more easily achieved in settings where participants share common objectives, than in environments in which resource allocation is being debated and stakeholders stand to gain by the outcome, at the expense of others.

Operating models are constructed for underlying processes. The models are used to evaluate different control rules and other heuristics that may be used on a routine basis by managers. Only those rules that work effectively over the full suite of scenarios are retained.

Most applications have been in the management of large corporations. Bennett *et al.* (2003) argued for greater attention to ecological issues in global environmental scenario development. Buckland *et al.* (1998, in Harwood 2000) used scenario planning to explore the consequences of different red deer (*Cervus elaphas*) culling programmes in Scotland. Scenario analysis has the potential for supporting the development of alternative conceptual models in environmental risk analysis. The models may form the basis of an on-going dialogue between stakeholders. They may provide a framework for data acquisition, model testing and monitoring, all components of the risk management cycle.

12.5.2 Multi-criteria decision analysis

Evaluating management decisions and performance measures may require a detailed examination of the values and perceptions of participants. Multi-criteria decision analysis (MCDA) measures preferences by eliciting and ordering judgements from people affected by a decision. The objectives may include arriving at a decision that reflects social values, identifying opportunities, and improving understanding among stakeholders. It can work when problems are complex and include monetary and nonmonetary values (Keeney and Raiffa 1976).

MCDA works by defining all criteria against which an action may be evaluated, and identifying a preference scale or some other measure of performance. A 'coherent' set of criteria has the following properties (Roy 1999):

1. Stakeholders are indifferent to alternative actions if they rank them the same against all criteria (implying the set is exhaustive).
2. An action will be preferred to others if it is substantially better than all others on one criterion and equal on all others (implying the set is cohesive).
3. The set is understood and accepted by all stakeholders.
4. The conditions above may be violated if any criterion is omitted (implying the set contains no redundant elements).

If the endpoints of preference scales are meaningful to stakeholders, they may be evaluated in terms of costs and benefits using multi-attribute utility functions (Borsuk *et al.* 2001a).

Successful MCDA depends on an effective social process. The analytical process should be preceded by rationalization of aims and scope, and by identification of stakeholders and experts who may contribute usefully. The rules governing selection, interactions between participants and aggregation of opinions (Chapter 4) largely determine the acceptability of outcomes.

Saaty (1992) described a process to structure and map the opinions of experts, generating a cohesive picture of areas of agreement and disagreement. Many factors influence stakeholder choices. An analytical hierarchy (a decision tree; see Chapter 8) may be used to order thinking about

---

Box 12.4 · *Basic steps in multi-criteria decision analysis (see Dodgson et al. 2001)*

1. Establish criteria that may affect a decision and outline options (scenario analysis may be helpful).
2. Classify criteria under broad headings (social, political, economic, landscape, ecological). Criteria are used to assess the performance of options, arranged under a hierarchy of objectives. The endpoints of the decision tree are the criteria (e.g. Figure 12.13).
3. Assign weights (each person assigns scores or preferences) at each branch of the tree that reflect the importance in determining outcomes.
4. Assign weights to each option (by experts or by group consensus).
5. Aggregate weighted scores (usually these are linear combinations, assuming that preference scores on one criterion are unaffected by scores on other criteria).
6. Present scores (anonymously) to the group and discuss differences of opinion. Participants revise their scores and generate new ranks for each criterion.
7. Aggregate the performance of each option across all criteria. Most approaches use linear numerical aggregation, adding the products of scores and weights for each criterion.
8. Conduct a sensitivity analysis (vary scores and weights, delete criteria in turn and create new options).

The value tree represents a model of values that affect the decision over management priorities (Belton 1999). The weights given to different criteria define acceptable tradeoffs.

---

these factors, and to provide a means for quantifying the priorities. The approach documents the diversity of opinions in a group that allows opinions to be examined and revised subsequently.

The gambling analogy introduced in Chapter 9 may be used to elicit preferences for nonmonetary values. Starmer (1998) provided an example of the method to elicit attitudes towards a state of health. Begin by selecting two extreme health states (they become endpoints for the utility scale). In this example, they are 'perfect health' and 'death'. Then the participant imagines a choice between two alternatives (Figure 12.12). Alternative 2 is to experience a state of health, H_i, for sure. Alternative 1 offers a chance, p, of perfect health and a chance, $1-p$, of death. The

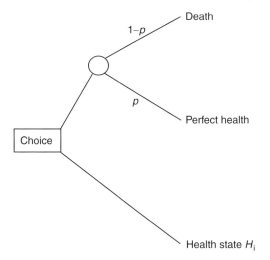

Death

1−p

p

Perfect health

Choice

Health state H_i

Figure 12.12. Choice strategy for elicitation of preferences for nonmonetary values (from Starmer 1998).

objective is to find a value for p such that the participant is indifferent to the choice of alternatives. Strategies such as those outlined in Chapter 9 (eliciting 'buying' and 'selling' prices) usually will result in a range of values for p.

At the 'point' of indifference, the expected utility of health state H_i, is $E(H_i) = p \times$ utility (perfect health) $+ (1 - p) \times$ utility (death).

If we define the utility of perfect health to be 1 and the utility of death to be 0, then $E(H_i) = p$ (or $[p_1, p_2]$ if an interval has been elicited).

The method may be used to elicit preferences for any aspiration or environmental decision. Sometimes, it is useful to ask for utilities directly (on a 0–1 scale) and compare the ranks with those derived from the alternative choice strategy.

Another way to elicit preferences is to ask participants to score criteria on a scale 0–100. Then, they are asked to compare how much a change from 0 to 100 on one criterion compares to a 0 to 100 change on another. This process identifies the most important criterion (in the eyes of this participant) and other criteria may be weighted relative to it.

Saaty (1980, 1994) described the analytical hierarchy process (AHP), a method for converting subjective judgements of importance into scores. It depends on a numerical interpretation of answers to questions such as, 'How important is (criterion) A compared to (criterion) B?'. Responses are coded by an analyst who uses a nine-point scale (Table 12.5).

Table 12.5. *Code and scale for AHP responses (after Saaty 1980)*

How important is A relative to B	Index
Equally important	1
Moderately more important	3
Strongly more important	5
Very strongly more important	7
Overwhelmingly more important	9

The pairwise comparisons between criteria form a matrix of preferences. Saaty (1980) used the elements of the first (dominant) eigenvector of the matrix as the weights for each criterion.

Usually, criteria are clustered in a value tree. The preference matrix is formed for each set in the tree, and then between sections at a higher level in the hierarchy.

The same strategy is used to calculate relative performance scores for each management option. Matrices are formed from comparisons between pairs of options. Options are judged in terms of their contribution to each criterion. Thus, for m options and n criteria, there are n separate $m \times m$ matrices.

There is no firm basis to link the verbal responses to the numerical scale (Table 12.5). Sometimes, new management options are added to an analysis once it is complete. New options can disrupt the relative ranks of original options. Such outcomes sometimes prompt people to develop alternative management strategies that have improved properties over a range of criteria (see Goodwin and Wright 1998).

Fernandes *et al.* (1999) used MCDA to explore acceptable alternatives to the management of a marine park. They stratified stakeholders by their interest in the issue (hotels, government regulators, NGOs and so on). They assessed the relative importance of each pair of objectives at each level in the hierarchy. Comparisons were limited to groups of subcriteria within a criterion (Table 12.6).

The assessment was determined by asking participants how important each component was for achieving the parent objective immediately above them in the hierarchy. Comparisons were made on a nine-point scale ranging from 'equally important' to 'overwhelmingly more important'. At the highest level in the hierarchy, the relative importance of each criterion was assessed independently by a single representative from each

Table 12.6. *MCDA analytical results showing assessments by stakeholders of the relative importance of the management planning objectives in contributing to the overall goal of preserving the marine resources of the reef for people in perpetuity (each row sums to 1) (a subset of the values reported by Fernandes et al. 1999)*

Stakeholders	Objectives (Criteria)			
	Ecological sustainability	Economic benefits	Social acceptability	Global model
Local government	0.309	0.388	0.249	0.054
Saba Conservation	0.476	0.093	0.360	0.071
Recreational fishers	0.525	0.168	0.177	0.129
Hotels / restaurants	0.293	0.285	0.260	0.163
Dive shops	0.683	0.111	0.150	0.056
Educators	0.600	0.230	0.057	0.114
Developer	0.283	0.529	0.095	0.093
Median value	0.344	0.168	0.207	0.111

stakeholder group. These values were used to generate weights for each criterion.

Surveys of stakeholders were also used to frame five different reef management options. Experts evaluated the extent to which each of the high-level objectives would be achieved under each management option. The weight of each management option depended on the weights assigned to each objective.

The analysis resulted in consensus among all stakeholders about the need for a managed park. It rejected options for elimination of the park and its no-fishing zones.

In MCDA, a preference may be expressed for all pairwise comparisons between actions [$n(n-1)/2$ comparisons], against each criterion. For instance, the following questions accommodate preference, indifference and incomparability (Bana e Costa and Vansnick 1999):

Is one of the actions (x or y) more attractive than the other?
(YES / NO / DON'T KNOW)
If yes, which is the more attractive?

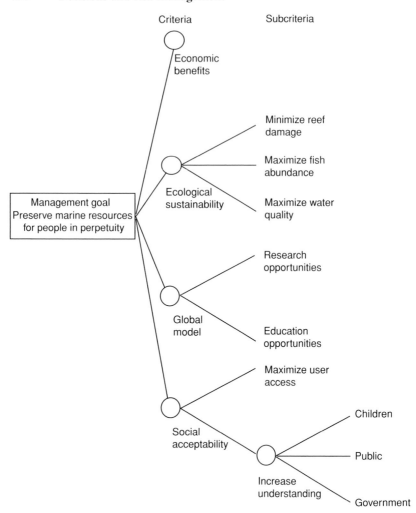

Figure 12.13. A MCDA value tree (a criteria hierarchy) for a decision about coral reef management on Saba, a small Caribbean island, involving economic, ecological and social tradeoffs. It provides for an assessment of decision alternatives against criteria (after Fernandes *et al.* 1999). The 'global model' criterion reflected the desire by stakeholders to have a reef management system that serves as an international standard. Economic benefits included several subcriteria such as tourism revenue from recreational divers, hotels and restaurants, development opportunities, fishing revenue and so on.

'Outranking' methods are used when preferences cannot be expressed as a unique numerical function, when at least one criterion is not quantitative or when compensating gains and losses among criteria are not clear; the 'Electre' method has been broadly applied in this context (see Roy and Vanderpooten 1996, Vincke 1999).

The structure of MCDA makes it relatively easy to assess decision alternatives against criteria at any point in the hierarchy (e.g. Figure 12.13). Visual devices such as the value tree may be used to inform stakeholders about the importance of various preferences in determining the outcome, and about the differences between them and other stakeholders. It represents a model of the values that weigh on a decision.

Preferences are always uncertain. MCDA is useful in exploring sensitivities interactively with stakeholders. The risk analyst may adjust the weights and relative ranks allocated to various subcriteria by stakeholders, to illustrate how much the overall outcome depended on various judgements.

The composition of the stakeholder group may affect the outcome. Even given a single group with well-defined preferences, there are many ways to define and arrange criteria. These uncertainties should be explored so that important differences in opinion are identified and communicated, and regions of uncertainty are understood. The best way to achieve this is through a sensitivity analysis.

The sensitivity analysis could answer the question, 'Given uncertainty in preferences or the structure of the criteria, what preferences and structures within the limits of uncertainty will result in each alternative being considered the best against each criterion?'. It could answer, 'How does the best alternative, or the rank of all alternatives, change when inputs to the MCDA process change?'.

Sometimes, one alternative will dominate most or all criteria for all regions of uncertainty. More often, the best solution will depend sensitively on the choice of preference scales, weights and structures (groupings) for the criteria. To communicate subjective uncertainties in the preferences of stakeholders, Fernandes *et al.* (1999) reported the ranges of relative values of management options (Figure 12.14).

Fernandes *et al.* (1999) used experts to assess the utility of management scenarios for coral reef management. These judgements could have been supported by more explicit models. For instance, the importance of no-fishing zones could have been explored by building models of the fish populations and evaluating fish abundances with and without restricted fishing. While there are some explicit guidelines for performing

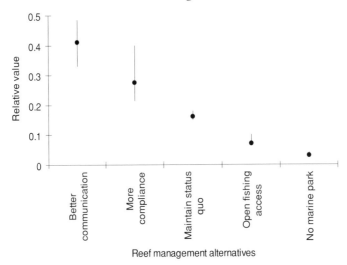

Figure 12.14. Relative value of management options to stakeholders under status quo conditions. The range for each option represents the range of preferences expressed by stakeholders combined with the degrees to which those objectives could be achieved. The midpoints are the median values for 15 stakeholders assessed (Fernandes *et al.* 1999).

sensitivity analysis (Tanino 1999), there is no general formula and the details of the sensitivity analysis should be guided by the needs of the specific circumstances of the case.

In MCDA, the person or group that defines the context is influential. The choices of participants, objectives, criteria, weights and aggregation rules are subjective. Either the group controlling the agenda or the participants need to decide if tradeoffs between criteria are acceptable. If they are, good performance on one compensates for a poor performance on another. Constraints on compensation can substantially limit the effectiveness of MCDA (Dodgson *et al.* 2001). Sensitivity analysis can alleviate some of the influence of arbitrary choices. If results are insensitive to differences of opinion, the approach can resolve difficult social differences.

12.5.3 Multi-criteria mapping

Multi-criteria mapping (MCM) takes the philosophical position that risks are socially constructed and (mostly) difficult to measure. Instead of using quantitative risk analyses, MCM uses the preferences of individuals to explore alternative management options (Stirling 1997).

Participants representing different social strata choose 'options', alternative scenarios that reflect a different decision about how best to manage a hazard. Participants then list 'criteria': all the things they would want taken into account to evaluate how best to fulfil a particular objective. Participants are asked to weight the criteria by ranking them, from most to least important. Participants then judge how well the options perform in the light of each evaluation criterion, and present a range of uncertainty for each assessment. Like MCDA, the process involves establishing context and available options, and uses subjective estimates for scores and weights for criteria.

The opinions and uncertainties of participants are shown to other participants, together with the social stratum that the individual represents.

Stirling (1999) outlined the application of MCM to choices about farming conventional and genetically modified (GM) crops in Europe. A panel of 12 people, all knowledgeable but representing a broad range of interests (expert stakeholders; Chapter 4), ranked six ways that oilseed rape might be grown on farms in Britain including:

1. Organic agriculture (no GM crops).
2. Integrated pest management without GM crops.
3. Conventional agriculture without GM crops.
4. Segregation and labelling of the GM produce.
5. GM crops with post-release monitoring.
6. GM crops with voluntary controls on areas of cultivation.

The criteria for evaluating options included social, economic, health, agricultural and environmental issues (Figure 12.15)

Alternative views may be created by varying weights. People were interviewed separately. They provided preferences that were communicated to other participants. The study identified a set of options that worked 'tolerably' well for all participants. The results were submitted to government prior to decisions about the use of genetically modified crops.

Stirling (1999) argued that the method works even in a hotly disputed controversy. It accommodates a broad range of perspectives. Its transparency helps builds trust. Uncertainty is acknowledged and carried through any calculations and summaries.

However, there is potential for ambiguity in definitions and concepts. In the absence of group meetings, some of these ambiguities will probably remain undiscovered and unacknowledged. The method is also time consuming, mainly because it relies on individual, face-to-face elicitation

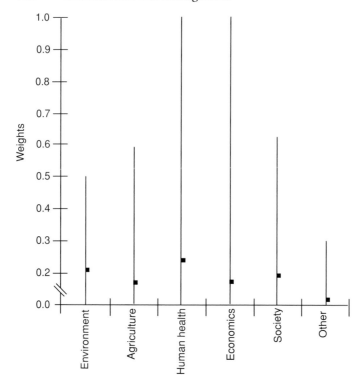

Figure 12.15. Multi-criteria mapping of attitudes (average and range over all participants) towards six issues affecting the evaluation of agricultural options in Britain (after Stirling and Mayer 1999).

procedures (Stirling 1999). Its primary value is that it makes the values and attitudes transparent to all participants, facilitating reconciliation and compromise decisions (Chapter 4).

MCDA and related techniques are useful when consensus, ownership and cooperation are sought. They can be used to achieve behavioural aggregation (Chapter 4). In other circumstances, the analyst may seek to design an MCDA that implements a management option, despite the conflicting self-interests of individual participants. The strategies for these different contexts will affect the details of the techniques selected. MCM differs in that it seeks primarily to elicit and communicate the values and preferences of participants. It is much closer to a communication device.

12.6 Risk communication

Risk communication is an interactive process involving the exchange of information and opinion between individuals, groups and institutions.

Warren-Hicks and Moore (1995) provided rules that they argued should guide risk communication. Their scheme represents a large proportion of risk communication strategies employed over the last few decades. They advised that analysts should:

- involve the public,
- plan for inclusiveness,
- listen to 'the audience',
- be honest, open and frank,
- collaborate with other 'credible' sources,
- meet the needs of the media, and
- follow up commitments.

Despite its positive attributes, this list carries an implicit us-and-them structure in which the audience listens to experts, and is listened to. 'Involving the public' means creating a process for working through devices such as public comment and response phases. This keeps stakeholders at a comfortable arm's length. It ensures that analysts can respond without experiencing critical cross-examination. In honest and complete risk assessments, stakeholders must have direct ownership of the process through direct involvement in all stages, including the design of protocols for interactions between interested parties.

Bier (2001) noted that despite a voluminous literature on the psychology of risk perception and acceptance, there are no definitive guides on the creation of risk communication strategies. Age, socio-economic status, gender, language, religion and cultural background overlay individual differences in knowledge and attitude towards risk (e.g. Flynn *et al.* 1994; Chapter 1) and may cloud a message.

Bier (2001) reviewed the status of risk communication and concluded that communication efforts should involve:

1. Understanding legal requirements or policies that may limit the design of risk communication.
2. Determining the purpose of communication, to
   a. raise awareness,
   b. educate people,
   c. motivate people to take action,
   d. reach agreement on a controversial issue, or
   e. obtain people's trust.
3. Determining the characteristics of the target audience, including level of knowledge, mental models, beliefs, experience, receptiveness and concerns.

4. Selecting a strategy that serves the purpose at hand, including the use of explanatory tools, techniques of persuasion, the means of stakeholder involvement, and the choice of participatory and communication formats (workshops, meetings, interviews, questionnaires, press, books, papers, surveys).

Focus groups, sociological analysis and published demographic information can assist in forming an effective communication strategy. Issues of detail need to be considered carefully. For instance, verbal interpretations of risk such as 'highly likely' are prone to linguistic ambiguity and there is no simple way to translate them into natural language (Chapter 4). Bier (2001) suggested that risk communication strategies be systematically pre-tested, especially as responses are often context specific. Each stage of the planning process may involve ethical choices, making the process dependent on the values of the communicator (Johnson 1999a).

12.6.1 Communicating probabilities: medical cases and framing

Gigerenzer (2002) noted that linguistic ambiguities surrounding statements about risk are one of the primary causes of misunderstanding. Cases from medicine provide perhaps the clearest examples of the causes of miscommunication. Their contexts and consequences are common to all risk assessments.

Gigerenzer (2002) identified several ways of communicating risk, including:

1. *Single event probabilities*: 'the probability that X will happen is $Y\%$'. There is no indication of the reference class. If the event is unique, the percentage may be interpreted as a belief (at best).
2. *Relative risks*: the change in risk experienced by a group, relative to the risk they would have experienced had they belonged to the reference class.
3. *Number needed to treat*: a concept in medicine that gives the number of people that need to participate in a treatment to save one life.
4. *Conditional probabilities*: given in circumstances in which it is easy to confuse $p(A|B)$ with $p(B|A)$ or $p(A \cup B)$.

When information is presented as probabilities, high variation in judgements is related to the logical strategies people use to draw inferences. For example, from randomized trials of breast cancer screening, there were 4 deaths per 1000 people without screening and 3 deaths per 1000 people with screening, in women 40 years of age and older. This could be communicated as (Gigerenzer 2002):

- a 25% risk reduction,
- 1/1000 absolute risk reduction,
- 1000 patients needed to treat to obtain one effective life saved,
- 12-day increase in mean life expectancy, or
- a risk reduction equivalent to the risk of driving a car about 500 km.

Each would elicit a different response from both experts and the public. Furthermore, such statistics do not communicate the physical or monetary costs of breast cancer screening.

Gigerenzer (2002) argued that people confronted with information in this form are incapable of informed consent. In 1957, Justice Bray of the California Court of Appeals coined the term 'informed consent' in relation to discussions about risk. He argued, '. . . a certain amount of discretion must be employed consistent with the full disclosure of the facts necessary to an informed consent' (Gigerenzer 2002, p. 96).

These words are ambiguous. Gigerenzer (2002) argued that doctors should be considered expert in diagnostic tools and treatment options, and that patients should be considered expert in personal preferences, goals and values. That is, acceptability is a subjective term and decisions should be left in the hands of the people who experience the risks and the benefits.

Schwartz et al. (1999) made suggestions to improve risk communication. They used the example of tamoxifen, a drug used to treat cancer. They reviewed software used by a national cancer institute to inform women of the costs and benefits of a particular treatment strategy. The message was communicated by:

'Women [taking tamoxifen] had about 49% fewer diagnoses of invasive breast cancer.

The benefit of tamoxifen was only expressed as a relative risk reduction without an explicit statement about baseline risk. Physicians and patients find the benefit of an intervention more compelling when it is expressed as a relative risk reduction, even though this form of communication is difficult to understand.

The baseline risk (the chance per year of developing invasive breast cancer for women in the placebo arm of the test) was 68 cases per 10 000 women per year. Applying the 49% relative risk reduction yields a risk of 34 cases per 10 000 women per year in the tamoxifen arm.

In contrast, one of the more important potential harms of tamoxifen, an increased chance of developing uterine cancer, was presented using raw frequencies for each group by:

... annual rate of uterine cancer in the tamoxifen arm was 30 per 10 000 compared to 8 per 10 000 in the placebo arm.

The asymmetric framing tended to emphasize the benefit of tamoxifen while minimizing the appearance of harm. If the increased uterine cancer was expressed as a relative risk, the statement would read, '... 275% more uterine cancer' and would likely elicit a different feeling. Schwartz *et al.* recommended using consistent framing, and using raw frequencies.

To further enhance the message, they recommended acknowledging uncertainty. In the presentation of both disease risk and treatment benefit, only point estimates (e.g. 49% risk reduction) were provided. They suggested comments describing lower bounds, such as:

If 1000 women do not take tamoxifen, six will be diagnosed with invasive breast cancer in the next year. If these 1000 women all take tamoxifen, our best guess is that three of these six women will **not** get breast cancer. It is possible that tamoxifen actually prevents as few as two women or as many as four women from getting breast cancer.

To capture the uncertainty associated with extrapolating from populations to individuals they recommended qualifying statements, such as:

There is no way of knowing [beforehand] whether you will be one of the women who gets breast cancer. In addition, if you take tamoxifen, there is no way of knowing whether you will be one of the women who benefited from it.

All risk communication exercises should take these and similar steps to make the message clear and to avoid the worst psychological pitfalls. Woloshin *et al.* (2002) had these considerations in mind when they constructed 'risk charts' to communicate the relative risks from different sources as a function of behaviour (smoking) and demography (gender and age) (Table 12.7).

12.6.2 Communicating comparative risks

When one risk is not unambiguously larger than another, any choice may turn out to be wrong. Risk management is balancing probabilities and consequences in a way that reflects the preferences of stakeholders (Finkel 1996).

Table 12.7. Risk control chart for women who currently smoke (from Woloshin et al. 2002).[a]

Age (years)	Vascular disease		Cancer					Infection			Accidents	Any cause
	Heart attack	Stroke	Lung	Breast	Colon	Ovarian	Cervical	Pneumonia	Influenza	AIDS		
20	Fewer than 1 death										2	8
25											1	10
30	1	1	1	1						1	2	14
35	1	1	2	1						2	2	22
40	3	2	4	2	1					2	2	32
45	6	4	10	4	1	1	1	1		1	2	50
50	13	6	21	5	2	1	1	1		1	2	80
55	24	9	36	6	3	2	1	2		1	2	125
60	45	16	65	7	4	3	1	4			2	199
65	50	16	85	9	6	3	1	8			3	301
70	88	30	124	10	8	4	1	16			5	470
75	153	58	137	11	11	4	1	33	1		7	725
80	261	99	136	12	14	4	1	66	3		11	>950
85	375	137	103	12	15	3	1	105	6		14	>950
90	462	154	64	10	14	2	1	140	12		15	>950

[a] Calculations for this chart are based on data that use the standard Centers for Disease Control and Prevention (CDC) definition of a smoker: someone who has smoked at least 100 cigarettes in her lifetime and smokes any amount now. The numbers in each row do not add up to the chance of dying from any reason because there are many other causes of death in addition to the ones listed here. AIDS = acquired immunodeficiency syndrome. Find the line next to your age. The numbers next to your age tell how many of 1000 women will die in the next 10 years from each cause of death.

Finkel (1996) argued that risk assessment is essentially comparative. Even if a single hazard is evaluated, it involves comparing the hazard with and without management. To understand risks, we need to compare them with concepts that make intuitive sense (such as an annual fatality risk of 10^{-6} is less than the risk of being struck by lightning).

Pairwise comparisons of risks fell into disrepute because comparing dissimilar risks was seen to be misleading. It doesn't make sense to compare the risk of exposure to pesticides with the risk of abseiling. Comparing risks may be invalid because the metrics differ. Comparisons often are misleading because they ignore uncertainty and the costs of prevention, and they fail to account for involuntary and inequitable exposure (Silbergeld 1994). Thus, risk comparisons can be used to manipulate reactions, rather than inform. They are even more misleading when they report relative importance based on the subjective opinions of experts, as though they were wholly objective observations.

Finkel (1996) suggested that it is possible to make risk comparisons informatively. We compare risks every day, breaking options into component attributes, making judgements about the value of attributes and aggregating the assessments before making a decision. MCDA formalizes this process. The 'unmet' challenge, according to Finkel, is to describe risks in informative and nonmanipulative ways. The first step is to use language that communicates variability and incertitude. Most comparative risk statements use point estimates. 'Which risk is larger' should be replaced by 'How confident can we be as to which risk is larger, and by how much, and with what consequences if I'm wrong?'. To account for variation, the question may be, 'Where, when and for whom is the risk larger?'. The tools that better communicate medical risks will help in communicating comparative risks in a wide range of environmental contexts.

12.7 Adaptive management, precaution and stakeholder involvement

Adaptive management was developed for the environment by Holling (1978) and Walters (1986). It had its foundations in the economic theory of the expected value of information (Ludwig and Walters 2002).

Raiffa and Schlaifer (1961) defined the expected value of information as the difference between the current state of knowledge and what might be learned from a given strategy. The more a strategy informs the manager

> **Box 12.5** · *Basic elements of adaptive management (see Holling 1978, Walters 1986)*
>
> Adaptive management involves:
>
> - planned learning though active management,
> - probing the system to elicit a response,
> - having a diversity of alternative, potential and applied management actions,
> - selecting management options based on alternative conceptual models that represent understanding of the system,
> - including environmental, social and economic considerations, and
> - monitoring to inform managers about the state of the system and the likelihood of the alternative conceptual models.
>
> It relies on a closed loop of activity. In many ways, the risk management cycle is just the adaptive management cycle in which formal environmental risk assessments and multiple models are accentuated.

about the state of a system, the more valuable it is. Valuable strategies retain flexibility for management to adapt through time. Such systems demand tailored monitoring programmes and regular, planned revision of strategies (Borsuk *et al.* 2001a,b).

Adaptive management has been applied in natural resource management, mainly in fisheries management. Precautionary approaches to harvest levels may result in lost benefits. The benefits of novel approaches may remain unexplored. On the other hand, aggressive approaches risk the collapse of fisheries. Adaptive management in this context uses routine management prescriptions to make large-scale, ongoing experiments. They may employ a range of management (harvesting) intensities and alternative strategies and monitor with sufficient power to detect important changes.

Management policies are chosen partly on the basis of their ability to accelerate learning. Marrying adaptive management with precaution means using observations, ecological principles and knowledge from other species and systems to build conceptual models. Managers experiment with alternatives and monitor the outcomes with sufficient power to detect important changes, while ensuring that the worst outcome from any action has acceptable risks. It means discarding and refining models as information accumulates (Parma *et al.* 1998). In short, it means applying the risk management cycle (Chapter 3) with the additional condition that options have tolerable risks.

The cost of adaptive management is tied to the cost of monitoring and of trying different management strategies, some of which may be suboptimal. These are set against the potential environmental and economic benefits of improving knowledge and reducing uncertainty.

For example, Punt and Smith (1999) described the management system for a stock of gemfish (*Rexea solandri*). It commenced as an open access fishery in the 1960s. The populations appear to have experienced a substantial decline during the 1970s and 1980s. The bulk of the historical catch came from bottom trawls of the upper continental shelf, in depths from 300 m to 500 m. Quota management was introduced in 1988. The first assessments of resource were made in the late 1980s, indicating substantial declines. Trawl catch was banned from 1993 to 1996. Following relatively strong recruitment into the population, trawling was reopened with the condition that there be a 50% chance that biomass (measured as 5+ males and 6+ females) exceeded 40% of the biomass estimated to exist in 1979.

To resolve differences in attitude towards management constraints among the members of the management committee, Punt and Smith (1999) used models to explore a range of harvest strategies and procedures, using a set of different assumptions about the ecology of the recruitment of the species. One management procedure used best estimates of model parameters, and another used estimates based on quantiles of the distributions of each parameter to generate conservative judgements. One harvest strategy aimed at achieving maximum sustainable yield. The other aimed at providing at least 50% certainty of having at least 40% of the 1979 population.

The results demonstrated tradeoffs between the expected catch and expected levels of depletion of the stock. But performance of management (in terms of yield and depletion) was sensitive to assumptions about the underlying biology of recruitment. The results of the simulations showed that the economic benefits of monitoring greatly exceeded the costs of running the surveys. The information had substantial value in providing assurances that management objectives had been achieved, and in informing the development of future management strategies. The results were embedded in a set of monitoring strategies and harvesting recommendations that were to be updated in subsequent years.

12.7.1   Involving stakeholders

Perhac (1998) identified three distinct rationales that may justify stakeholder involvement in a risk assessment. Each implies a different motivation.

First, stakeholders may be necessary to ensure public support and acceptance. This imperative is strongest for localized decisions such as locating hazardous waste facilities. The definition of stakeholders becomes a question of whose acceptance is necessary for political viability. Stakeholders may provide views on acceptable alternatives, leading to endpoints and management objectives that are valued by the public (Keeney 1992, McDaniels et al. 1999). If stakeholders are used in this way, adaptive management strategies should then be sensitive both to scientific uncertainties and the uncertain and dynamic nature of public preferences.

Second, stakeholders may be considered to be the appropriate source of value judgements, making it important to canvas a full spectrum of viewpoints. Behavioural aggregation (see Chapter 4) may be necessary to reconcile disparate positions. The appropriate course of action depends on whether the question is ethical, valuational or evidentiary.

Third, stakeholders may possess important factual knowledge such as consumption patterns or responses to management.

Stakeholders may also be involved because of legal obligations. For example, the US Comprehensive Environmental Response, Compensation and Liability Act, aimed at cleaning up old waste sites, requires extensive 'interested party' participation in risk assessments. Stakeholders may mobilize public sentiment, reduce antagonism and prevent or reduce liability issues (Warren-Hicks and Moore 1995).

But political viability may be bought at the price of scientific rigor, leading to less efficient management of things such as public health (Finkel 1996, Perhac 1998). Once the motivation for stakeholder engagement is defined, the challenge is to develop a mode of involvement without compromising whatever independent evidence may be relevant to making a decision (Perhac 1998).

In managing the gemfish fishery, for example, Punt and Smith (1999) involved a management committee composed of fishery managers (regulators), representatives from the catching and processing industries, government and independent scientists, an economist and a 'conservation member'. Smith et al. (1999a) argued that even the most sophisticated risk assessment strategies fail if they cannot accommodate effective stakeholder involvement in all stages of the assessment. They recommended

Table 12.8. *Stakeholder-identified objectives and measures for developing a water management plan and monitoring its effects on the Alouette River in British Columbia (after McDaniels* et al. *1999)*

Objective	Measure(s)
Avoid adverse effects of flooding (on people, property, cultural resources and perceptions)	Flood frequency
Promote river ecological health	Hectares of high-quality fish habitat Shape of the river hydrograph
Avoid cost increases for provincial electricity supply	$ cost of lost generation capacity Environmental cost of increased generation
Promote recreational opportunities	Number and quality of opportunities
Promote flexibility, learning and adaptive management	Learning opportunities Flexibility of management structure

stakeholder involvement in stock assessment, setting research priorities, enforcement and decision making, none of which are the traditional domains of nonexpert stakeholders.

McDaniels *et al.* (1999) used stakeholder values to define the objectives of ideal management of flows in the Alouette River in British Columbia, Canada. They also employed an 'informative' decision rule in which each participant was asked what alternatives they could support, a process termed 'approval voting'. The final decision was not based on majority opinion. Instead, the outcomes of discussions and approval voting were used to form a report on tolerable alternatives to an appointed decision maker.

The process identified five broad objectives and a set of measures (endpoints) for each objective (Table 12.8). Having begun with a somewhat contentious social environment, after 15 meetings, they reported complete consensus (approval) on all major issues for at least one management option.

There may be more than one mental model, and more than one quantitative model. Hilborn and Mangel (1997) argued that it is entirely sensible to have multiple descriptions of nature. They are, after all, metaphorical. Several may be consistent with the data. Hilborn and Mangel advised that we should avoid acting as though one model is true and the others false. We would do better to weigh the costs and consequences of each against the likelihood that it is correct.

Experience in fisheries management suggests that useful predictive models often are simpler than intuition suggests. But if models are used to explore the consequences of decisions, more complexity may be needed. Certainly, the simplest way of dealing with irreconcilable differences of opinion among stakeholders is to retain them in the analysis as different alternative models and parameter values. Alternatives may be discarded and others constructed as information accumulates.

12.7.2 Equity and efficiency

Decisions involve criteria on which a final judgement will be made. It may be to minimize cost, to elect the option that has greatest support and that also satisfies management objectives, or to select the 'best' option on some other grounds.

The outcomes of decision processes may lead to the inequitable distribution of risk. It can result in what Wigley and Shrader-Frechette (1996) call 'environmental racism'. In the USA, minorities disadvantaged by education, income and occupation bear disproportionate environmental risk. In most circumstances they have assumed the risks involuntarily. It is likely that the same kinds if inequities exist in other countries, as risk management decisions are influenced by political and social context.

Risk equity is defined as the distribution of risk among individuals in a population. Risk efficiency is defined as the magnitude of risk reduction obtained under an option. Efficiency might also consider the magnitude of risk reduction within a stratum of a population, against the background risks they already carried. Good decisions should consider equity and efficiency. Risk reduction strategies could favour those who already carry the largest individual risks from all sources.

Both risks and risk-reduction decision options should be ranked. A 90% reduction of the third largest risk may be more valuable than a 5% reduction in the largest risk. Solutions may affect more than one risk. Ranking decision options can sidestep the issue of how to aggregate risks for the purpose of ranking.

12.8 Conclusions

The effectiveness of risk management depends on trust, more easily destroyed than created (Slovic 1999). The asymmetry is caused by things such as the visibility of negative events, and that sources of bad news are seen as more credible than sources of good news. The public is aware that special interest groups use experts selectively to find solutions that are best for the interest group, often in an adversarial legal system.

Risk communication and management work best when those affected by the risks are involved closely and continuously in the risk assessment process. The best risk assessments involve stakeholders and experts in an iterative process, making a marriage of the technical and social dimensions of risk. The best we can hope for is decisions informed by honest and complete risk assessments that seek tolerable outcomes that are robust to as broad a range of uncertainty as possible.

Adams (1995), O'Brien (2000), Fischer (2000) and many others have been critical of technical risk assessments. O'Brien characterized risk analysis as a process which makes the illegitimate exercise of power by scientists and vested interests acceptable. She saw it as a collection of selective assumptions, data and assertions that justify business as usual. Risk assessments may create an aura of unassailable authority (Walton 1997, Fischer 2000). Much of the evidence concerning the importance of psychology in expert judgement supports their position.

O'Brien's (2000) solution was to require businesses and government to explore publicly, in understandable language, options for causing least environmental damage. Stakeholders should formulate the options. To do so implies that preferences are shaped by context and that there is no single, best vantage point (Fischer 2000).

I agree with this position. It is beyond the capabilities of technical risk analysts to anticipate the full range of legitimate points of view and ideas about cause and effect. The role of risk assessment is to cross-examine the ideas of stakeholders, to ensure internal consistency, to eliminate linguistic uncertainty and other sources of arbitrary disagreement, to clarify the implications of assumptions, and to leave honest disagreements in plain view. This role depends on partnerships between stakeholders and scientists (Poff *et al.* 2003) in which the position of the analyst is subordinate to the stakeholder. In this context, quantitative risk assessment has an essential role to assist people to critically examine their ideas and to understand the ideas and values of those with whom they disagree.

Glossary

acceptable daily intake (ADI) Termed a reference dose by the EPA, it is the amount of a food ingested every day without experiencing much likelihood of an adverse effect.

acute (effect) Short term, usually 96 hours.

adaptive sampling The sample size or the interval between samples depends on the position of the current sample. The relative costs of 'false alarms' versus 'failure to act' should contribute to the allocation of sampling effort such that warning thresholds equalize the relative costs of the two kinds of errors.

advection The transportation of a chemical from one medium to another by a carrier unrelated to the presence of the chemical. Such vehicles include dust, rainfall, food, or sediment particles suspended in a water column.

ALARP principle The idea that risks should be reduced to a level called 'as low as reasonably practicable'.

algorithm A logical arithmetical or computational step by step procedure that, if correctly applied, ensures the solution of a problem.

alternative hypothesis *see* Null hypothesis.

ambiguity

Uncertainty arising from the fact that a word can have more than one meaning and it is not clear which meaning is intended.

analytical hierarchy process (AHP)

A method for converting subjective judgements of importance into scores. It depends on a numerical interpretation of answers to questions such as; 'how important is (criterion) A compared to (criterion) B?' It may be used to quantify priorities.

anchoring

The tendency to be influenced by initial estimates – people will be drawn to the guesses made by others, and will defer their judgements to people they believe have greater authority.

Anderson–Darling (A-D) statistic

Sums the squared vertical distances between two cumulative distributions. The A–D statistic is useful in risk analysis because it pays greater attention to the tails of the distribution than the Kolmogorov–Smirnoff (K–S) statistic.

arbitrary evidence

Describes the situation where there is no evidence common to all subsets, even though some subsets have elements in common

area under the curve (AUC)

Measures the performance of a measure over a range of thresholds, equivalent to a Mann–Whitney statistic

assessment endpoints

Formal expressions of the environmental values to be protected. They provide a means by which management goals may be identified, measured, audited and evaluated.

atrazine	A chemical used to control weeds in crops.
autocorrelation (temporal)	If the value of a variable, x, at time t, depends in some way on the value of x at time $t-1$, the variables will be temporally autocorrelated.
average run length (ARL)	For in-control processes, it is defined as the average number of samples between points that result in false alarms. The ARL for out-of-control processes is defined as the average number of samples between true alarms, the average delay before an alarm is raised.
averaging	*see* Mixing or averaging probabilities.
backing	Background assumptions or foundations that support a warrant, including axioms, theory and formal principles.
Bayesian analysis	Provides a mechanism for combining knowledge from subjective sources with current information to produce a revised estimate of a parameter.
Bayesian credible intervals	The shortest interval that contains a specified amount of a (posterior) probability distribution, or the amount of a probability distribution contained within specified bounds.
Bayesian networks	(also called probability networks, influence networks and Bayes' belief nets) Graphical models that represent relationships among uncertain variables, in which probabilities may be estimated subjectively and updated using Bayes' theorem.
behavioural aggregation	Uses behavioural strategies, rather than quantitative ones, to arrive at a

belief

combined estimate (*see* closure, Delphi technique and resolution). The degree to which a proposition is judged to be true, often reported on an interval (0, 1) or percent scale creating an analogy with 'chance'.

belief function

see Dempster–Shafer structures.

benchmark dose

The dose that corresponds to a predetermined level of response, such as the dose at which, say, 10% of the population exhibit an effect.

benefit–cost analysis

Examination, usually in economic terms, of the advantages and disadvantages of a particular course of action.

beta (betaPert) distribution

Defined by three parameters, usually used to represent the probability of a random event in a series of trials. It can take on a wide variety of shapes, including both symmetric and asymmetric forms (either left or right skewed), and horseshoe shapes.

binomial distribution

A statistical distribution giving the probability of obtaining a number of successes in a specified number of Bernoulli trials.

bioconcentration factor

The steady-state ratio of chemical concentration in organisms relative to the concentration of the chemical in the media in which the organisms live.

'bounded rationality'

Perception and uncertainty limit the ability of people to achieve goals. People construct simple models to deal with these difficulties. The key idea is that people 'satisfice' rather than optimize.

bounds (statistical)

Limits within which we are sure (to some extent) the truth lies.

	Generally, upper and lower bounds are intended to provide an envelope that brackets the true value and the majority of possible outcomes.
Brier score	Summarizes one of the elements of calibration between judgement and outcome. It is the weighted average of the mean squared differences between the proportion correct in each category and the probability associated with the category.
***c*-charts**	*see* Control (statistical process)
calibration	The likelihood that the expert's probabilities correspond with a set of repeated experimental results, the probability that the difference between the expert's judgement and the observed values have arisen by chance.
calibration curves	Expert judgements plotted against reality.
calibration of models	Adjusting model parameters, structures and assumptions so that they fit available data and intuition, i.e. refinement of ideas.
carrying capacity	The maximum number of individuals that a given environment can support indefinitely, usually determined by the organism's resource requirements.
cause-consequence diagram	Another name for a logic tree.
central tendency	The tendency of the values of a random variable to cluster around the mean, median and mode.
chance	The frequency of a given outcome, among all possible outcomes of a random process, or within a given time frame.

charts	*see* Control (statistical process): charts.
chi-square test	Compares the numbers of sample observations in discrete classes with those expected under the proposed distribution.
chronic effects	Long-term effects.
claim	An assertion or proposition, usually the end result of an argument, but neither necessarily certain or true.
closure	*see* Sound argument, consensus, negotiation, natural death, procedures guide.
coefficient of variation	A measure of the relative variation of a distribution independent of the units of measurement; the standard deviation divided by the mean.
cognitive availability	The tendency to judge the probability of an event by the ease with which examples are recalled.
common failure mode	Occurs when events influence several branches of a logic tree simultaneously, so that responses are correlated, making the probabilities on the tree wrong.
conceptual models	Verbal models, diagrams, logic trees, or sets of mathematical equations representing components in a system, including input and output, flows, cycles, system boundaries, causal links and so on.
conditional lethality	Deaths per 100 000 from a disease given the disease is diagnosed.
conditional probabilities (statistics)	The probability of occurrence of an event given the occurrence of another conditioning event.
confidence	The degree to which we are sure that an estimate lies within some distance of the truth.

confidence intervals	In the long run, someone who computes 95% confidence intervals will find that the true values of parameters lie within the computed intervals 95% of the time.
confusion matrix	A two-by-two classification table showing the number of true and false positive and negative predictions.
congruential method	The most popular algorithm for generating pseudorandom numbers.
conjunctive pooling	(A∩B) retains only those opinions that are common to all experts.
connectedness	There are preferences for all outcomes.
consensus	A means of achieving closure in which the experts agree that a particular position is 'best'.
consensus tree	Also known as an average tree, it summarizes the common properties of a set of trees linking the same set of objects (vertices).
consequence curve	The cumulative distribution of LC50s for a large set of species.
consistent evidence	At least some evidence is common to all of the subsets.
consonant evidence	Can be represented by nested subsets. The content of the smallest set is contained within the next largest, and so on.
context	The setting of the problem at hand.
contingent valuation	A method for valuing intangibles in benefit–cost analysis. It uses questionnaires and/or interviews to elicit preferences and demand functions for environmental goods and services.
contraction (of info–gap models)	The nominal value, $\tilde{v}$ is the only possible value in the absence of uncertainty.

convergence	Behavioural consensus techniques in which participants agree to negotiate to resolve conflict.
co-optation	A means of achieving resolution in which experts acknowledge that the conflict is 'resolvable', and sound argument, consensus or negotiation may bring closure.
correlation (statistical)	The extent of correspondence between the ordering of two variables.
correlation coefficient	A statistic measuring the degree of correlation between two variables.
cost-benefit analysis	*see* Benefit-cost analysis.
cotton pyrethroid insecticides	Chemicals used to control insects. They interfere with ion channels in insect nervous systems.
credibility	The believability of detail in a narrative or model (acceptance of ideas based on the skill of the communicator, the trust placed in a proponent).
credible intervals	*see* Bayesian credible intervals.
cumulative probabilities	A cumulative probability distribution gives the probability, p, that the random variable X will be less than or equal to some value x. It sums the value of the probability distribution from left to right.
CUSUM charts	Make use of sequential information and are more sensitive to correlated processes and small changes than are the other kinds of control charts. CUSUM charts accumulate deviations above the target in one statistic ($C+$) and deviations below the target in another ($C-$).
decision	A choice between two or more acts, each of which will produce one of several outcomes.

decision tables	Tables that link acts, states and outcomes. The acts refer to the decision alternatives, the states refer to the relevant possible states of the system, and the outcomes refer to what will occur if an act is implemented in a given state.
decision trees	Event trees in which one or more of the branch-points are decisions; a graphical representation of decision pathways.
Delphi technique	A form of behavioural aggregation that consists of questionnaires, elicitation, aggregation of results, review of combined results by experts and iteration of feedback until consensus is achieved.
demographic variation	The chance events in the births and deaths of a population.
demographic stochasticity (with reference to risk assessment)	The variation in the average chances of survivorship and reproduction that occur because of the demographic structure of populations made up of a finite, integer number of individuals; random sampling of distributions for variables which must logically take a discrete integer value.
Dempster's rule	A generalization of Bayes' theorem that applies the conjunctive AND to combine evidence. More formally, the probability density of the combined evidence is limited to the interval over which experts agree.
Dempster-Shafer structures	Sets of plausible values that the available evidence does not distinguish. The lower bound is termed a 'belief' function. The upper bound is termed a 'plausibility' function.

density dependence

This is when survival or fecundity is a function of the difference between the total number of adults and the carrying capacity of the environment, creating a feedback between population size, and the rate at which the population grows.

dependency of a variable (statistical)

Implies that variation in one variable contributes to or causes the values in another variable.

detectable effects sizes (*d*)

If data x are not statistically significantly different from H_0 (the null hypothesis), and the power to detect effect d is high, then x constitutes good evidence that the actual effect is no greater than d. Conversely, if data x are not statistically significantly different from H_0, and the power to detect effect d is low, then x constitutes poor evidence that the actual effect is no greater than d.

deterministic model

A model in which there is no representation of variability.

deterministic sensitivity

If a parameter is changed by a small amount in the region of the best estimate, it is the magnitude of change we see in model output, relative to the amount of change in the parameter.

dichotomy

Division into two parts or classifications, especially when they are sharply distinguished or opposed.

diminishing returns

The utility resulting from any small increase in wealth is inversely proportional to the quantity of goods already possessed.

disagreement	*see* alternative hypothesis, epistemic uncertainty, information disagreement, preference disagreement, semantic disagreement.
disjoint evidence	Implies that any two subsets have no elements in common with any other subset.
disjunctive pooling	(A∪B) retains the full breadth of opinion offered by all experts.
dispersal (with reference to PVAs)	The movement of individuals among spatially separate patches of habitat, including all immigration and emigration events.
EC $_{50}$	Median effective concentration, the concentration at which 50% of individuals exhibit the effects of exposure to a contaminant within a specified time.
ecosystem services	The processes through which natural ecosystems sustain human life.
ecotoxicology	A basis for assessing whether chemicals are likely to have adverse effects on ecosystems.
effect size	The level of impact that is required to be detected by a study, given that there is an effect; the magnitude of change that we wish confidently to detect.
Electre method (for preference ranking)	*see* 'Outranking'.
empirical	Derived from or relating to experiment and observation rather than theory.
endpoints	*see* Assessment endpoints, management goals, measurement endpoint.

environmental aspect A human activity, product or service that can interact with the environment (such as emissions, chemical handling and storage, road construction).

environmental effect Any change to the environment, whether adverse or beneficial.

'environmental racism' Occurs when minorities disadvantaged by education, income and occupation bear disproportionate environmental risk, in most cases through involuntary risks.

environmental uncertainty (in PVAs) The effects of the environment on vital rates, usually represented by time-dependent stochastic survivorships and fecundities.

enzyme Biological catalyst, usually a protein, which increases the rate of a reaction.

epistemic uncertainty Reflects incomplete knowledge, including measurement error, systematic error, natural variation, model uncertainty; and subjective judgement.

equipossible / equiprobable Different events occur with equal probabilities. When uncertainties are unknown, it is common practice to assign equal probabilities to different events. The logic is that when there is no evidence to the contrary, the best assumption is that all events are equipossible, and therefore equiprobable.

'equivalence' region The discrepancy between the null case and a response that may be considered unimportant or tolerable.

event tree A form of logic tree, an event tree begins with a triggering event and follows all possible outcomes to their final consequences (event tree analysis).

evidence Direct experimental observation of cause and effect, probability or frequency.

expected dose The average daily rate of intake, ADRI.

expected utility The magnitude of an anticipated gain, discounted by the chance that the outcome will be achieved.

expected value of information The difference between the current state of knowledge and what might be learned from a given strategy.

expert Someone who has knowledge, skill, experience, training or education about an issue at an appropriate level of detail and who is capable of communicating their knowledge. *See also* substantive expertise and normative expertise.

exponential distribution The time between random, successive events, sometimes called the negative exponential distribution.

exposure pathway The way a chemical reaches a target, usually an assessment endpoint.

failure modes Describe the way in which a product or process could fail to perform its desired function, defined in terms of the needs, wants and expectations of people (shareholders, customers or stakeholders).

failure modes and effects analysis (FMEA) A systematic process for identifying failures before they occur, with the intent to eliminate them or minimize the risk associated with them.

false negative/ **false positive/false alarm**	*see* type I and type II errors.
fault tree	A form of logic tree, linking chains of events to the outcome (fault tree analysis).
fecundity (with reference to PVAs)	The number of offspring born per adult, and alive at the time of the next census.
final acute value (FAV)	A value summarizing the susceptibility of a group of organisms to a toxic substance. Typically, a model is selected to represent the variation in LC_{50} values over all species. Its parameters are estimated from the sample data.
fixed probability	Method of asking a series of questions to elicit points on a distribution in which values of the variable that bound specified quantiles were elicited. Answers to these questions approximate points on a cumulative density function (a 'cdf'). The assessor concentrates on medians, quartiles and extremes (such as the 1% and 99% limits).
fixed value	Method of asking a series of questions to elicit points on a distribution in which the assessor asks experts to judge the probability that the variable lies within a specified interval. The answers approximate points on a probability density function (a 'pdf').
free-riding	When people receive the benefits without bearing risk or making a contribution.
frequentist statistics	See probabilities as relative frequencies (as opposed to Bayesian statistics).

fugacity	A measure of a chemical's tendency to escape from the phase it's in (*see* Partitioning coefficient).
fuzzy numbers	Essentially stacks of intervals, each level of which represents a different degree of surety about the boundary.
goodness of fit	The probability that the fitted distribution could have produced the data.
hazard	A situation that in particular circumstances could lead to harm.
hazard index	The sum of the hazard quotients for all of the substances to which an individual is exposed, and that act by a similar mechanism.
hazard matrices	A matrix of interactions between human activities and valued components of the environment that may be affected by the actions.
hazard and operability analysis (HAZOP)	A kind of structured, expert brainstorming session that uses conceptual models and influence diagrams together with guide words such as 'more of', 'less of' and 'reverse flow' to prompt the thinking of a small team of experts.
hazard probability rank	A verbal description of the relative likelihood of the event (the hazard), ranging from frequent or continuous to improbable.
hazard quotient	The estimated exposure to a chemical divided by a toxicity threshold, i.e. the calculated daily dose divided by the reference value representing a tolerable daily dose.
'hedonic' pricing	Using market valuations such as house prices for a benefit-cost analysis.

hierarchical holographic models (HHM) — An approach that recognizes that more than one conceptual (or mathematical) model is possible for any system, and tries to capture the intuition and perspectives embodied in different conceptual and mathematical models of the same system.

hormesis — Beneficial responses observed to some toxicants at very low exposures because response systems have beneficial effects until the system becomes overloaded at higher concentrations.

hydrolysis — The term given to chemical reactions in which organic compounds react with water to produce other compounds.

hypothesis — The error rates, α and β, define acceptance or rejection of the null hypothesis. They may be adjusted so that their ratio equals the ratio of the 'costs' of making a type I error versus a type II error. *See also* null hypothesis, type I/type II errors.

'illusory certainty' — The belief that scientific tests (such as mammograms) are infallible or highly reliable.

incertitude — Lack of knowledge about parameters or models (including parameter and model uncertainty).

indeterminacy — Future usage of terms is not completely fixed by past usage.

indicators — Biological entities whose interactions with an ecosystem make them especially informative about communities and ecosystem processes.

influence diagrams — A visual representation of the functional components and dependencies of a system. Shapes (ellipses, rectangles) represent variables, data and parameters. Arrows link the elements, specifying causal relations and dependencies.

information disagreement   Experts differ in their views on measurements, the validity of experiments, the methods used to obtain observations, or the rules of inference.

interval arithmetic   Arithmetic operations defined for ranges.

interval probabilities   Represent an expert's degree of belief, where lower and upper bounds encompass the range of beliefs.

Kent scales   Translate the linguistic interpretations of uncertainty into quantitative values.

Kolmogorov–Smirnoff (K–S) statistic   Compares the maximum distance between an empirical cumulative distribution and a theoretical, fitted distribution, with a table of critical values.

LCxx   Lethal threshold concentration at which xx% of individuals die within a specified time. L stands for 'lethal', C is 'concentration'.

likelihood   The extent to which a proposition or model explains available data (the relation between hypothesis and evidence). *See also* maximum likelihood.

linguistic uncertainty   Arises because language is not exact, including vagueness, context dependence, ambiguity, indeterminacy and underspecificity.

LOAEL/LOEL   Lowest observed adverse effect level, the smallest dose at which a

LOEC

Lowest observed effect concentration, *see* LOAEL/LOEL.

logic trees

Diagrams that link all the processes and events that could lead to, or develop from, a hazard. They are sometimes called cause–consequence diagrams. *See also* fault tree and event tree.

management goals

Statements that embody broad objectives, such as clean water or a healthy ecosystem.

Markov Chain Monte Carlo (MCMC)

Useful in analysing large and complicated data sets with some form of hierarchical structure such as chemical speciation and exposure relationships, and life history parameters in natural populations of plants and animals. It is used often (although not exclusively) in conjunction with Bayesian prior distributions to estimate parameters from data in circumstances in which ordinary methods are too complicated.

MaxiMin strategy

Given an ordinal scale for outcomes, in which larger numbers represent greater utility, the procedure is to identify the minimal outcome associated with each act, and select the act with the largest minimal value.

measurement endpoint

Quantitative physical and biological responses, things that we can measure, such as toxic effects on survival and fecundity, operational definitions of assessment endpoints that are, in turn, conceptual representations of management goals.

statistically significant effect has been demonstrated.

measurement error	Error caused by imprecise and inaccurate instruments and operators.
median	The point that divides an ordered set of data into two equal parts.
metapopulation	A set of local populations which interact via individuals moving between local populations.
minimum expected population size (in PVA)	The average minimum population size from a set of forecasts, summarizing the chances of a population falling below a lower threshold within a specified time period.
minimum viable population	The smallest isolated population having an acceptable chance of surviving for the foreseeable future.
mixing or averaging probabilities	Probabilities associated with belief may be combined as weighted linear combinations of opinions.
model averaging	Combines the predictions of a set of plausible models into a single expectation in which individual weights reflect the degree to which each model is trusted.
model uncertainty	Uncertainty arising from the fact that, often, many alternative assumptions and models could be constructed that are consistent with data and theory, but which would generate different predictions.
monitoring	Sampling and analysis to determine compliance with a standard or deviation from a target or prediction, or to measure the state and response of the system to management strategies.
Monte Carlo analysis	Uses statistical distributions to represent different kinds of uncertainty, combining them to generate estimates of a risk.

motivational bias Biased assessments that arise because the results benefit the people who make the assessments.

multi-attribute utility A utility-based criterion for decision making in which a utility function that incorporates the outcomes of all important attributes is used in place of economic value.

multicriteria decision analysis (MCDA) Measures preferences by eliciting and ordering judgements from people affected by a decision.

multicriteria mapping (MCM) Takes the philosophical position that risks are socially constructed and (mostly) difficult to measure and uses the preferences of individuals to explore alternative management options.

natural death (in behavioural aggregation) A means of achieving closure in which the conflict declines gradually and is resolved by ignoring it, usually because it turns out to be unimportant.

natural variation Environmental change (with respect to time, space or other variables) that is difficult to predict.

'negative' null hypothesis An a priori assumption that there is no effect of a treatment, or no impact of an activity.

negotiation A means of achieving closure in which an arranged resolution is reached that is acceptable to the participating experts and that is 'fair' rather than correct.

nesting (of info-gap models) A family of nested sets of possible values of the uncertain entity.

no observed effect level (NOEL, or concentration, NOEC) The highest amount of a substance for which no statistically significant effect was found in a statistical test between a treatment and a control.

nonmetric multidimensional scaling (NMDS)	A nonparametric ordination technique designed to represent multivariate information in two or three dimensions efficiently, so that it can be visualized.
normal distribution	Parameters that result from the sum of a large number of independent random processes.
normative expertise	The ability to communicate, including interpersonal skills, flexibility and impartiality.
normative theories (of rational consensus)	Expert judgement takes the form of degrees of belief about facts and in order to estimate an uncertain quantity, an analyst may combine the distributions provided by more than one expert. There is an underlying assumption that there is a fact and the job of the experts is to estimate it.
number needed to treat	A concept in medicine that gives the number of people that need to participate in a treatment to save one life.
numerical aggregation	Uses quantitative strategies, rather than behavioural ones, to arrive at a combined estimate. *see* Bayesian analysis, Dempster's rule.
odds ratio	Expresses the probability of one outcome relative to the probability of its opposite.
'outranking' (multicriteria decision analysis)	Methods used when preferences cannot be expressed as a unique numerical function, when at least one criterion is not quantitative or when compensating gains and losses among criteria are not clear; the 'Electre' method has been broadly applied in this context.

p-bounds

A modelling method which does not require specific guesses about distributional shape; 'p-bounds' calculations bound arithmetic operations, making only those assumptions about dependencies, distribution shapes, moments of distributions, or logical operations that are justified by the data.

p-box

'Sure' bounds on cumulative distribution functions.

p-value

The probability associated with the observed data under the assumption that the null hypothesis is true – if the null hypothesis is true, and if an experiment is repeated many times, the *p*-value is the proportion of experiments that would give less support to the null than the experiment that was performed.

partitioning coefficient

The ratio of concentrations of chemical in the two phases at equilibrium.

paternalistic model for stakeholder involvement

Governments or regulators invite participation under strictly controlled conditions such as public comment phases.

photolysis

Reactions caused by light, such as sunlight photolysis of organic chemicals in surface waters, on soil and in the atmosphere.

plausibility

The relative chance or relative degree of belief (the rank) assigned to elements of a set of possible outcomes.

plausibility function

see Dempster-Shafer structures.

point estimates

A measurement or estimate of the value of a parameter that ignores uncertainty.

Poisson distribution	Models the number of occurrences of an event that are likely to occur within a time, t, when the probability of an event occurring per interval of time is constant, and independent of any other events that may have occurred. The distribution is used to estimate the number of failures in a repetitive process such as manufacturing, and the distribution of births per female in natural populations of plants and animals.
Poisson process	The occurrence of independent, random events.
population viability analysis (PVA)	The use of population models to estimate extinction risk and to compare management options in terms of the risk of decline of a population or metapopulation.
possibility	The set of things (events, outcomes, states, propositions) that could be true, to which some (nonzero) degree of belief or relative frequency might be assigned.
posterior probabilities	Estimates of the prior probabilities (probabilities, distributions) are combined with Bayes' theorem to give posterior probabilities, the updated degrees of belief that the hypotheses are true.
power (statistical)	The probability of detecting a given true difference between two 'populations' (data sets) when using a statistical test.
power curve (statistical)	The relationship between number of samples and the power of a test plotted on a graph.
precautionary principle	Where there are threats of serious or irreversible damage, lack of full

scientific certainty shall not be used as a reason for postponing cost-effective measures to prevent environmental degradation.

prediction intervals Provide an estimate of the confidence interval within which you expect the next single observation to lie.

preference disagreement Experts have different preferences for methods and standards used to evaluate claims.

preference matrix The result of pairwise comparisons of preferences between criteria. The elements of the first (dominant) eigenvector of the matrix can be used as the weights for each criterion.

principle of insufficient reason Suggests that when making decisions under uncertainty each state is equally likely because there are no data on which to discriminate among them.

prior probabilities Represent the probability that the data would be observed, if the various hypotheses were true. It requires the experts to specify their subjective belief in a distribution, prior to the analysis, even if no data are available.

probabilistic benefit-cost A utility-based criterion for decision making in which benefits and costs of alternatives are estimated in economic terms and expected value (weighted by risk) is used to find the option with the greatest net benefit.

probabilistic logic trees The AND and OR statements that make up fault trees are subject to ordinary probabilistic calculus.

probability	The statistical frequency (or relative frequency) with which an event is expected to occur, or the degree of belief warranted by evidence; *see also* belief, bounds, chance, confidence, credibility, cumulative probabilities, likelihood, plausibility, possibility, risk, posterior probabilities, prior probabilities, tendency.
probability density function	The distribution of relative frequencies of different kinds of events (outcomes).
probability networks	*see* Bayesian networks.
problem formulation	Identification of the scope of the problem, including the ecological, social, geographic, temporal and political limits.
procedures guide (for expert elicitation)	A method for achieving aggregation of expert opinion, including weighted combination based on an expert's performance against known standards.
process model	A mathematical representation of a conceptual model, such as a population model, an expert system, a logic tree or any other quantitative model.
'prosecutor's fallacy'	In law, the probability of an event (p(event)) is sometimes confused with the probability that someone is guilty, given the event (p(guilty | event)).
pseudorandom numbers	Sequences of numbers that satisfy statistical tests for a random number sequence, generated by algorithms; *see also* congruential method.
public participation	Deliberation on the issues by those affected by a decision.

QSAR (quantitative structure activity relationship) Equations that describe the relationship between the structure of a compound and its toxicity.

quantiles A point in the ordered data set below which a specified percentage of the data lie.

quartiles The point below which 25%, 50% or 75% of the data of an ordered set lie. The third quartile is the point below which 75% of the data lie. The interquartile range is the interval between the first and third quartiles and it encloses 50% of the data.

range (statistical) The interval between the smallest and largest values of an ordered set.

rank correlations *see* Spearman rank correlation.

rational consensus *see* normative theories (of rational consensus).

rational subgroups (in control limits) Those that maximize the effects of 'assignable causes' between groups and minimize their effects within groups.

'real' risks Objective, analytical, evidenced-based risks.

reasonable maximum exposure (RME) A conservative exposure case (i.e. well above the average case) that is still within the range of possible outcomes.

rebuttal The conditions under which and the reasons why the claim does not necessarily hold, and which may apply to the warrant or the claim.

reference dose (RfD) The dose that accounts for uncertainty in interspecies variability (UF_A), intraspecies variability (UF_H), uncertainty in the duration of exposure and the duration of the study (UF_S), the use of a LOAEL, incomplete databases, age dependence of thresholds, and additional modifying factors (MF) such that the true threshold dose of an ingested substance is lower for the most sensitive members of a population.

reference lottery
A tree-based method to elicit a judgement for a probability, p, that event E will occur.

regret
The difference between the result of the best action and the result of the action taken.

relative risk
The chance of an event in an exposed population relative to the chance in the unexposed population, i.e. the change in risk experienced by a group, relative to the risk they would have experienced had they belonged to the reference class.

rights–based criteria
Consideration of process, allowed actions and equity independent of benefits and costs.

risk
The chance, within a time frame, of an adverse event with specific consequences.

risk analysis
Evaluation and communication of the nature and extent of uncertainty.

risk assessment
Completion of all stages of the risk management cycle, a marriage of risk analysis methods, adaptive management, decision tools, monitoring and validation.

risk aversion
When people prefer to have a smaller reward with greater certainty, than a larger reward with less certainty.

risk charts
Communicate the relative risks from different sources as a function

	of behaviour (e.g. smoking) and demography (e.g. gender and age).
risk efficiency	The magnitude of risk reduction obtained under an option. Efficiency might also consider the magnitude of risk reduction within a stratum of a population, against the background risks they already carry.
risk equity	The distribution of risk among individuals in a population.
risk priority number (RPN)	The product of three quantities, severity, occurrence and detection, used to set priorities for action on hazards and to identify elements that require additional planning.
risk ranking	Risk assessment that relies on qualitative, usually subjective estimates of likelihoods and consequences to rank hazards. *See also* risk assessment tables, AS/NZS 4360 1999, failure modes and effects analysis (FMEA), hazard and operability analysis (HAZOP), hierarchical holographic models (HHM), comparative risk, relative risk.
robust Bayesian analyses	Estimate a set of posterior distributions for a quantity, based on prior distributions and likelihoods selected from classes believed to be plausible by the analyst.
robust decisions	Decisions that provide a satisfactory outcome, despite uncertainty, thereby avoiding unacceptable outcomes.
robust satisficing	A strategy that maximizes the reliability of an adequate outcome.

ROC (receiver operating characteristic) curves	Used to judge the effectiveness of predictions for repeated binary decisions (act / don't act, present / absent, diseased / healthy, impact / no impact), a special case of operating characteristic curves in which a binary decision is judged against a threshold for a continuous variable (an indicator), built around confusion matrices that summarize the frequencies of false and true positive and negative predictions, of various values of a prediction threshold.
scenario analysis	An approach to creating alternatives for problem formulation, constructed and communicated with a story line in which events unfold through time through a series of imagined causes and effects.
scenarios	Shared, agreed mental models, internally consistent descriptions of possible futures created in structured brainstorming exercises.
second order (two-dimensional) Monte Carlo	The stochastic parameters are themselves drawn from statistical distributions that reflect uncertainty about true values.
semantic disagreement	Misunderstand the meanings of words.
sensitivity analysis (for models)	An analysis of how a model's output responds to changes in a variable or an assumption.
Shannon entropy index	Measures the 'information' contained in a distribution including expert opinions.
sigmoidal curve	An 'S' shaped curve which is typical when the logarithms of chemical concentrations are plotted against

the percentage of organisms exhibiting an effect, i.e. dose–effect curve.

snowballing
Generating lists of potential participants in elicitation exercises by asking people to recommend other people.

Spearman rank correlation
Measures the similarity in the rank order of objects (samples) in two lists. When objects have the same ordering, they have a rank correlation of 1. When the orders are reversed, the rank correlation is −1. When the orders are random with respect to one another, the rank correlation is 0.

stakeholder
In law, someone who has custody of the possessions of other people. Generally, in risk assessment, it is anyone who has an interest in an issue, or anyone who shares the burden of the risk of a wrong decision. In a social context, a stakeholder usually is an individual or a representative of a group affected by or affecting the issues in question.

stakeholder expert
An expert retained by a stakeholder or group to evaluate technical information and represent the stakeholders in technical deliberations.

stochastic dominance
The extent to which a given decision is 'best', depending on the source or magnitude of uncertainty.

stochastic dynamic programming
A Markov chain that gives the transition probabilities among possible states, maximizing the chances of a desired outcome.

stochastic models

A model in which at least some of the parameters are drawn from statistical distributions, or in which there is some other explicit recognition of uncertainty.

strategic risk management

Using risk assessment to determine organizational activities, the process of deciding what actions to take in response to a risk.

stressors

The elements of a system that precipitate an unwanted outcome.

structure activity relationship (SAR)

Establishing the kinds of tests and the expected levels for safety of a new chemical by inferring its mode of action and toxicity from similarly configured chemicals.

subjective belief

Personal judgement in the truth of a proposition.

substantive expertise

Technical experience and training.

supremacy

A means of achieving resolution in which expert disagreements are tested to determine the 'correct' position; experts agree on the grounds upon which 'supremacy' will be based, and what evidence would cause them to alter their position.

sure bounds

With reference to interval arithmetic – a range encompassing 100% of the data.

surprise index (surprise frequency)

The percentage of true outcomes that lie outside expert confidence regions.

systematic error

Errors that occur when measurements are biased; the difference between the true value of a parameter and the value to which the mean of the measurements converges as sample sizes increase.

tendency
The physical properties or traits of a system or a test that result in stable long-run frequencies with repeated trials, or that yield one of several possible outcomes.

theory-based inference
Relationships and causal mechanisms inferred from an understanding of physical or ecological principles.

transformation
The set of processes and reactions by which chemicals change composition, including physical processes such as oxidation and biologically mediated transformations.

triangular distribution
Defined by three parameters (lower bound, best guess (central tendency), upper bound), based on providing the maximum and minimum 'plausible' values for a parameter, knowing that more extreme outcomes are possible, if unlikely. It has no theoretical basis but is used particularly to represent expert judgement or belief.

type I, type II errors
Monitoring systems should: (1) tell us there *is* a serious problem when one exists (thus avoiding overconfidence, called 'false negatives' or type II errors) and (2) tell us there *is not* a serious problem when there isn't one (thus avoiding false alarms, called 'false positives' or type I errors).

underspecificity
Occurs when there is unwanted generality in language.

uniform distribution
Defined by an upper and lower bound, within which all values are equally likely, used for parameters for which few data are available, but

firm bounds are known, and as a model for independent random variation from which other distributions may be constructed.

utilities
A measure of the total benefit or cost resulting from each of a set of alternative actions (decisions), a scale of preferences among outcomes.

utility function
A continuous representation of utilities. Calculations depend on a probability associated with each state.

utility-based criteria
Decisions based on the valuation of outcomes; for example, probabilistic benefit-cost, maximizing multi-attribute utility, or maximizing/minimizing chances of extreme outcomes.

vagueness
Arises because language permits borderline cases.

validation/verification (of model)
Comparing independent field observations with predictions, i.e. testing ideas.

variability
Naturally occurring, unpredictable change, differences in parameters attributable to 'true' heterogeneity or diversity in a population.

volatilization
Processes in which a chemical changes from a solid or liquid form to a vapour.

warrant
The causal laws, inference mechanisms, models or rules that link the data to the claim.

Weibull distribution
Used to model time to failure, material strength and related properties.

zero risk strategies
Strategies that eliminate risks entirely, irrespective of costs and benefits.

Bibliography

Wilkes, G. A. and Krebs, W. A. (eds.) 1982. *The New Collins Concise English Dictionary*. Australian Edition. Sydney: Collins.

Wordsworth Editions 1995. *Dictionary of Science and Technology*. Hertfordshire: Wordsworth Editions Ltd. First published in 1988 as *Chambers Science and Technology Dictionary*. Edinburgh: W&R Chambers Ltd.

References

Adams, J. 1995. *Risk*. London: UCL Press.

ADD. 1995. *Submission to the Commonwealth Commission of Inquiry, East Coast Armament Complex Point Wilson, Victoria*. Canberra: http://www.environment. gov.au/epg/eianet/casestudies/studies/cs1.html. Australian Department of Defence.

Akçakaya, H. R. 1990. A method for simulating demographic stochasticity. *Ecological Modelling* **54**, 133–6.

Akçakaya, H. R. and Atwood, J. L. 1997. A habitat-based metapopulation model of the Californian Gnatcatcher. *Conservation Biology* **11**, 422–34.

Akçakaya, H. R., Ferson, S. and Root, W. T. 1999. *RAMAS Red List: Threatened Species Classification under Uncertainty*. Version 2. New York: Applied Biomathematics, Setauket.

Akçakaya, H. R., Ferson, S., Burgman, M. A., Keith, D. A., Mace, G. M. and Todd, C. R. 2000. Making consistent IUCN classifications under uncertainty. *Conservation Biology* **14**, 1001–13.

ALRC. 1985. *Evidence*. Australian Law Reform Commission Report 26. Volume 1. Canberra: Australian Government Publishing Service. http://www.austlii.edu. au/au/other/alrc/publications/reports/26/.

  2000. *Managing Justice: a Review of the Federal Civil Justice System*. Australian Law Reform Commission Report 89. Canberra: http://www.austlii.edu.au/ au/other/alrc/publications/reports/89/. Australian Government Publishing Service.

Ames, B. N. and Gold, L. S. 1989. Pesticides, risks, and applesauce. *Science* **244**, 755.

Andelman, S. J., Beissinger, S., Cochrane, J. F., Gerber, L., Gomez-Priego, P., Groves, C., Haufler, J., Holthausen, R., Lee, D., Maguire, L., Noon, B., Ralls, K. and Regan, H. 2001. *Scientific Standards for Conducting Viability Assessments under the National Forest Management Act: Report and Recommendations of the NCEAS Working Group*. Santa Barbara, California: National Center for Ecological Analysis and Synthesis.

Anderson, C. 1998a. *Caution: Precautionary Principle at Work*. J. E. Cummins, OBE Memorial Oration to The Royal Society of Victoria, Melbourne. Tuesday, 22nd September, 1998.

Anderson, D. R., Burnham, K. P. and Thompson, W. L. 2000. Null hypothesis testing: problems, prevalence, and an alternative. *Journal of Wildlife Management* **64**, 912–23.

Anderson, J. L. 1998b. Embracing uncertainty: the interface of Bayesian statistics and cognitive psychology. *Conservation Ecology [Online]* **2**(1), 2. http://www.consecol.org/vol2/iss1/art2.

Anonymous. 1995. *Biodiversity: the UK Steering Group Report.* Volume 2. Action Plans. London: HMSO.

ANZECC / ARMCANZ. 2000. *Australian Guidelines for Water Quality Monitoring and Reporting.* Australian and New Zealand Environment and Conservation Council / Agriculture and Resource Management Council of Australia and New Zealand. Canberra: Australian Government Publishing Service.

Arrow, K. J. 1950. A difficulty in the concept of social welfare. *Journal of Political Economy* **58**, 328–46.

1971. *Essays in the Theory of Risk-bearing.* Chicago, Illinois: Markham Publishing.

Arrow, K. J. and Fischer, A. C. 1974. Environmental preservation, uncertainty and irreversibility. *Quarterly Journal of Economics* **88**, 312–19.

AS/NZS 4360. 1999. *Australian/ New Zealand Standard 4360. Risk Management.* Strathfield, NSW: Standards Association of Australia.

Ayyub, B. M. 2001. *Elicitation of Expert Opinions for Uncertainty and Risks.* Boca Raton, Florida: CRC Press.

Baker, S. R. 1996. Regulating and managing risks: impact of subjectivity on objectivity. In C. R. Cothern (ed.) *Handbook for Environmental Risk Decision Making: Values, Perceptions and Ethics.* Boca Raton, Florida: CRC Lewis Publishers, pp. 83–92.

Bana e Costa, C. M. and Vansnick, J.-C. 1999. Preference relations and MCDM. In T. Gal, T. J. Stewart and T. Hanne (eds.) *Multicriteria Decision-Making: Advances in MCDM Models, Algorithms, Theory and Applications.* Boston, Massachusetts: Kluwer, pp. 4/1–4/23.

Baran, N. 2000. Effective Survey Methods for Detecting Plants. MSc Thesis. University of Melbourne: School of Botany.

Barry, T. M. 1996. Recommendations on testing and use of pseudo-random number generators used in Monte Carlo analysis for risk assessment. *Risk Analysis* **16**, 93–105.

Bartell, S. M. 1990. Ecosystem context for estimating stress-induced reductions in fish populations. *American Fisheries Society Symposium* **8**, 167–82.

Bartell, S. M., Lefebvre, G., Kaminski, G., Carreau, M. and Campbell, K. R. 1999. An ecosystem model for assessing ecological risks in Quebec rivers, lakes and reservoirs. *Ecological Modelling* **124**, 43–67.

Beer, T. 2003. Environmental risk and sustainability. In T. Beer and A. Ismail-Zadeh (eds.) *Risk Science and Sustainability.* Dordrecht: Kluwer, pp. 39–61.

Beer, T. and Ziolkowski, F. 1995. *Environmental Risk Assessment: an Australian Perspective.* Supervising Scientist Report 102. Canberra: Office of the Supervising Scientist.

Beissinger, S. R. 2002. Population viability analysis: past, present, future. In S. R. Beissinger and D. R. McCullough (eds.) *Population Viability Analysis.* Chicago, Illinois: University of Chicago Press, pp. 5–17.

Beissinger, S. R. and McCullough, D. R. (eds.) 2002. *Population Viability Analysis.* Chicago, Illinois: University of Chicago Press.

Beissinger, S. R. and Westphal, M. I. 1998. On the use of demographic models of population viability in endangered species management. *Journal of Wildlife Management* **62**, 821–41.

Bell, D. E. and Schleifer, A. 1995. *Risk Management*. Cambridge, Massachusetts: Course Technology.

Belton, V. 1999. Multi-criteria problem structuring and analysis in a value theory framework. In T. Gal, T. J. Stewart, and T. Hanne (eds.) *Multicriteria Decision-Making: Advances in MCDM Models, Algorithms, Theory and Applications*. Boston, Massachusetts: Kluwer, pp. 12/1–12/32.

Ben-Haim, Y. 2001. *Information-Gap Decision Theory: Decisions Under Severe Uncertainty*. San Diego, California: Academic Press.

Bennett, E. M., Carpenter, S. R., Peterson, G. D., Cumming, G. S., Zurek, M. and Pingali, P. 2003. Why global scenarios need ecology. *Frontiers in Ecology and the Environment* **1**, 322–9.

Bernard, H. B. 1988. *Research Methods in Cultural Anthropology*. London: Sage Publications.

Bernstein, P. L. 1996. *Against the Gods: the Remarkable Story of Risk*. New York: Wiley.

Berry, G. 1986. Statistical significance and confidence intervals. *The Medical Journal of Australia* **144**, 618–19.

Bier, V. M. 2001. On the state of the art: risk communication to the public. *Reliability Engineering and System Safety* **71**, 139–50.

Bier, V. 2002. *Implications of the Research on Overconfidence for Challenge Problem Solution Strategies*. Paper presented to the SANDIA Conference on Uncertainty, Albuquerque, New Mexico, August, 2002.

  2004. Should the model for regulation be game theory rather than decision theory? *Risk Analysis* (in press)

Bier, V. M., Haimes, Y. Y., Lambert, J. H., Matalas, N. C. and Zimmerman, R. 1999. A survey of approaches for assessing and managing the risks of extremes. *Risk Analysis* **19**, 83–94.

Bingham, G., Bishop, R., Brody, M., Bromley, D., Clark, E., Cooper, W., Costanza, R., Hale, T., Hayden, G., Kellert, S., Norgaard, R., Norton, B., Payne, J., Russell, C. and Suter, G. 1995. Issues in ecosystem valuation: improving information for decision making. *Ecological Economics* **14**, 73–90.

Borsuk, M. E., Clemen, R. T., Maguire, L. A. and Reckhow, K. H. 2001a. Stakeholder values and scientific modelling in the Neuse River watershed. *Group Decision and Negotiation* **10**, 355–73.

Borsuk, M. E., Stow, C. A., Luettich, R. A., Paerl, H. W. and Pinckney, J. L. 2001b. Modelling oxygen dynamics in an intermittently stratified estuary: estimation of process rates using field data. *Estuarine Coastal and Shelf Science* **52**, 33–49.

Borsuk, M. E., Stow, C. A. and Reckhow, K. H. 2003. An integrated approach to TMDL development for the Neuse River Estuary using a Bayesian probability network model (Neu-BERN). *Journal of Water Resources Planning and Management* **129**, 271–82.

Bottom, W., Lakha, K. and Miller, G. J. 2002. Propagation of individual bias through group judgment: error in the treatment of asymmetrically informative signals. *Journal of Risk and Uncertainty* **25**, 147–63.

Box, G. and Luceno, A. 1997. *Statistical Control: by Monitoring and Feedback Adjustment.* New York: Wiley.

Boyce, M. S. 1992. Population viability analysis. *Annual Review of Ecology and Systematics* **23**, 481–506.

Brereton, S., Lane, M., Smith, C. and Yatabe, J. 1998. Risk management plan for the National Ignition Facility. In A. Mosleh and R. A. Bari (eds.) *Probabilistic Safety Assessment and Management.* PSAM 4. Proceedings of the 4th International Conference on Probabilistic Safety Assessment and Management, 13–18 September 1990, New York City, USA, pp. 2807–12.

Breyer, S. 1993. *Breaking the Vicious Circle: Towards Effective Regulation.* Cambridge, Massachusetts: Harvard University Press.

Briggs, J. D. and Leigh, J. H. 1996. *Rare or Threatened Australian Plants.* Melbourne: CSIRO.

Brook, B. W., O'Grady, J. J., Chapman, A. P., Burgman, M. A., Akçakaya, H. R. and Frankham, R. 2000. Predictive accuracy of population viability analysis in conservation biology. *Nature* **404**, 385–7.

Brook, B. W., Burgman, M. A., Akçakaya, H. R., O'Grady, J. J. and Frankham, R. 2002. Critiques of PVA ask the wrong questions: throwing the heuristic baby out with the numerical bathwater. *Conservation Biology* **16**, 262–3.

Brown, D. A. 1996. The urgent need to integrate ethical considerations into risk assessment procedures. In C. R. Cothern (ed.) *Handbook for Environmental Risk Decision-Making: Values, Perceptions and Ethics.* Boca Raton, Florida: CRC Lewis Publishers, pp. 115–30.

Brunton, R. 1995. The perils of the precautionary principle. *Australasian Biotechnology* **5**, 236–8.

Bui, E. 2000. Risk assessment in the face of controversy: tree clearing and salinization in North Queensland. *Environmental Management* **26**, 447–56.

Burgman, M. A. 1999. Are Australian standards for risk analysis good enough? *Australian Biologist* **12**, 125–37.

  2000. Population viability analysis for bird conservation: theory, practice, models and psychology. *Emu* **100**, 347–53.

  2001. Flaws in subjective assessments of ecological risks and means for correcting them. *Australian Journal of Environmental Management* **8**, 219–26.

  2002. Are listed threatened plant species actually at risk? *Australian Journal of Botany* **50**, 1–13.

Burgman, M. A. and Lindenmayer, D. B. 1998. *Conservation Biology for the Australian Environment.* Chipping Norton, Sydney: Surrey Beatty.

Burgman, M. A. and Possingham, H. P. 2000. Population viability analysis for conservation: the good, the bad and the undescribed. In A. G. Young and G. M. Clarke (eds.) *Genetics, Demography, and Viability of Fragmented Populations.* Cambridge: Cambridge University Press, pp. 97–112.

Burgman, M. A., Ferson, S. and Akçakaya, H. R. 1993. *Risk Assessment in Conservation Biology.* London: Chapman and Hall.

Burgman, M. A., Davies, C., Morgan, D. and Maillardet, R. 1997. *Statistical Power and the Design of Flora and Fauna Survey and Monitoring Programs.* Flora Section of the Victorian Department of Conservation and Natural Resources, Melbourne. Unpublished report.

Burgman, M. A., Maslin, B. R., Andrewartha, D., Keatley, M. R., Boek, C. and McCarthy, M. 2000. Inferring threat from scientific collections: power tests and application to Western Australian Acacia species. In S. Ferson and M. A. Burgman (eds.) *Quantitative Methods for Conservation Biology*. New York: Springer-Verlag, pp. 7–26.

Burgman, M. A., Breininger, D. R., Duncan, B. W. and Ferson, S. 2001. Setting reliability bounds on habitat suitability indices. *Ecological Applications* **11**, 70–8.

Burmaster, D. E. and Harris, R. H. 1993. The magnitude of compounding conservatisms in superfund risk assessments. *Risk Analysis* **13**, 131–4.

Burnham, K. P. and Anderson, D. R. 2002. *Model Selection and Multimodel Inference: A Practical Information-Theoretic Approach*, 2nd edition. New York: Springer-Verlag.

Byrd, D. M. and Cothern, C. R. 2000. *Introduction to Risk Analysis. A Systematic Approach to Science-Based Decision Making*. Rockville, Maryland: Government Institutes.

Calabrese, E. J. and Baldwin, L. A. 2001. The frequency of U-shaped dose responses in the toxicological literature. *Toxicological Sciences* **62**, 330–8.

Calow, P. and Forbes, V. E. 2003. Does ecotoxicology inform ecological risk assessment? *Environmental Science and Technology* April 1, 146–51.

Campbell, L. M. 2002. Science and sustainable use: views of marine turtle conservation experts. *Ecological Applications* **12**, 1229–46.

Capen, E. C. 1976. The difficulty of assessing uncertainty. *Journal of Petroleum Technology*, August 1976, 843–50.

Caputi, N., Brown, R. S. and Chubb, C. F. 1995. Regional prediction of the western rock lobster, *Panulirus cygnus*, commercial catch in Western Australia. *Crustaceana* **68**, 245–56.

Carlon, C., Critto, A., Marcomini, A. and Nathanail, P. 2001. Risk based characterization of contaminated industrial site using multivariate and geostatistical tools. *Environmental Pollution* **111**, 417–27.

Carpenter, S. R., Bolgrien, D., Lathrop, R. C., Stow, C. A., Reed, T. and Wilson, M. A. 1998. Ecological and economic analysis of lake eutrophication by non-point pollution. *Australian Journal of Ecology* **23**, 68–79.

Carpenter, S., Brock, W. and Hanson, P. 1999. Ecological and social dynamics in simple models of ecosystem management. *Conservation Ecology* [*online*] **3**(2):4. http://www.consecol.org/vol3/iss2/art4.

Carson, R. 1962. *Silent Spring*. Second edition, 2002. Boston: Houghton, Mifflin.

Cartwright, N. 2003. What is wrong with Bayes nets? In H. E. Kyburg and M. Thalos (eds.) *Probability is the Very Guide to Life: the Philosophical Uses of Chance*. Chicago, Illinois: Open Court. pp. 253–75.

Casella, G. and Berger, R. L. 1990. *Statistical Inference*. Belmont, California: Duxbury Press.

Caswell, H. 2001. *Matrix Population Models*, 2nd edition. Sunderland, Massachusetts: Sinauer.

Cauchi, S. 2003. New fears raised about GM plants. *The Age*, February 6 http://www.theage.com.au/articles/2003/02/05/1044318670302.html.

Cavieres, M. F., Jaeger, J. and Porter, W. 2002. Developmental toxicity of a commercial herbicide mixture in mice: I. Effects on embryo implantation and litter size. *Environmental Health Perspectives* **110**, 1081–5.

Chalmers, A. 1999. *What is This Thing Called Science?* 3rd edition. Brisbane: University of Queensland Press.

Chapman, M. 1995. The expert in France. *Arbitration* **61**, 264.

Chee, Y. E. 2004. An ecological perspective on the valuation of ecosystem services. *Biological Conservation* (in press).

Chesson, P. 1978. Predator – prey theory and variability. *Annual Review of Ecology and Systematics* **9**, 323–47.

Christensen-Szalanski, J. and Bushyhead, J. 1981. Physicians use of probabilistic information in a real clinical setting. *Journal of Experimental Psychology: Human Perception and Performance* **7**, 928–35.

CIA. 1977. *A Guide to Hazard and Operability Studies.* London: Chemical Industries Association.

Clark, A. J. 1933. *The Mode of Action of Drugs on Cells.* Baltimore, Maryland: Williams and Wilkins.

Clemen, R. T. 1996. *Making Hard Decisions: an Introduction to Decision Analysis.* 2nd edition. California: Duxbury, Pacific Grove.

Clemen, R. T. and Winkler, R. L. 1999. Combining probability distributions from experts in risk analysis. *Risk Analysis* **19**, 187–203.

Clyde, M. A. 2000. Model uncertainty and health effect studies for particulate matter. *Environmetrics* **11**, 745–63.

Cogliano, V. J. 1997. Plausible upper bounds: are their sums plausible? *Risk Analysis* **17**, 77–84.

Cohen, J. 1988. *Statistical Power Analysis for the Behavioural Sciences.* 2nd edition. New Jersey: Lawrence Erlbaum Associates.

Cohen, J. T., Lampson, M. A. and Bowers, T. S. 1996. The use of two-stage Monte Carlo simulation techniques to characterize variability and uncertainty in risk analysis. *Human and Ecological Risk Assessment* **2**, 939–71.

Colyvan, M. 2004. Is probability the only coherent approach to uncertainty? *Risk Analysis* (in press).

Colyvan, M., Regan, H. M. and Ferson, S. 2003. Is it a crime to belong to a reference class? In H. E. Kyburg and M. Thalos (eds.) *Probability is the Very Guide to Life.* Chicago, Illinois: Open Court, pp. 331–47.

Cooke, R. M. 1991. *Experts in Uncertainty: Opinion and Subjective Probability in Science.* Oxford: Oxford University Press.

Cooke, R. M. and Goossens, L. H. J. 2000. Procedures guide for structured expert judgement in accident consequence modelling. *Radiation Protection and Dosimetry* **90**, 303–9.

Cooke, R. and Kraaikamp, C. 2000. Risk analysis and jurisprudence; a recent example. In W. M. Doerr (ed.) *Safety Engineering and Risk Analysis* (SERAS), Volume 10. New York: American Society of Mechanical Engineers, pp. 67–72.

Cooke, R. and Kraan, B. 2000. Processing expert judgements in accident consequence modelling. *Radiation Protection Dosimetry* **90**, 311–15.

CPR. 1999. *The English Civil Procedure Rules.* CPR (UK) R 35.7. Civil Procedure Rules 1999 (UK).

Crawford-Brown, D. J. 1999. *Risk-Based Environmental Decisions: Method and Culture.* Boston, Massachusetts: Kluwer Academic.

Crawley, M. J., Brown, S. L., Hails, R. S., Kohn, D. D. and Rees, M. 2001. Biotechnology: transgenic crops in natural habitats. *Nature* **409**, 682–3.

Cross, F. B. 1996. Paradoxical perils of the precautionary principle. *Washington and Lee Law Review* **53**, 851–925.

Crossley, S. J. 2000. Joint FAO / WHO Geneva consultation – acute dietry intake methodology. *Food Additives and Contaminants* **17**, 557–62.

Cullen, A. C. 1994. Measures of compounding conservatism in probabilistic risk assessment. *Risk Analysis* **14**, 389–93.

Cumming, G. and Finch, S. 2004. Inference by eye: confidence intervals, and how to read pictures of data (submitted).

Cushing, D. 1995. *Population Production and Regulation in the Sea: a Fisheries Perspective.* Cambridge: Cambridge University Press.

Daily, G. C. 1997. Introduction: what are ecosystem services? In G. C. Daily (ed.) *Nature's Services: Societal Dependence on Natural Ecosystems.* Washington, DC: Island Press, pp. 1–10.

Daily, G. C. 2000. Management objectives for the protection of ecosystem services. *Environmental Science and Policy* **3**, 333–9.

Davies, J. C. (ed.) 1996. Comparative risk analysis in the 1990s: the state of the art. *Comparing Environmental Risks.* Washington, DC: Resources for the Future, pp. 1–8.

de Finetti, B. 1974. *Theory of Probability.* New York: Wiley.

Dennis, B., Brown, B. E., Stage, A. R., Burkhart, H. E. and Clark, S. 1985. Problems of modeling growth and yield of renewable resources. *American Statistician* **39**, 374–83.

Deville, A. and Harding, R. 1997. *Applying the Precautionary Principle.* Sydney: The Federation Press.

Dewey, J. 1927. *The Public and its Problems.* New York: Swallow.

Dixon, P. M. 1998. Assessing effect and no effect with equivalence regions. In M. C. Newman and C. L. Strojan (eds.) *Risk Assessment: Logic and Measurement.* Chelsea, Michigan: Ann Arbor Press, pp. 275–301.

Dodgson, J., Spackman, M., Pearman, A. and Phillips, L. 2001. *Multicriteria Analysis: a Manual.* London: Department for Transport, Local Government and the Regions. http://www.dtlr.gov.uk/about/multicriteria/17.htm.

Donaldson, T., and Preston, L. E. 1995. The stakeholder theory of the corporation. *Academy of Management Review* **20**, 65–91.

Draper, D. 1995. Assessment and propagation of model uncertainty. *Journal of the Royal Statistical Society Series B-Methodological* **57**, 45–97.

Draper, D., Pereira, A., Prado, P., Saltelli, A., Cheal, R., Eguilior, S., Mendes, B. and Tarantola, S. 1999. Scenario and parametric uncertainty in GESAMAC: a methodological study in nuclear waste disposal risk assessment. *Computer Physics Communications* **117**, 142–55.

Drechsler, M., Burgman, M. A. and Menkhorst, P. W. 1998. Uncertainty in population dynamics and its consequences for the management of the orange-bellied parrot *Neophema chrysogaster. Biological Conservation* **84**, 269–81.

Efron, B. and Tibshirani, R. 1991. Statistical data analysis in the computer age. *Science* **253**, 390–5.

Elith, J. 2000. Quantitative methods for modeling species habitat: comparative performance and an application to Australian plants. In S. Ferson and M. Burgman (eds.) *Quantitative Methods for Conservation Biology.* New York: Springer, pp. 39–58.

Ellner, S. P., Frieberg, J., Ludwig, D. and Wilcox, C. 2002. Precision of population viability analysis. *Conservation Biology* **16**, 258–61.

Engelhardt, H. and Caplan, H. 1986. Patterns of controversy and closure: the interplay of knowledge, values, and political forces. In H. Engelhardt and H. Caplan (eds.) *Scientific Controversies: Case Studies in the Resolution and Closure of Disputes in Science and Technology*. New York: Cambridge University Press, pp. 1–23.

Erdfelder, E., Faul, F. and Buchner, A. 1996. GPOWER: a general power analysis program. *Behavior Research Methods, Instruments and Computers* **28**, 1–11.

Evans, P. D. 1999. An evolution in risk management standards. *InDepth*, September 1999. Sydney: Freehill, Hollingdale and Page, pp. 1–2.

Fairweather, P. G. 1991. Statistical power and design requirements for environmental monitoring. *Australian Journal of Marine Freshwater Research* **42**, 555–67.

FAO / WHO. 2001. Call for experts for the Joint FAO / WHO risk assessment activities in the areas of *Campylobacter* in broilers and *Vibrio* in seafood. *Joint Expert Consultations on Risk Assessment of Microbiological Hazards in Food*. Rome, Italy: Food and Nutrition Division, Food and Agriculture Organization of the United Nations.

Fernandes, L., Ridgley, M. A. and van't Hof, T. 1999. Multiple criteria analysis integrates economic, ecological and social objectives for coral reef managers. *Coral Reefs* **18**, 393–402.

Ferson, S. 1996a. *Reliable calculation of probabilities: accounting for small sample size and model uncertainty.* Paper presented to Intelligent Systems: a semiotic perspective. NIST, October 1996. http://gwis2.circ.gwu.edu/~joslyn/sem96.

1996b. What Monte Carlo methods cannot do. *Human and Ecological Risk Assessment* **2**, 990–1007.

2002. *RAMAS RiskCalc Version 4.0. Software: Risk Assessment with Uncertain Numbers*. Boca Raton, Florida: Lewis.

Ferson, S. and Ginzburg, L. R. 1996. Different methods are needed to propagate ignorance and variability. *Reliability Engineering and Systems Safety*. 54: 133–144.

Ferson, S. and Moore, D. R. J. 2004. Bounding uncertainty analysis. In A. Hart (ed.) Proceedings from a workshop on the application of uncertainty analysis to ecological risks of pesticides. Pensacola, Florida: Society for Environmental Toxicology and Chemistry.

Ferson, S., Root, W. and Kuhn, R. 1999. *RAMAS Risk Calc: Risk Assessment with Uncertain Numbers*. Setauket, NY: Applied Biomathematics.

Ferson, S., Kreinovich, V., Ginzburg, L., Myers, D. and Sentz, K. 2003. *Constructing Probability Boxes and Dempster-Shafer Structures*. SAND2002–4015. Albuquerque, New Mexico: Sandia National Laboratories.

Finizio, A. and Villa, S. 2002. Environmental risk assessment for pesticides: a tool for decision making. *Environmental Impact Assessment Review* **22**, 235–48.

Finkel, A. M. 1995. A second opinion on an environmental misdiagnosis: the risky prescriptions of breaking the vicious circle. *Environmental Law Journal* **3**, 295–381.

1996. Comparing risks thoughtfully. *Risk: Health, Safety and Environment* **7**, 349. www.piercelaw.edu/risk/vol7/fall/.

Fischer, F. 2000. *Citizens, Experts, and the Environment*. Durham, North Carolina: Duke University Press.

Fischhoff, B. 1994. Acceptable risk: a conceptual proposal. *Risk: Health, Safety and Environment* **1**, 1–28.

  1995. Risk perception and communication unplugged: twenty years of progress. *Risk Analysis* **15**, 137–45.

Fischhoff, B., Lichtenstein, S., Slovic, P., Derby, S. L. and Keeney, R. L. 1981. *Acceptable Risk*. Cambridge: Cambridge University Press.

Fischhoff, B., Slovic, P. and Lichtenstein, S. 1982. Lay foibles and expert fables in judgements about risk. *American Statistician* **36**, 240–55.

Flynn, J., Slovic, P. and Merta, C. K. 1994. Gender, race, and perception of environmental health risks. *Risk Analysis* **14**, 1101–8.

FMEAInfoCenter. 2002. Links to information resources and general information. http://www.fmeainfocentre.com/introductions.htm.

Frank, M. J. 1979. On the simultaneous associativity of $F(x, y)$ and $x + y + F(x, y)$. *Aequationes Mathematicae* **19**, 194–226.

Frank, M. J., Nelson, R. B. and Schweizer, B. 1987. Best-possible bounds for the distribution of a sum – a problem of Kolmogorov. *Probability Theory and Related Fields* **74**, 199–211.

Franklin, J. 2002. Enhancing a regional vegetation map with predictive models of dominant plant species in chaparral. *Applied Vegetation Science* **5**, 135–46.

Freckelton, I. 1995. The challenge of junk psychiatry, psychology and science: the evolving role of the forensic expert. In H. Selby (ed.) *Tomorrow's Law*. Sydney: Federation Press, pp. 58–9.

Freeman, A. M. III. 1993. *The Measurement of Environmental and Resource Values*. Washington, DC: Resources for the Future.

Freeman, R. E. 1984. *Strategic Management: a Stakeholder Approach*. Boston, Massachusetts: Pitman.

French, S. 1986. *Decision Theory: an Introduction to the Mathematics of Rationality*. Chichester: Ellis Horwood.

Freudenburg, W. R. 1992. Heuristics, biases, and the not-so-general publics: expertise and error in the assessment of risks. In S. Krimsky and D. Golding (eds.) *Social Theories of Risk*. Westport, Connecticut: Praeger Publishing, pp. 229–49.

  1996. Risky thinking: irrational fears about risk and society. *Annals AAPSS* **545**, 44–53.

  1999. Tools for understanding the socioeconomic and political settings for environmental decision making. In V. H. Dale and M. R. English (eds.) *Tools to Aid Environmental Decision Making*. New York: Springer, pp. 94–125.

Freudenburg, W. R. and Rursch, J. A. 1994. The risks of 'putting the numbers in context'; a cautionary tale. *Risk Analysis* **14**, 949–58.

Freudenburg, W. R., Coleman, C.-L., Gonzales, J. and Helgeland, C. 1996. Media coverage of hazard events: analyzing the assumptions. *Risk Analysis* **16**, 31–42.

Frey, H. C. and Rhodes, D. S. 1996. Characterizing, simulating, and analyzing variability and uncertainty: an illustration of methods using an air toxic emissions example. *Human and Ecological Risk Assessment* **2**, 762–97.

Gabbay, D. M. and Smets, P. (eds.) 1998. *Handbook of Defeasible Reasoning and Uncertainty Management Systems*, Volume 3. Dordrecht: Kluwer.

Garrod, G. and Willis, K. G. 1999. *Economic Valuation of the Environment*. Cheltenham: Edward Elgar.

Gigerenzer, G. 2002. *Calculated Risks: How to Know when Numbers Deceive You*. New York: Simon and Schuster.

Gigone, D. and Hastie, R. 1997. Proper analysis of the accuracy of group judgments. *Psychological Bulletin* **121**, 149–67.

Gilbert, R. O. 1987. *Statistical Methods for Environmental Pollution Monitoring*. New York: Van Nostrand Reinhold.

Gilpin, M. E. and Soulé, M. E. 1986. Minimum viable populations: Processes of species extinctions. In M. E. Soulé. (ed.) *Conservation Biology: the Science of Scarcity and Diversity*. Sunderland, Massachusetts: Sinauer Associates, pp. 19–34.

Glicken, J. 2000. Getting stakeholder participation 'right': a discussion of participatory processes and pitfalls. *Environmental Science and Policy* **3**, 305–10.

Goklany, I. M. 2001. *The Precautionary Principle: a Critical Appraisal of Environmental Risk Assessment*. Washington DC: CATO Institute.

Goldring, J. 2003. An introduction to statistical 'evidence'. *Australian Bar Review* **23**, 239–62.

Gollier, C., Jullien, B. and Treich, N. 2000. Scientific progress and irreversibility: an economic interpretation of the 'precautionary principle'. *Journal of Public Economics* **75**, 229–53.

Good, I. J. 1959. Kinds of probability. *Science* **129**, 443–7.

Goodwin, P. and Wright, G. 1998. *Decision Analysis for Management Judgement*. Chichester: Wiley.

Goossens, L. H. J. and Cooke, R. M. 2001. Expert judgement elicitation in risk assessment. In I. Linkov and J. Palma-Oliveira (eds.) *Assessment and Management of Environmental Risks*. Dordrecht: Kluwer Academic, pp. 411–26.

Goossens, L. H. J., Cooke, R. M. and Kraan, B. C. P. 1998. Evaluation of weighting schemes for expert judgement studies. In A. Mosleh and R. A. Bari (eds.) *Probabilistic Safety Assessment and Management. PSAM* **4**, *1935–1942*. New York: Springer Verlag.

Graham, J. D. and Hammitt, J. K. 1996. Refining the CRA framework. In J. C. Davies (ed.) *Comparing Environmental Risks*. Washington, DC: Resources for the Future, pp. 93–109.

Groombridge, B. (ed.) 1994. *Biodiversity Data Sourcebook*. World Conservation Monitoring Centre, Biodiversity Series No 1. Cambridge: World Conservation Press.

Gumbel, E. J. 1958. *Statistics of Extremes*. New York: Columbia University Press.

Gustafson, J., Jehl, J., Kohfield, B., Laabs, L. and La Berteaux, D. 1994. Inyo California towhee. In C. Thelander (ed.) *Life on the Edge*, Volume 1: *Wildlife*. Santa Cruz, California: Biosystems Books, pp. 218–19.

Haas, C. N. 1997. Importance of distributional form in characterizing inputs to Monte Carlo risk assessments. *Risk Analysis* **17**, 107–13.

Hacking, I. 1975. *The Emergence of Probability: a Philosophical Study of Early Ideas about Probability, Induction and Statistical Inference*. Cambridge: Cambridge University Press.

Haight, R. G. 1995. Comparing extinction risk and economic cost in wildlife conservation planning. *Ecological Applications* **5**, 767–75.

Haimes, Y. Y. 1998. *Risk Modeling, Assessment and Management*. New York: Wiley.

Haimes, Y. Y., Lambert, J. H. and Mahoney, B. B. 2000. *Risk Modeling, Assessment and Management of Interdependent Critical Structures.* Sixteenth Annual Security Technology Symposium, Center for Risk Management of Engineering Systems, 29 June 2000. Williamsburg, Virginia: University of Virginia.

Hammersley, J. M. and Handscomb, D. C. 1964. *Monte Carlo Methods.* London: Methuen.

Hanski, I. 2002. Metapopulations of animals in highly fragmented landscapes and population viability analysis. In S. R. Beissinger and D. R. McCullough (eds.) *Population Viability Analysis.* Chicago, Illinois: University of Chicago Press, pp. 86–108.

Hanski, I. and Gilpin, M. 1991. Metapopulation dynamics: brief history and conceptual domain. *Biological Journal of the Linnean Society* **42**, 3–16.

Hanski, I. and Gyllenberg, M. 1993. Two general metapopulation models and the core-satellite hypothesis. *The American Naturalist* **142**, 17–41.

Hanson, M. L. and Solomon, K. R. 2002. New technique for estimating thresholds of toxicity in ecological risk assessment. *Environmental Science and Technology* **36**, 3257–64.

Harremoes, P., Gee, D., MacGarvin, M., Stirling, A., Keys, J., Wynne, B. and Vaz, S. G. 2001. *Late Lessons from Early Warnings: the Precautionary Principle 1896–2000.* European Environmental Agency, Environmental Issue Report 22. Copenhagen, Denmark: European Environmental Agency.

Harrison, S. 1991. Local extinction in a metapopulation context: an empirical evaluation. *Biological Journal of the Linnean Society* **42**, 73–88.

Hart, A. 1986. *Knowledge Acquisition for Expert Systems.* New York: McGraw-Hill.

Hart, B. T., Maher, B. and Lawrence, I. 1999. New generation water quality guidelines for ecosystem protection. *Freshwater Biology* **41**, 347–59.

Hart, B. T., Lake, P. S., Webb, J. A. and Grace, M. R. 2003. Ecological risk to aquatic systems from salinity increases. *Australian Journal of Botany* **51**, 689–702.

Hart, M. K. and Hart, R. F. 2002. *Statistical Process Control for Health Care.* Pacific Grove, California: Duxbury / Thomson Learning.

Harwood, J. 2000. Risk assessment and decision analysis in conservation. *Biological Conservation* **95**, 219–26.

Harwood, J. and Stokes, K. 2003. Coping with uncertainty in ecological advice: lessons from fisheries. *Trends in Ecology and Evolution* **18**, 617–22.

Hattis, D. 1990. Three candidate 'laws' of uncertainty analysis. *Risk Analysis* **10**, 11.

Hattis, D. and Burmaster, D. E. 1994. Assessment of variability and uncertainty distributions for practical risk analyses. *Risk Analysis* **14**, 713–29.

Haviland. 2002. *Failure Modes and Effects Analysis (FMEA).* Michigan: The Haviland Consulting Group. http://www.fmeca.com/ffmethod/methodol.htm.

Hayakawa, H., Fischbeck, P. S. and Fischhoff, B. 2000. Traffic accident statistics and risk perceptions in Japan and the United States. *Accident Analysis and Prevention* **32**, 827–35.

Hayes, K. R. 1997. *A Review of Ecological Risk Assessment Methodologies.* CSIRO CRIMP Technical Report Number 13, Hobart, Australia: CSIRO Division of Marine Research.

  2002a. *Best Practice and Current Practice in Ecological Risk Assessment for Genetically Modified Organisms.* KRA Project 1: Robust methodologies for ecological risk assessment [unpublished report] Hobart, Australia: CSIRO. Division of Marine Research.

2002b. Identifying hazards in complex ecological systems. Part 2. Infections modes and effects analysis for biological invasions. *Biological Invasions* **4**, 251–61.

Hayes, K. R., Gregg, P. C., Jessop, R., Lonsdale, M., Sindel, B., Stanley, J., Vadakutta, G. and Williams, C. K. 2004. Identifying hazards in complex ecological systems. Part 3. Hierarchical holographic model for herbicide tolerant Canola. *Environmental Biosafety Review* (in review).

Hilborn, R. 1987. Living with uncertainty in resource management. *North American Journal of Fisheries Management* **7**, 1–5.

Hilborn, R. and Mangel, M. 1997. *The Ecological Detective: Confronting Models with Data.* Monographs in Population Biology 28. Princeton, New Jersey: Princeton University Press.

Hill, R. A., Chapman, P. M., Mann, G. S. and Lawrence, G. S. 2000. *Marine Pollution Bulletin* **40**, 471–7.

Hoeting, J. A., Madigan, D., Raftery, A. E. and Volinsky, C. T. 1999. Bayesian model averaging: a tutorial. *Statistical Science* **14**, 382–401.

Hoffman, E. O. and Hammonds, J. S. 1994. Propagation of uncertainty in risk assessments: the need to distinguish between uncertainty due to lack of knowledge and uncertainty due to variability. *Risk Analysis* **14**, 707–12.

Hoffman, F. O. and Kaplan, S. 1999. Beyond the domain of direct observation: how to specify a probability distribution that represents the 'state of knowledge' about uncertain inputs. *Risk Analysis* **19**, 131–4.

Hoffman, F. O. and Thiessen, K. M. 1996. The use of Chernobyl data to test model predictions for interindividual variability of ^{137}Cs concentrations in humans. *Reliability Engineering and System Safety* **54**, 197–202.

Holling, C. S. 1978. *Adaptive Environmental Assessment and Management.* New York: Wiley.

Hone, J. and Pech, R. 1990. Disease surveillance in wildlife with emphasis on detecting foot and mouth disease in feral pigs. *Journal of Environmental Management* **31**, 173–84.

Hora, S. C. 1993. Acquisition of expert judgment: examples from risk assessment. *Journal of Energy Engineering* **118**, 136–48.

Horst, H. S., Dijkhuizen, A. A., Huirne, R. B. M., and De Leeuw, P. W. 1998. Introduction of contagious animal diseases into the Netherlands: elicitation of expert opinions. *Livestock Production Science* **53**, 253–64.

Hughes, G. and Madden, R. V. 2003. Evaluating predictive models with application in regulatory policy for invasive weeds. *Agricultural Systems* **76**, 755–74.

ICE/FIA. 1998. *RAMP: Risk Analysis and Management for Projects.* Institution of Civil Engineers and the Faculty and Institute of Actuaries. London: Thomas Telford.

Ignizio, J. P. 1985. *Introduction to Linear Goal Programming.* New York: Sage Publications.

Iman, R. L. and Conover, W. J. 1982. A distribution-free approach to inducing rank order correlation among input variables. *Communications in Statistics – Simula Computata* **B11**, 311–34.

Imwinkelried, E. J. 1993. The Daubert decision: Frye is dead: long live the Federal Rules of Evidence. *Trial* **29**, 60–5.

IRC. 2002. *Western Rock Lobster Ecological Risk Assessment*. IRC Environment, Project JOO-207. Report to the Western Australian Department of Fisheries, Perth, Western Australia.

ISO 14971–1. 1998. International Standard 14971–1. *Medical Devices – Risk Management. Part 1: Application of Risk Analysis*. Geneva: International Organisation for Standardization.

IUCN. 1994. IUCN *Red List of Threatened Animals*. Gland, Switzerland: International Union for the Conservation of Nature.

  2001. *IUCN Red List Categories*. IUCN Species Survival Commission, The World Conservation Union. Gland, Switzerland: IUCN.

Jasanoff, S. 1993. Bridging the two cultures of risk analysis. *Risk Analysis* **13**, 123–9.

Jaynes, E. T. 1976. Confidence intervals vs Bayesian intervals. In W. L. Harper and C. A. Hooker (eds.) *Foundations of Probability Theory, Statistical Inference, and Statistical Theories of Science*, Volume 2. Dordrecht, the Netherlands: D. Reidel Publishing, pp. 175–257.

Jennings, M. M. 1999. *Stakeholder Theory: Letting Anyone Who's Interested Run the Business – No Investment Required*. Paper presented to the Conference on Corporate Governance: Ethics Across the Board. Houston, Texas, April 1999. http://www.stthom.edu/cbes/conferences/marianne_jennings.html.

Johansson, P.-O. 2002. On the definition and age-dependency of the value of a statistical life. *The Journal of Risk and Uncertainty* **25**, 251–63.

Johnson, B. B. 1999a. Ethical issues in risk communication. *Risk Analysis* **19**, 335–48.

Johnson, D. H. 1999b. The insignificance of statistical significance testing. *Journal of Wildlife Management* **63**, 763–72.

Johnson, L. E., Ricciardi, A. and Carlton, J. T. 2001. Overland dispersal of aquatic invasive species: a risk assessment of transient recreational boating. *Ecological Applications* **11**, 1789–99.

Johnson, N. L., Kotz, S. and Kemp, A. W. 1992. *Univariate Discrete Distributions*, 2nd edition. New York: Wiley.

Johnson, N. L., Kotz, S. and Balakrishnan, N. 1994. *Continuous Univariate Distributions*, Volume 1, 2nd edition. New York: Wiley.

Johnson, N. L., Kotz, S. and Balakrishnan, N. 1995. *Continuous Univariate Distributions*, Volume 2, 2nd edition. New York: Wiley.

Johnson, R. H. and Blair, J. A. 1983. *Logical Self-defense*, 2nd edition. Toronto: McGraw-Hill Ryerson.

Jones, S. 1997. Detecting *Aprasia parapulchella*. Paper presented to the Ecological Society of Australia, Albury, Victoria, October, 1997.

Jonzen, N., Cardinale, M., Gardmark, A., Arrhenius, F. and Lundberg, P. 2002. Risk of collapse in the eastern Baltic cod fishery. *Marine Ecology Progress Series* **240**, 225–33.

Kahn, H. and Wiener, A J. 1967. *The Year 2000: a Framework for Speculation*. New York: Macmillan.

Kahneman, D. and Tversky, A. 1979. Prospect theory: an analysis of decision under risk. *Econometrica* **47**, 263–91.

  1984. Choices, values, and frames. *American Psychologist* **39**, 342–7.

Kaiser, J. 2002. Software glitch threw off mortality estimates. *Science* **296**, 1945–6.

Kalof, L., Dietz, T., Guagnano, G. and Stern, P. C. 2002. Race, gender and environmentalism: the atypical values and beliefs of white men. *Race, Gender and Class* **9**, 1–19.

Kammen, D. M. and Hassenzahl, D. M. 1999. *Should We Risk It? Exploring Environmental, Health, and Technological Problem Solving*. Princeton, New Jersey: Princeton University Press.

Kang, S. H., Kodell, R. L., and Chen, J. J. 2000. Incorporating model uncertainties along with data uncertainties in microbial risk assessment. *Regulatory Toxicology and Pharmacology* **32**, 68–72.

Kanji, G. K. 2000. Quality improvement methods and statistical reasoning. In S. H. Park and G. G. Vining (eds.) *Statistical Process Monitoring and Optimization*. New York: Marcel Dekker, pp. 35–43.

Kaplan, S. 1992. 'Expert opinion' versus 'expert opinions.' Another approach to the problem of eliciting / combining / using expert opinion in PRA. *Reliability Engineering and System Safety* **35**, 61–72.

  1997. The words of risk analysis. *Risk Analysis* 17, 407–17.

Kaplan, S. and Garrick, B. 1981. On the quantitative definition of risk. *Risk Analysis* **1**, 11–27.

Kaufmann, A. and Gupta, M. M. 1985. *Introduction to Fuzzy Arithmetic: Theory and Applications*. New York: Von Nostrand Reinhold.

Keeney, R. 1992. *Value-focused Thinking: a Path to Creative Decision-making*. Cambridge, Massachusetts: Harvard University Press.

Keeney, R. L. and Raiffa, H. 1976. *Decisions with Multiple Objectives: Preferences and Value Tradeoffs*. New York: Wiley.

Keith, D. A. 1998. An evaluation and modification of World Conservation Union Red List Criteria for classification of extinction risk in vascular plants. *Conservation Biology* **12**, 1076–90.

Kelly, E. J. and Campbell, K. 2000. Separating variability and uncertainty in environmental risk assessment – making choices. *Human and Ecological Risk Assessment* **6**, 1–13.

Kerr, R. 1996. A new way to ask the experts: rating radioactive waste risks. *Science* **274**, 913–14.

Kershaw, K. A. 1964. *Quantitative and Dynamic Ecology*. London: Edward Arnold.

Keynes, J. M. 1921. *A Treatise on Probability*. London: Macmillan.

  1936. *The General Theory of Employment, Interest and Money*. New York: Harcourt, Brace.

Kirsch, E. W. 1995. Daubert v. Merrell Dow Pharmaceuticals: active judicial scrutiny of scientific evidence. *Food and Drug Law Journal* **50**, 213–34.

Klaassen, C. D. (ed.) 1996. *Casarett and Doull's Toxicology: the Basic Science of Poisons*, 5th edition. New York: McGraw-Hill.

Kletz, T. A. 1999. *Hazop and Hazan*. Philadelphia: Taylor and Francis.

Klir, G. J. and Harmanec, D. 1997. Types and measures of uncertainty. In J. Kacprzyk, H. Nurmi and M. Fedrizzi (eds.) *Consensus under Fuzziness*. Boston, Massachusetts: Kluwer, pp. 29–51.

Klir, G. and Wierman, M. J. 1998. *Uncertainty-based Information: Elements of Generalized Information Theory*. Heidelberg: Physica-Verlag.

Knuth, D. E. 1981. *The Art of Computer Programming*, Volume 2, *Seminumerical Algorithms*, (2nd edition). Reading, Massachusetts: Addison–Wesley.

Kolar, C. S. and Lodge, D. M. 2002. Ecological predictions and risk assessment for alien fishes in North America. *Science* **298**, 1233–6.

Korb, K. B. and Nicholson, A. E. 2003. *Bayesian Artificial Intelligence*. New York: Chapman and Hall, Boca Raton, Florida: CRC Press.

Krinitzsky, E. L. 1993. Earthquake probability in engineering. Part 1: the use and misuse of expert opinion. *Engineering Geology* **33**, 257–88.

Kumamoto, H. and Henley, E. J. 1996. *Probabilistic Risk Assessment and Management for Engineers and Scientists*, 2nd edition. New York: IEEE Press.

Laabs, D. M., Allaback, M. L. and LaPre, L. F. 1995. Census of the Inyo California towhee in the eastern third of its range. *Western Birds* **26**, 189–96.

Landres, P. B., Verner, J. and Thomas, J. W. 1988. Ecological uses of vertebrate indicator species: a critique. *Conservation Biology* **2**, 316–28.

Laskowski, R. 1995. Some good reasons to ban the use of NOEC, LOEC and related concepts in ecotoxicology. *Oikos* **73**, 140–4.

Lathrop, R. C., Carpenter, S. R., Stow, C. A., Soranno, P. A. and Panuska, J. C. 1998. Phosphorus loading reductions needed to control blue-green algal blooms in Lake Mendota. *Canadian Journal of Fisheries and Aquatic Sciences* **55**, 1169–78.

Lawson, J. D. 1900. *The Law of Expert and Opinion Evidence* (2nd edition). Chicago, Illinois: T. H. Flood.

Lehrer, K. 1997. Consensus, negotiation and mediation. In J. Kacprzyk, H. Nurmi and M. Fedrizzi (eds.) *Consensus under Fuzziness*. Boston, Massachusetts: Kluwer, pp. 3–14.

Lehrer, K. and Wagner, C. 1981. *Rational Consensus in Science and Society*. Dordecht: Reidel.

Leon, A. A. and Bonano, E. J. 1998. Legal admissibility vis-à-vis scientific acceptability of experts judgements in environmental management. In A. Mosleh and R. A. Bari (eds.) *Probabilistic Safety Assessment and Management, PSAM 4*. Proceedings of the 4th International Conference on Probabilistic Safety Assessment and Management, 13–18 September 1990, New York City, pp. 1943–8.

Levins, R. 1966. The strategy of model building in population biology. *American Scientist* **54**, 421–31.

Lichtenstein, S. and Newman, J. R. 1967. Empirical scaling of common verbal phrases associated with numerical probabilities. *Psychonometric Science* **9**, 563–4.

Lihou. 2002. HAZOP. Version 4.0. An integrated system for the recording, reporting and analysis of hazard and operability studies and other safety reviews. Birmingham: Lihou Technical and Software Services.

Lindblad, M. 2001. Development and evaluation of a logistic risk model: predicting fruit fly infestation in oats. *Ecological Applications* **11**, 1563–72.

Link, W. A., Cam, E., Nichols, J. D. and Cooch, E. G. 2002. Of BUGS and birds: Markov Chain Monte Carlo for hierarchical modeling in wildlife research. *Journal of Wildlife Management* **66**, 277–91.

Lonsdale, W. M. and Smith, C. S. 2001. Evaluating pest-screening systems – insights from epidemiology and ecology. In R. H. Groves, J. G. Virtue and F. D. Panetta (eds.) *Weed Risk Assessment*. Melbourne: CSIRO Publishing, pp. 52–60.

Lord Woolf. 1996. *Access to Justice*. Draft civil proceedings rules, HMSO London 1996 (Woolf rules) 32.1–32.9. London: HMSO.

Ludwig, D. and Walters, C. J. 2002. Fitting population viability analysis into adaptive management. In S. R. Beissinger and D. R. McCullough (eds.) *Population Viability Analysis*. Chicago, Illinois: University of Chicago Press, pp. 511–20.

MacLean, D. 1996. Environmental ethics and human values. In C. R. Cothern (ed.) *Handbook for Environmental Risk Decision Making: Values, Perceptions and Ethics*. Boca Raton, Florida: CRC Lewis Publishers, pp. 177–93.

Maguire, L. A. 1986. Using decision analysis to manage endangered species populations. *Journal of Environmental Management* **22**, 345–60.

Mapstone, B. D. 1995. Scaleable decision rules for environmental impact studies: effect size, type I and type II errors. *Ecological Applications* **5**, 401–10.

Marvier, M. 2002. Improving risk assessment for nontarget safety of transgenic crops. *Ecological Applications* **12**, 1119–24.

Mathews, R. A. J. 1997. Decision-theoretic limits on earthquake prediction. *Geophysics Journal International* **131**, 526–9.

May, R. M., Lawton, J. H. and Stork, N. E. 1995. Assessing extinction rates. In J. H. Lawton and R. M. May (eds.) *Extinction Rates*. Oxford: Oxford University Press, pp. 1–24.

Mayo, D. 1985. Behavioristic, evidentialist, and learning models of statistical testing. *Philosophy of Science* **52**, 493–516.

Mayo, D. G. 1996. *Error and the Growth of Experimental Knowledge*. Chicago, Illinois: University of Chicago Press.
  2003. Severe testing as a guide to inductive learning. In H. E. Kyburg and M. Thalos (eds.) *Probability is the Very Guide to Life: the Philosophical Uses of Chance*. Chicago, Illinois: Open Court, pp. 89–117.

Mayo, D. G. and Spanos, A. 2004. A severe-testing interpretation of Neyman-Pearson methods. (in prep.)

McArdle, B. H. 1990. When are rare species not there? *Oikos* **57**, 276–7.

McCarthy, M. A., and Thompson, C. 2001. Expected minimum population size as a measure of threat. *Animal Conservation* **4**, 351–5.

McCarthy, M. A., Keith, D., Tietjen, J., Burgman, M. A., Maunder, M., Master, L., Brook, B., Mace, G., Possingham, H. P., Medellin, R., Andelman, S., Regan, H., Regan, T. and Ruckelshaus, M. 2004. Comparing predictions of extinction risk using models and subjective judgement. *Acta Oecologica* (in press).

McDaniels, T. L., Gregory, R. S. and Fields, D. 1999. Democratizing risk management: successful public involvement in local water management decisions. *Risk Analysis* **19**, 497–510.

Meer, E. 2001. *Using Comparative Risk to Set Pollution Prevention Priorities in New York State: a Formula for Inaction: a Critical Analysis of the New York State Department of Environmental Conservation Comparative Risk Project*. A report from the New York State Assembly Legislative Commission on Toxic Substances and Hazardous Wastes. New York: Assemblyman Steve Englebright, Chair.

Metropolis, N., Rosenbluth, A. W., Rosenbluth, M. N., Teller, A. H. and Teller, E. 1953. Equations of state calculations by fast computing machines. *Journal of Chemical Physics* **1**, 1087–91.

Metz, C. E. 1978. Basic principles of ROC analysis. *Seminars in Nuclear Medicine* **8**, 283–98.

Meyer, M. A. and Booker, J. M. 1990. *Eliciting and Analyzing Expert Judgment: a Practical Guide.* Washington, DC: Office of Nuclear Regulatory Research, Division of Systems Research. US Nuclear Regulatory Commission.

Mills, L. S. and Lindberg, M. S. 2002. Sensitivity analysis to evaluate the consequences of conservation actions. In S. R. Beissinger and D. R. McCullough (eds.) *Population Viability Analysis.* Chicago, Illinois: University of Chicago Press, pp. 338–66.

Montgomery, D. C. 2001. *Introduction to Statistical Quality Control.* New York: Wiley.

Montgomery, D. C. and Mastrangelo, C. M. 2000. Process monitoring with auto-correlated data. In S. H. Park and G. G. Vining (eds.) *Statistical Process Monitoring and Optimization.* New York: Marcel Dekker, pp. 139–60.

Moore, R. E. 1966. *Interval Analysis.* Englewood Cliffs, New Jersey: Prentice-Hall.

Morgan, M. G. 1990. Choosing and managing technology-induced risk. In T. S. Glickman and M. Gough (eds.) *Readings in Risk.* Washington, DC: Resources for the Future, pp. 17–29.

1993. Risk analysis and management. *Scientific American* **269**, 32–42.

Morgan, M. G. and Henrion, M. 1990. *Uncertainty: a Guide to Dealing with Uncertainty in Quantitative Risk and Policy Analysis.* Cambridge: Cambridge University Press.

Morgan, M. G., Fischhoff, B., Lave, L. and Fischbeck, P. 1996. A proposal for ranking risk within Federal agencies. In J. C. Davies (ed.) *Comparing Environmental Risks.* Washington, DC: Resources for the Future, pp. 111–47.

Morley, R. S. 1993. A model for the assessment of the animal disease risks associated with the importation of animals and animal products. *Scientific and Technical Review, Office Internationale des Epizooties* **12**, 1055–92.

Morris, P. A. 1977. Combining expert judgments: a Bayesian approach. *Management Science* **20**, 1233–41.

Morris, W. F. and Doak, D. F. 2002. *Quantitative Conservation Biology: Theory and Practice of Population Viability Analysis.* Sunderland, Massachusetts: Sinauer.

Murphy, A. H. and Winkler, R. L. 1977. Can weather forecasters formulate reliable probability forecasts of precipitation and temperature? *National Weather Digest* **2**, 2–9.

Murtaugh, P. A. 1996. The statistical evaluation of ecological indicators. *Ecological Applications* **6**, 132–9.

Myers, N. 1993. Biodiversity and the precautionary principle. *Ambio* **22**, 74–9.

Nabholz, J. V., Clements, R. G. and Zeeman, M. G. 1997. Information needs for risk assessment in EPA's Office for Pollution Prevention and Toxics. *Ecological Applications* **7**, 1094–8.

Nelson, L. S. 1984. The Shewhart Control Chart – tests for special cases. *Journal of Quality Technology* **16**, 237–9.

NEPC 2000. National Environment Protection Council. Environmental Health Risk Assessment. April 2000. Public Consultation Draft. Commonwealth Government, Canberra, Australia. www.nepc.gov.au.

Neumaier, A. 1990. *Interval Methods for Systems of Equations.* Cambridge: Cambridge University Press.

Newman, M. C. and McCloskey, J. T. 1996. Time-to-event analyses of ecotoxicity data. *Ecotoxicology* **5**, 187–96.

Newman, M. C., Ownby, D. R., Mezin, L. C. A., Powell, D. C., Christensen, T. R. L., Lerberg, S. B. and Anderson, B.-A. 2000. Applying species-sensitivity distributions in ecological risk assessment: assumptions of distribution type and sufficient numbers of species. *Environmental Toxicology and Chemistry* **19**, 508–15.

NFPRF 2002. *Comparative Risk Estimates.* Recommendations of the Research Advisory Council on Post-fire Analysis. Foundation Report Appendix C. Quincy, Massachusetts: The Fire Protection Research Foundation.

Nichols, A. L. and Zeckhauser, R. J. 1988. The perils of prudence: how conservative risk assessments distort regulation. *Regulatory Toxicology and Pharmacology* **8**, 61–75.

Noss, R. F. 1990. Indicators for monitoring biodiversity: a hierarchical approach. *Conservation Biology* **4**, 355–64.

Oakes, M. 1986. *Statistical Inference: A Commentary for the Social and Behavioural Sciences.* Chichester: J. Wiley & Sons, Inc.

O'Brien, M. 2000. *Making Better Environmental Decisions: an Alternative to Risk Assessment.* Cambridge, Massachusetts: MIT Press.

Olson, L. J., Erickson, B. J., Hinsdill, R. D., Wyman, J. A., Porter, W. P., Binning, L. K., Bidgood, R. C. and Nordheim, E. V. 1987. Aldicarb immunomodulation in mice: an inverse dose-response to parts per billion levels in drinking water. *Archives of Environmental Contamination and Toxicology* **16**, 433–9.

Oosthuizen, J. 2001. Environmental health risk assessments: how flawed are they? A methyl-mercury case. *Environmental Health* **1**, 11–17.

Parma, A. M. and the NCEAS Working Group on Population Management (1998). What can adaptive management do for our fish, forests, food and biodiversity? *Integrative Biology* 1998, 16–26.

Parris, K. M., Norton, T. W. and Cunningham, R. B. 1999. A comparison of techniques for sampling amphibians in the forests of south-east Queensland, Australia. *Herpetologica* **55**, 271–83.

Pastorok, R. A., Akçakaya, H. R., Regan, H., Ferson, S. and Bartell, S. 2003. Role of ecological modeling in risk assessment. *Human and Ecological Risk Assessment* **9**, 939–72.

Paté-Cornell, M. E. 1998. Risk comparison: uncertainties and ranking. In A. Mosleh and R. A. Bari (eds.) *Probabilistic Safety Assessment and Management, PSAM4, 1991–1996.* New York: Springer.

Pearl, J. 1988. *Probabilistic Reasoning in Intelligent Systems: Networks of Plausible Inference.* San Mateo, California: Morgan Kaufmann.

  2000. *Causality: Models, Reasoning, and Inference.* Cambridge: Cambridge University Press.

Perhac, R. M. 1998. Comparative risk assessment: where does the public fit in? *Science, Technology, and Human Values* **23**, 221–41.

Peterman, R. M. 1990. Statistical power analysis can improve fisheries research and management. *Canadian Journal of Fisheries and Aquatic Sciences* **47**, 2–15.

Pheloung, P. 1995. *Determining the Weed Potential of New Plant Introductions to Australia.* Agriculture Protection Board Report, Western Australian Department of Agriculture, Perth. 26 pp.

Pheloung, P. C., Williams, P. A. and Halloy, S. R. 1999. A weed risk assessment model for use as a biosecurity tool evaluating plant introductions. *Journal of Environmental Management* **57**, 239–51.

Philip, M. S. 1994. *Measuring Trees and Forests*. Wallingford: CAB International.

Philippi, T. E., Dixon, P. M. and Taylor, B. E. 1998. Detecting trends in species composition. *Ecological Applications* **8**, 300–8.

Pidgeon, N., Hood, C., Jones, D., Turner, B. and Gibson, R. 1992. Risk perception. In *Risk: Analysis Perception and Management*. Report of a Royal Society for the Prevention of Accidents Study Group. London: The Royal Society.

Piegorsch, W. W. and Bailer, A. J. 1997. *Statistics for Environmental Biology and Toxicology*. London: Chapman and Hall.

Plous, S. 1993. *The Psychology of Judgment and Decision Making*. New York: McGraw-Hill.

Poff, N. L., Allan, J. D., Palmer, M. A., Hart, D. D., Richter, B. D., Arthington, A. H., Rogers, K. H., Meyer, J. L. and Stanford, J. A. 2003. River flows and water wars: emerging science for environmental decision making. *Frontiers in Ecology and the Environment* **1**, 298–306.

Pollard, S. J., Yearsley, R., Reynard, N., Meadowcroft, I. C., Duarte-Davidson, R. and Duerden, S. 2002. Current directions in the practice of environmental risk assessment in the United Kingdom. *Environmental Science and Technology* **36**, 530–8.

Possingham, H. P. 1996. Decision theory and biodiversity management: how to manage a metapopulation. In R. B. Floyd, A. W. Sheppard and P. J. De Barro (eds.) *The Proceedings of the Nicholson Centenary Conference 1995*. Canberra: CSIRO Publishing, pp. 391–8.

Possingham, H. P., Lindenmayer, D. B. and T. W. Norton. 1993. A framework for the improved management of threatened species based on population viability analysis. *Pacific Conservation Biology* **1**, 39–45.

Possingham, H. P., Andelman, S. J., Burgman, M. A., Medellén, R. A., Master, L. L. and Keith, D. A. 2002a. Limits to the use of threatened species lists. *Trends in Ecology and Evolution* **17**, 503–7.

Possingham, H. P., Lindenmayer, D. B. and Tuck, G. N. 2002b. Decision theory for population viability analysis. In S. R. Beissinger and D. R. McCullough (eds.) *Population Viability Analysis*. Chicago, Illinois: University of Chicago Press, pp. 470–89.

Potter, M. E. 1996. Risk assessment terms and definitions. *Journal of Food Protection Supplement*, 6–9.

Power, M. and McCarty, L. S. 1998. A comparative analysis of environmental risk assessment / risk management frameworks. *Environmental Science and Technology* 1 May, 224–31.

Preston, B. J. 2003. Science and the law: evaluating evidentiary reliability. *Australian Bar Review* **23**, 263–95.

Preston, L. E. 1998. Agents, stewards, and stakeholders. *Academy of Management Review* **23**, 9.

Punt, A. E. and Hilborn, R. 1997. Fisheries stock assessment and decision analysis: the Bayesian approach. *Reviews in Fish Biology and Fisheries* **7**, 35–63.

Punt, A. E. and Smith, A. D. M. 1999. Harvest strategy evaluation for the eastern stock of gemfish (*Rexea solandri*). *ICES Journal of Marine Science* **56**, 860–75.

Raftery, A. E. 1996. Approximate Bayes factors and accounting for model uncertainty in General Linear Models. *Biometrika* **83**, 251–66.

Raftery, A. E., Madigan, D. and Volinsky, C. T. 1996. Accounting for model uncertainty in survival analysis improves predictive performance. In J. Bernardo, J. Berger, A. Dawid and A. Smith (eds.) *Bayesian Statistics*. Oxford: Oxford University Press, pp. 323–49.

Raiffa, H. and Schlaifer, R. 1961. *Applied Statistical Decision Theory*. Cambridge, Massachusetts: Harvard University Press.

Ralls, K. and Starfield, A. M. 1995. Choosing a management strategy: two structured decision-making methods for evaluating the predictions of stochastic simulation models. *Conservation Biology* **9**, 175–81.

Ralls, K., Beissinger, S. R. and Cochrane, J. F. 2002. Guidelines for using population viability analysis in endangered-species management. In S. R. Beissinger and D. R. McCullough (eds.) *Population Viability Analysis*. Chicago, Illinois: University of Chicago Press, pp. 521–50.

Reckhow, K. H. 1994. Importance of scientific uncertainty in decision making. *Environmental Management* **18**, 161–6.

Reed, J. M. 1996. Using statistical probability to increase confidence of inferring species extinction. *Conservation Biology* **10**, 1283–5.

Reed, J. M., Mills, L. S., Dunning, J. B. Jr., Menges, E. S., McKelvey, K. S., Frye, R., Beissinger, S. R., Anstett, M.-C. and Miller, P. 2002. Emerging issues in population viability analysis. *Conservation Biology* **16**, 7–19.

Regan, H. M., Colyvan, M. and Burgman, M. A. 2000. A proposal for fuzzy IUCN categories and criteria. *Biological Conservation* **92**, 101–8.

Regan, H. M., Lupia, R., Drinnan, A. N. and Burgman, M. A. 2001. The currency and tempo of extinction. *American Naturalist* **157**, 1–10.

Regan, H. M., Colyvan, M. and Burgman, M. A. 2002a. A taxonomy and treatment of uncertainty for ecology and conservation biology. *Ecological Applications* **12**, 618–28.

Regan, H. M., Hope, B. K. and Ferson, S. 2002b. Analysis and portrayal of uncertainty in a food-web exposure model. *Human and Ecological Risk Assessment* **8**, 1757–77.

Regan, H. M., Auld, T. D., Keith, D. A. and Burgman, M. A. 2003. The effects of fire and predators on the long-term persistence of an endangered shrub, *Grevillea caleyi*. *Biological Conservation* **109**, 73–83.

Regan, H. M., Ben-Haim, Y., Lundberg, P., Langford, B., Andelman, S., Wilson, W. and Burgman, M. 2004. Using information gap theory to make robust decisions: applications to decision tables. (in review)

Renner, R. 2002. The K_{OW} controversy. *Environmental Science and Technology*, November, 411–13.

Resnik, M. 1987. *Choices: an Introduction to Decision Theory*. Minnesota: University of Minnesota Press.

Reynolds, M. and Rinderknecht, S. 1993. The expert witness in England and Germany: a comparative study. *Arbitration* **59**, 118–19.

Reynolds, M. R. and Stoumbos, Z. G. 2000. Some recent developments in control charts for monitoring a proportion. In S. H. Park and G. G. Vining (eds.) *Statistical Process Monitoring and Optimization*. New York: Marcel Dekker, pp. 117–38.

Ripley, B. D. 1987. *Stochastic Simulation*. New York: Wiley.

Robb, C. A. and Peterman, R. M. 1998. Application of Bayesian decision analysis to management of a sockeye salmon (*Oncorhynchus nerka*) fishery. *Canadian Journal of Fisheries and Aquatic Sciences* **55**, 86–98.

Robbins, C. S., Dawson, D. K. and Dowell, B. A. 1989. Habitat area requirements of breeding forest birds of the middle Atlantic states. *Wildlife Monographs* **103**, 1–34.

Rohrmann, B. 1994. Risk perception of different societal groups: Australian findings and cross-national comparisons. *Australian Journal of Psychology* **46**, 150–63.

1998. The risk notion: epistemological and empirical considerations. In M. G. Stewart and R. E. Melchers (eds.) *Integrated Risk Assessment*. Rotterdam: Balkema.

Roy, B. 1999. Decision-aiding today: what should we expect? In T. Gal, T. J. Stewart and T. Hanne (eds.) *Multicriteria Decision-Making: Advances in MCDM Models, Algorithms, Theory and Applications*. Boston, Massachusetts: Kluwer, pp. 1/1–1/35.

Roy, B. and Vanderpooten, D. 1996. The European school of MCDA: emergence, basic features and current works. *Journal of Multi-Criteria Decision Analysis* **5**, 22–37.

Royal Society. 1983. *Risk Assessment: Report of a Royal Society Study Group*. London: The Royal Society.

Ruckelshaus, M., Hartway, C. and Karieva, P. 1997. Assessing the data requirements of spatially explicit dispersal models. *Conservation Biology* **11**, 1298–306.

Ruckelshaus, M. H., Levin, P., Johnson, J. B. and Kareiva, P. M. 2002. The Pacific Salmon wars: what science brings to the challenge of recovering species. *Annual Review of Ecology and Systematics* **33**, 665–706.

Saaty, T. L. 1980. *The Analytic Hierarchy Process*. New York: McGraw Hill.

1992. *Multicriteria Decision Making – the Analytical Hierarchy Process*. Pittsburg: RWS Publications.

1994. Highlights and critical points in the theory and application of the analytic hierarchy process. *European Journal of Operational Research* **74**, 426–47.

Salsburg, D. 2001. *The Lady Tasting Tea: how Statistics Revolutionized Science in the Twentieth Century*. New York: Freeman.

Savage, L. J. 1972. *The Foundations of Statistics*. New York: Dover Publications.

Schwartz, L. M., Woloshin, S. and Welch, H. G. 1999. Risk communication in clinical practice: putting cancer in context. *Journal of the National Cancer Institute Monographs* **25**, 124–33.

Seiler, F. A. and Alvarez, J. L. 1996. On the selection of distributions for stochastic variables. *Risk Analysis* **16**, 5–18.

Sentz, K. and Ferson, S. 2002. Combination of evidence in Dempster-Shafer theory. *SAND Report, SAND2002–0835*. Albuquerque, New Mexico: Sandia National Laboratories.

Shafer, G. 1976. *A Mathematical Theory of Evidence*. Princeton, New Jersey: Princeton University Press.

Shaffer, M. L. 1981. Minimum population sizes for species conservation. *Bioscience* **31**, 131–4.

　1987. Minimum viable populations: coping with uncertainty. In M. E. Soulé (ed.) *Viable Populations for Conservation*. Cambridge: Cambridge University Press, pp. 59–68.

　1990. Population viability analysis. *Conservation Biology* **4**, 39–40.

Sharpe, V. A. 1996. Ethical theory and the demands of sustainability. In C. R. Cothem (ed.) *Handbook for Environmental Risk Decision Making: Values, Perceptions, and Ethics*. New York: American Academy of Environmental Engineers.

Shavell, S. 1987. *Economic Analysis of Accident Law*. Washington, DC: Harvard University Press.

Shewhart, W. A. 1931. *Economic Control of Quality of Manufactured Product*. Princeton, New Jersey: Van Nostrand Reinhold.

Shine, J. P., Trapp, C. J. and Coull, B. A. 2003. Use of Receiver Operating Characteristic curves to evaluate sediment quality guidelines for metals. *Environmental Toxicology and Chemistry* **22**, 1642–8.

Shipley, B. 2000. *Cause and Correlation in Biology*. Cambridge: Cambridge University Press.

Shlyakhter, A. I. 1994. Uncertainty estimates in scientific models: lessons from trends in physical measurements, population and energy projections. In B. M. Ayyub and M. M. Gupta (eds.) *Uncertainty Modelling and Analysis: Theory and Applications*. Amsterdam: Elsevier Science, pp. 477–96.

Shrader-Frechette, K. 1996a. Methodological rules for four classes of scientific uncertainty. In J. Lemons (ed.) *Scientific Uncertainty and Environmental Problem Solving*. Cambridge, Massachusetts: Blackwell, pp. 12–39.

　1996b. Value judgments in verifying and validating risk assessment models. In C. R. Cothern (ed.) *Handbook for Environmental Risk Decision Making: Values, Perceptions and Ethics*. Boca Raton, Florida: CRC Lewis Publishers, pp. 291–309.

　2001. Non-indigenous species and ecological explanation. *Biology and Philosophy* **16**, 507–19.

Shrader-Frechette, K. and McCoy, E. 1992. Statistics, costs, and rationality in ecological inference. *Trends in Ecology and Evolution* **7**, 96–9.

Silbergeld, E. K. 1994. The risks of comparing risks. *New York University Environmental Law Journal* **3**, 405–30.

Simberloff, D. 1998. Flagships, umbrellas and keystones – is single-species management *passe* in the landscape era. *Biological Conservation* **83**, 247–57.

Simberloff, D. and Dayan, T. 1991. The guild concept and the structure of ecological communities. *Annual Review of Ecology and Systematics* **22**, 115–43.

Simon, H. A. 1959. Theories of decision making in economics and behavioral science. *American Economic Review* **49**, 253–83.

Slovic, P. 1999. Trust, emotion, sex, politics, and science: surveying the risk-assessment battlefield. *Risk Analysis* **19**, 689–701.

Slovic, P., Fischhoff, B. and Lichtenstein, S. 1979. Rating the risks. *Environment* **2**, 14–20, 36–9.

1984. Perception and acceptability of risk from energy systems. In W. R. Fruendenburg and E. A. Rosa (eds.) *Public Reactions to Nuclear Power: are there Critical Masses?* Boulder, Colorado: AAAS/Westview, pp. 115–35.

Slovic, P., Monahan, J. and MacGregor, D. G. 2000. Violence risk assessment and risk communication: the effects of using actual cases, providing instruction, and employing probability versus frequency formats. *Law and Human Behavior* **24**, 271–96.

Smith, A. D. M., Sainsbury, K. J. and Stevens, R. A. 1999a. Implementing effective fisheries-management systems – management strategy evaluation and the Australian partnership approach. *ICES Journal of Marine Science* **56**, 967–79.

Smith, C. S., Lonsdale, W. M. and Fortune, J. 1999b. When to ignore advice: invasion predictions and decision theory. *Biological Invasions* **1**, 89–96.

Sokal, R. R. and Rohlf, F. J. 1995. *Biometry*, 3rd edition. San Francisco, California: Freeman.

Solomon, K. R., Giddings, J. M. and Maund, S. J. 2001. Probabilistic risk assessment of cotton pyrethroids: I. Distributional analyses of laboratory aquatic toxicity data. *Environmental Toxicology and Chemistry* **20**, 652–9.

Solow, A. R. 1993. Inferring extinction from sighting data. *Ecology* **74**, 962–4.

Solow, A. R. and Roberts, D. L. 2003. A nonparametric test for extinction based on a sighting record. *Ecology* **84**, 1329–32.

Soulé, M. E. (ed.) 1987. *Viable Populations for Conservation*. Cambridge: Cambridge University Press.

Spencer, M. and Ferson, S. 1998. *RAMAS / Ecotoxicology. Ecological Risk Assessment for Food Chains and Webs.* Version 1.0. User's Manual, Volume 1. New York: Applied Biomathematics.

Standards Australia. 1999. *Risk Management.* AS/NZS 4360: 1999. Strathfield, Australia: Standards Association of Australia.

Starfield, A. M. and Bleloch, A. L. 1992. *Building Models for Conservation and Wildlife Management*. Edina: Burgess International Group.

Starmer, C. V. 1998. The economics of risk. In P. Calow (ed.) *Handbook of Environmental Risk Assessment and Management*. Oxford: Blackwell Science, pp. 319–44.

Stephens, M. E., Goodwin, B. W. and Andres, T. H. 1993. Deriving parameter probability density functions. *Reliability Engineering and System Safety* **42**, 271–91.

Stewart, A. 1993. Environmental risk assessment: the divergent methodologies of economists, lawyers and scientists. *Environment and Planning Law Journal* **10**, 10–18.

Stewart, M. G. and Melchers, R. E. 1997. *Probabilistic Risk Assessment of Engineering Systems*. London: Chapman and Hall.

Stigler, S. M. 1986. *The History of Statistics: the Measurement of Uncertainty before 1900*. Cambridge, Massachusetts: Belknap Press.

Stirling, A. 1997. Multicriteria mapping: mitigating the problems of environmental valuation? In J. Foster (ed.) *Valuing Nature: Economics, Ethics and Environment*. London: Routledge.

1999. *On 'Science' and 'Precaution' in the Management of Technological Risk.* Report to the EU Forward Studies Unit, IPTS, Sevilla. EUR 19056.

Stirling, A. and Mayer, S. 1999. *Rethinking Risk: a Pilot Multicriteria Mapping of a Genetically Modified Crop and Agricultural Systems in the UK*. Science Policy Research Unit, Sheffield University, Report 21. Sheffield, UK.

Stow, C. A. and Borsuk, M. E. 2003. Enhancing causal assessment of estuarine fishkills using graphical models. *Ecosystems* **6**, 11–19.

Sugihara, G., Grenfell, B. and May, R. M. 1990. Distinguishing error from chaos in ecological time. *Philosophical Transactions of the Royal Society, London, Series B* **330**, 235–51.

Suter, G. W. 1993. *Ecological Risk Assessment*. Boca Raton, Florida: Lewis.

   1995. Introduction to ecological risk assessment for aquatic toxic effects. In G. M. Rand (ed.) *Fundamentals of Aquatic Toxicology: Effects, Environmental Fate and Risk Assessment*, 2nd edition. Washington: Taylor and Francis, pp. 802–25.

Swaay, C. A. M. van and Warren, M. S. 1999. *Red Data Book of European butterflies* (Rhopalocera). Nature and Environment, No. 99. Strasbourg: Council of Europe Publishing.

Swartout, J., Price, S. and Dourson, M. 1998. A probabilistic framework for the reference dose (Probabilistic RfD). *Risk Analysis* **18**, 271.

Swets, J. A., Dawes, R. M. and Monahan, J. 2000. Better decisions through science. *Scientific American*, October 2000, 82–7.

Tanino, T. 1999. Sensitivity analysis in MCDM. In T. Gal, T. J. Stewart and T. Hanne (eds.) *Multicriteria Decision-making: Advances in MCDM Models, Algorithms, Theory and Applications*. Boston, Massachusetts: Kluwer, pp. 7/1–7/29.

Tarplee, B. 2000. Atrazine – re-evaluation by the FQPA Safety Factor Committee. US EPA FQPA Safety Factor Committee, Memorandum to C. Eiden, Health Effects Division. HED Doc. No. 014375.

Taylor, B. L. 1995. The reliability of using population viability analysis for risk classification of species. *Conservation Biology* **9**, 551–8.

Taylor, B. L. and Gerrodette T. 1993. The uses of statistical power in conservation biology: the Vaquita and Northern Spotted Owl. *Conservation Biology* **7**, 489–500.

Tengs, T. O., Adams, M. E., Pliskin, J. S., Safran, D. G., Siegel, J. E., Weinstein, M. and Graham, J. D. 1995. Five hundred life-saving interventions and their cost effectiveness. *Risk Analysis* **15**, 369–90.

Thaler, R. H. 1991. *Quasi-rational Economics*. New York: Russell Sage Foundation.

Thompson, J. R. and Koronacki, J. 2002. *Statistical Process Control: the Deeming Paradigm and Beyond*. London: Chapman and Hall / Boca Raton, Florida: CRC Press.

Toulmin, S. 1958. *The Uses of Argument*. Cambridge: Cambridge University Press.

Trumbo, C. W. and McComas, K. A. 2003. The function of credibility in information processing for risk perception. *Risk Analysis* **23**, 343–53.

Tufte, E. 1997. *Visual Explanations: Images and Quantities, Evidence and Narrative*. Cheshire, Connecticut: Graphics Press.

Tukey, J. W. 1991. The philosophy of multiple comparisons. *Statistical Sciences* **6**, 100–16.

Tversky, A. and Kahneman, D. 1971. Belief in the law of small numbers. *Psychological Bulletin* **76**, 105–10.

1974. Judgement under uncertainty: heuristics and biases. *Science* **211**, 453–8.

1981. The framing of decisions and the psychology of choice. *Science* **211**, 453–8.

1982a. Causal schemas in judgments under uncertainty. In D. Kahneman, P. Slovic, A. Tversky (eds.) *Judgment under Uncertainty: Heuristics and Biases.* Cambridge: Cambridge University Press.

1982b. Belief in the law of small numbers. In D. Kahneman, P. Slovic and A. Tversky (eds.) *Judgement under Uncertainty: Heuristics and Biases.* Cambridge: Cambridge University Press, pp. 23–30.

Ulam, S. M. 1976. *Adventures of a Mathematician.* New York: Charles Scribner.

UN. 1992. *Report of the United Nations Conference on Environment and Development.* Rio de Janeiro, 3–14 June 1992. United Nations Environment Programme.

Underwood, A. J. 1997. *Experiments in Ecology.* Cambridge: Cambridge University Press.

US EPA. 1989. *Human Health Evaluation Manual.* Washington, DC: US EPA.

1992. Framework for ecological risk assessment. EPA/630/R-92/ 001. US Environmental Protection Agency, Risk Assessment Forum. Washington, DC: US EPA.

1997a. US Environmental Protection Agency. *Policy for the Use of Probabilistic Risk Analysis.* (15 May, 1997). http://www.epa.gov/ncea/mcpolicy.htm.

1997b. US Environmental Protection Agency. *Guiding Principles for Monte Carlo analysis.* EPA/630/r-97/001. http://www.epa.gov/ncea/monteabs.htm.

1997c. US Environmental Protection Agency, Office of Air Quality Planning and Standards, Office of Research and Development. *Mercury Study, Report to Congress*, Volume 3. Springfield, Virginia: National Technical Information Service, pp. 205–402.

1998. Guidelines for ecological risk assessment. US Environmental Protection Agency, Washington, DC Federal Register, Volume 63, no. 93, May 14, 26846–26924.

US FDA. 1998. *Design Control Inspection Results, 6/1/97–6/1/98.* Center for Devices and Radiological Health. http://www.fda.gov/cdrh/dsma/dcisresults.html.

US Federal Register, undated. Volume 65, no. 218, Section 219.36.

US NRC. 1983. *Risk Assessment in the Federal Government: Managing the Process.* Washington, DC: US National Research Council.

Valverde, L. J. 2001. Expert judgment resolution in technically-intensive policy disputes. In I. Linkov and J. Palma-Oliveira (eds.) *Assessment and Management of Environmental Risks.* Dordrecht, the Netherlands: Kluwer Academic, pp. 221–38.

Vanackere, G. 1999. Minimizing ambiguity and paraconsistency. *Logique et Analyse* **165–166**, 139–160.

Van der Heijden, K. 1996. *Scenarios: the Art of Strategic Conversation.* Chichester: Wiley.

Varis, O. and Kuikka, S. 1999. Learning Bayesian decision analysis by doing: lessons from environmental and natural resource management. *Ecological Modelling* **119**, 177–95.

Venables, W. M. and Ripley, B. D. 1997. *Modern Applied Statistics with S-Plus*, 2nd edition. New York: Springer-Verlag.

Verdonck, F. A. M., Aldenberg, T., Jaworska, J. and Vanrolleghem, P. A. 2003. Limitations of current risk characterization methods in probabilistic environmental risk assessment. *Environmental Toxicology and Chemistry* **22**, 2209–13.

Vincke, P. 1999. Outranking approach. In T. Gal, T. J. Stewart and T. Hanne (eds.) *Multicriteria Decision-making: Advances in MCDM Models, Algorithms, Theory and Applications*. Boston, Massachusetts: Kluwer, pp. 11/1–11/29.

Viscusi, W. K. 2000. Corporate risk analysis: a reckless act? *Harvard Law School Discussion Paper 304* (11/2000), http://www.law.harvard.edu/programs/olin_center/.

Vose, D. 1996. *Quantitative Risk Analysis: a Guide to Monte Carlo Simulation Modelling.* Chichester: Wiley.

Wade, P. R. 2000. Bayesian methods in conservation biology. *Conservation Biology* **14**, 1308–16.

Walker, S. F. and Marr, J. W. 2001. *Stakeholder Power: a Winning Plan for Building Stakeholder Commitment and Driving Corporate Growth.* Boulder, Colorado: Perseus Publishing.

Walley, P. 1991. *Statistical Reasoning with Imprecise Probabilities*. London: Chapman and Hall.

  2000a. Coheret upper and lower provisions. Imprecise Probabilities Project, http://www.sipta.org/.

  2000b. Towards a unified theory of imprecise probability. *International Journal of Approximate Reasoning* **24**, 125–48.

Walley, P. and DeCooman, G. 2001. A behavioral model for linguistic uncertainty. *Information Sciences* **134**, 1–37.

Walters, C. J. 1986. *Adaptive Management of Renewable Resources.* New York: MacMillan.

Walton, D. 1997. *Appeal to Expert Opinion: Arguments from Authority.* Pennsylvania: Pennsylvania State University Press.

Walton, D. N. and Batten, L. M. 1984. Games, graphs and circular arguments. *Logique et Analyse* **106**, 133–64.

Warren-Hicks, W. J. and Moore, D. R. J. 1995. (eds.) *Uncertainty Analysis in Ecological Risk Assessment.* Pensacola, Florida: SETAC Press.

Watson, P. 1998. A process for estimating geological risk of petroleum exploration prospects. *APPEA Journal* 1998, 577–82.

Watts, T. 1998. The dangers of ignoring public ire. *Business Review Weekly* August 31 1998, 60–1.

Weaver, J. C. 1995. Indicator species and the scale of observation. *Conservation Biology* **9**, 939–42.

Weber, E. U., Blais, A.-R. and Betz, N. E. 2002. A domain-specific risk-attitude scale: measuring risk perceptions and risk behaviors. *Journal of Behavioral Decision Making* **15**, 263–90.

Weitzman, M. 1998. The Noah's Ark problem. *Econometrica* **66**, 1279–98.

Wen-Qiang, B. and Keller, L. R. 1999. Chinese and Americans agree on what is fair, but disagree on what is best in societal decisions affecting health and safety risks. *Risk Analysis* **19**, 439–52.

WHO 1994. *Environmental Health Criteria: 170 Mercury.* Geneva: International Program on Chemical Safety. World Health Organization, pp. 23–56.

Wigley, D. C. and Shrader-Frechette, K. S. 1996. Environmental racism and biased methods of risk assessment. *Risk: Health, Safety and Environment* **55**, 55–88.

Williamson, R. and Downs, T. 1990. Probabilistic arithmetic, I: Numerical methods for calculating convolutions and dependency bounds. *International Journal of Approximate Reasoning* **4**, 89–158.

Wilson, R. 1979. Analyzing the daily risks of life. *Technology Review* **81**, 41–6.

Windschitl, P. D. 2002. Judging the accuracy of a likelihood judgment: the case of smoking risk. *Journal of Behavioral Decision Making* **15**, 19–35.

Windsor, D. 1998. *The Definition of Stakeholder Status.* Proceedings of the International Association for Business and Society (IABS) Annual Conference, Kona-Kailua, Hawaii (June 1998), pp. 537–42.

Wintle, B. A., McCarthy, M., Volinsky, C. T. and Kavanagh, R. P. 2003. The use of Bayesian Model Averaging to better represent uncertainty in ecological models. *Conservation Biology* **17**, 1579–90.

Wolfgang, P. 2002. Witness 'conferencing'. *Arbitration International* **18**, 47–58.

Woloshin, S., Schwartz, L. M. and Welch, H. G. 2002. Risk charts: putting cancer in context. *Journal of the National Cancer Institute* **94**, 799–804.

Wright, G., Bolger, F. and Rowe, G. 2002. An empirical test of the relative validity of expert and lay judgments of risk. *Risk Analysis* **22**, 1107–22.

Yamada, K., Ansari, M., Harrington, R., Morgan, D. and Burgman, M. 2004. Sindh Ibex in Kirthar National Park, Pakistan. In H. R. Akçakaya, M. Burgman, O. Kindvall, C. C. Wood, P. Sjogren-Gulve, J. Hatfield and M. McCarthy (eds.) *Species Conservation and Management: Case Studies Using RAMAS GIS.* New York: Oxford University Press.

Zadeh, L. A. 1986. A simple view of the DempsterShafer theory of evidence and its implication for the rule of combination. *The AI Magazine* 1986, 85–90.

Zaunbrecher, M. 1999. BHP's environmental risk assessment process. Unpublished presentation, School of Botany, March 1999. Melbourne, Australia: University of Melbourne.

Zweig, M. H. and Campbell, G. 1992. Receiver-Operating Characteristic (ROC) plots: a fundamental evaluation tool in clinical medicine. *Clinical Chemistry* **39**, 561–77.

Index

Page numbers in **bold** type refer to tables; those in *italic* type refer to figures and boxes.